Pvt. Ltd.

Biofuels

Advances and Perspectives

Geetanjali Kaushik
Shivani Chaturvedi
Satish Patil
Arvind Chel

2018

Studium Press (India) Pvt. Ltd.

Biofuels

Advances and Perspectives

ISBN: 978-93-85046-22-3

Published by:

Studium Press (India) Pvt. Ltd.
4735/22, 2nd Floor, Prakash Deep Building
(Near Delhi Medical Association),
Ansari Road, Darya Ganj, New Delhi-110 002
Tel.: + 91-11-43240200-15 (15 lines); Fax: 91-11-43240215
E-mail: studiumpress@gmail.com

Printed at India

केन्द्रीय प्रदूषण नियंत्रण बोर्ड
CENTRAL POLLUTION CONTROL BOARD
पर्यावरण, वन एवं जलवायु परिवर्तन मंत्रालय, भारत सरकार
MINISTRY OF ENVIRONMENT, FORESTRY & CLIMATE CHANGE GOVT. OF INDIA

Dr. DIPANKAR SAHA
ADDITIONAL DIRECTOR
DIVISIONAL HEAD, AIR QUALITY DIVISION

Foreword

"Biofuels: Advances and Perspectives"

With the depletion of fossil fuels and the adverse impacts associated with their usage there has been a renewed interest in alternative sources of energy. Since the global fossil fuel reserves are limited, currently the research is focused on renewables and other non-polluting sources of energy. Biofuels provide a suitable alternative as energy sources of future and form the basis of sustainable development with regard to socio-economic and environmental concerns. Biodiesel, Bioethanol derived from plant and algal sources appear to be promising future energy sources. India's energy policies reflect the need to shift towards sustainable energy sources along with ensuring energy security. Currently the country meets 84.5% of its crude oil demand through imports, making the nation vulnerable to price fluctuations. Biofuels produced from agricultural by-products (such as molasses) and lignocellulosic biomass, offer alternate sources of income to farmers and also address food security concerns. Further increasing the expansion of biofuel processing and production facilities will further generate sustainable livelihood opportunities. Since biofuels are produced from renewable biological sources their usage has the potential

'परिवेश भवन' पूर्वी अर्जुन नगर, दिल्ली - 110032
Parivesh Bhawan, East Arjun Nagar, Delhi-110032
दूरभाष/Tel: 43102030, 22305792; वेबसाईट/Website: www.cpcb.nic.in

to reduce vehicular emissions while also ensuring significant foreign exchange savings, besides revitalizing the rural economy.

In this regard, this book is timely as it provides a comprehensive overview of India's biofuels policy and the Ethanol Blended Petrol Programme. The book entitled "Biofuels: Advances and Perspectives" with the use of data and case studies demonstrates the advantages of fuel blending particularly towards tackling the growing menace of air pollution in urban India. I congratulate Dr. Geetanjali Kaushik and her team in inviting manuscripts from noted researchers engaged in biofuel research across the world and bringing out this compilation. It is hoped that this book would be used as a point of reference for policy makers, academicians and other stakeholders.

31.08

(D. Saha)

'परिवेश भवन' पूर्वी अर्जुन नगर, दिल्ली - 110032
Parivesh Bhawan, East Arjun Nagar, Delhi-110032
दूरभाष/Tel: 43102030, 22305792; वेबसाईट/Website: www.cpcb.nic.in

Bhartiya Janta Party
Kishanchand Tanwani
BJP President, Sambhajinagar (Aurangabad)
(Ex-Mayor & Ex-MLC)

Dated: Oct 2, 2017

Prologue

Air Pollution is a global problem. As per TERI's recent report around 80% of Indian cities are unable to meet the prescribed air quality standards. Managing India's air pollution requires a comprehensive national action plan with a focus on city level management plans.

Accordingly to CPCB's latest observations there are total 17 non-attainment cities in the state of Maharashtra. Aurangabad, the historical city is also listed within the list with regards to particulate matter. I congratulate Dr. Geetanjali Kaushik & Arvind Chel (Mahatama Gandhi Mission, Jawaharlal Nehru Engineering College, Aurangabad) and other editors — Prof. Satish S. Patil (Dr Babasaheb Marathwada University, Aurangabad, Maharashtra) & Dr. Shivani Chaturvedi (Department of Chemistry, Indian Institute of Technology Delhi, New Delhi.) — for their efforts in conducting Air Quality Assessment and bringing out Clean Air Action Plan for Aurangabad and other cities of India.

Utilization of biofuels in vehicles will improve air quality of city and reduction in the fossil fuel consumption. This book "Biofuels: Advances and Perspectives" will be extremely useful for students, budding researchers and scientist in the expert area. This book contains the researchers, contribution as authors from Canada, USA, UK, Germany, Vietnam and Bangladesh, in addition to reputed Universities from India.

Shri Kishanchand Tanwani

Party Office : BJP Karayalay, Opp. SSC Board, Sambhajinagar (Aurangabad) Ph : 0240 2349011
Office Address : Jethliya Tower Gulmandi, Sambhajinagar, (Aurangabad)
Ph : 0240 2361555, 09823006358, **email : tanwanikl@gmail.com**

About the Editors

Dr. Geetanjali Kaushik is presently working as Associate Professor in MGM's Jawaharlal Nehru Engineering College, Aurangabad, Maharashtra. Dr. Geetanjali Kaushik has a Masters from TERI and a BSc from University of Delhi (Silver and Golden Jubilee Awardee). Dr. Geetanjali Kaushik completed her PhD in Environmental Sciences, from the prestigious Indian Institute of Technology Delhi with more than 45 papers in International Journals of repute. She has participated in International Conferences held in Germany (won first prize) and USA (received American Society of Nutrition travel grant). Subsequent to her PhD she has also worked with UC Berkley on an air quality monitoring project for over 40 locations across Delhi. She has also edited 4 International books. Her taught assignments include stints as Assistant Professor with the University of Petroleum and Energy Studies, Dehradun and Amity University, Noida. In 2012, Dr. Kaushik received full scholarship from School of Management, Swansea University UK to undertake MBA. She topped the MBA class and has worked in UK and Turkey in various positions. She returned to India in 2015 and worked as Associate Professor in Environmental Management before joining as an International Air Quality Consultant. She has worked with Clean Air Asia India and World Resources Institute.

Shivani Chaturvedi, born in 1972, Chaturvedi is a Post doctoral fellow under UGCWM at IIT Delhi. She has research experience of over thirteen years in fields of nutrition, biomass biology and enzyme & microbial biochemistry. She graduated from Ajmer University, Rajasthan in the year 1990. She then went on to obtain her Masters in Chemistry from Kumaon University, Nainital in 1995. She obtained her doctoral degree in Chemistry from Kumaon University in collaboration with D.R.D.O Pithoragarh in 2000. She also possesses more than 8 years experience of teaching at the Graduate and Post Graduate Levels. Between 2001 and 2017 she has worked under various capacities in National Botanical Research Institute, Lucknow and Indian Institute of Technology Delhi. Since 2015, she has also been associated with the UGC funded project of ePG Pathshala in subject material of Biochemistry being implemented by IIT Delhi. She has published many research papers and book chapters in international & national Journals of repute. She has also presented papers in various reputed international conferences. She has been awarded "Bioved Mahila Gaurav Samman" in 2013; "Vigyan Parishad Pryag Shatabdi Samaan" in 2013 and "Young Scientist Associate Award" at the 10th Indian Agricultural Scientists and Farmers Congress held in 2008. At present she is acting as reviewer of several international Journals. She has contributed in several international books and has authored four books. Her passion towards writing is reflected not only in scientific areas but also in the field of literature. Her multiple hobbies such astrology, gardening, miniature painting, teaching and yoga keeps her engaged in life.

Prof. Satish Patil is currently Prof. in Environmental Sciences in the Dept of Environmental Sciences & Director of Central Facility for Advanced Research and Training at Dr. Babasaheb Ambedkar Marathwada University, Aurangabad. Prof. Satish Patil has been serving in the field of teaching and research in Environmental Science for the past 27 years. He has more than 100 International/National publications and 6 books in this field. He also has 2 patents to his credit. His research has been highly supported by various government/non-government funding agencies for more than Rs 1 Cr. Till now. 22 students have been awarded PhD degree under his guidance and 8 students are undertaking research under his valuable guidance. In addition to teaching and research, Prof Patil has extraordinarily worked in administration. He has held the position of Director of Board of Colleges & University Development, Dr. Babasaheb Ambedkar Marathwada University, Aurangabad and currently working as Director of Central Facility for Advanced Research and Training at Dr. Babasaheb Ambedkar Marathwada University, Aurangabad. He has also worked as Incharge Officer on Special Duty for Academics & Planning of University as well as Head of the

Department of Environmental Science, Coordinator of Science & Technology Park, Member of Board of Studies in various universities. Maharashtra Pollution Control Board (MPCB), Government of Maharashtra has appointed him as Member of Action Plan 2020 and Vision 2020.

Dr. Arvind Chel is presently working as Associate Professor (Dept of Mechanical Engineering) MGM's Jawaharlal Nehru Engineering College Aurangabad Maharashtra. Dr. Chel has undertaken his post doctoral from Ghent University, Belgium. He completed his PhD from Indian Institute of Technology Delhi with more than 40 International Journal papers in high impact factor journals. He received the best paper award in International Conference Solaris 2007. Dr. Chel undertook his M.Tech in Energy Systems Engineering from IIT Bombay, and completed his B.E. in Mechanical Engineering from Govt. Engg. College, Aurangabad. For his innovative Solar Biomass Cookstove Dr. Chel received the best innovation award from IIT Kanpur. This cookstove has been published online by Indian Patent office. He has also undertaken research with UC Berkley and also published a paper with the Energy Resources Group in Atmospheric Environment. He has travelled to research labs at Florida, USA, Sweden, London, France, Germany and Belgium for presenting his papers in International Conferences. He has been an Invited as member by Chamber of Marathwada Industries and Agriculture for Energy and Environment Health Safety cell.

Preface

Owing to exhaustion of fossil fuels in coming decades, globally there is a serious lookout for greener sources of energy. In terms of identifying future resources of energy, biomass can be considered as a promising source of energy. Biomass is a non-exhaustible resource primarily derived from biomass. At present, solid, liquid and gaseous fuels can be generated from different biomass resources. Among liquid biofuels the most sought after biofuels, which can be successfully used as transport fuel are ethanol and biodiesel. Traditionally production of biofuels was sourced from sugary/starchy crops and oil containing plants. Cellulosic biomass is attractive source as it utilizes waste from various industrially important crops which are not rendered as food crops. So, they do not compete with food crops. Further, oil producing plants can be easily generated for production of diesel substitutes or blends. However, owing to their viscosity these cannot be directly used for energy production.

But, they could be easily blended with other fuels to produce energy. Chapter-1 presents an Overview to Biofuels and the methods of their production. Most extensively recognized biofuel is corn or sugar based ethanol, generated in the course offer mentation of plant products. Worldwide bioethanol production augmented from 4.4-46.2 barrels between year 1980 and 2005. Globally USA, Brazil, and China are the major bioethanol producers (almost 75% global biofuels). Chapter 2 discusses the environmental, economical and societal perspective of this major alternative energy resource. Chapter 3 is focused on the various feed stocks that can be used for the production of biofuels and the key policy and issues addressing the development of biofuels.

In 2003 Indian launched its Ethanol blended programme for 5% ethanol blended petrol. In 2009 Government of India introduced the National Policy on Biofuels which aimed to encourage biofuel usage with reduction in fossil fuel dependence along with addressing environmental and fuel efficiency concerns. It is evident that for India a gradual shift from fossil fuels to biofuels would reduce harmful emissions of pollutants and consequently improve the air quality in our cities. Chapter 4 discusses the potential role of biofuels in air pollution mitigation.

For achieving sustainable economic growth and for reducing the dependence on international oil to look for are new able alternative fuel sources, bioethanol blend petrol and octane could be attractive alternatives for Bangladesh. Chapter 5 outlines the prospects of producing bioethanol locally, possible usage of bioethanol blended petrol and octane as transportation fuel and the associated impacts on food sector, national economy and environment in the light of technical advances and regulatory requirements.

For developing India's biomass energy, funding from government and project financing are the two main sources. Projects directly related to farmers should have investment by government.To develop the biomass industries private financing is critical for securing self-sustaining financing cycles. Government should invest in the forms of grant subsidies and interest subsidies for centralized straw gasification plants and straw biogas plants; R&D; co-financing pilot and demonstration projects (at about 50%). There is a strong need of a broad partnership. To raise the financing necessary for the project international finance institutions, such as ADB, may play a key role. Coordination between national policy and planning is necessary to establish a comprehensive industrial policy, investment and taxation frame work and to establish goals for the development of rural biomass energy. Chapter 6 presents the scope and development of biomass energy in India.

Jatropha curcas is cultivated in tropical areas as commercial crop for liquid fuel production and also for many other applications. Its plantation has a potential to provide large-scale rural employment. The non-edible oil produced by *Jatropha* seeds does not compete with commercial oil seed crops, which reduces food *versus* fuel crisis. Chapter 7 highlights the utilization of *Jatropha* for biofuel production.

In chapter 8 castor (*Ricinus communis L.*) oil alkyl esters were found to be effective lubricity enhancer for low lubricity diesel fuel apart from being an alternative to diesel fuel.

Chapter 9 provides review of carbon sequestration using algae as one of the important aspect of algae cultivation. It also discusses the use of algal biodiesel in commercial transportation. In chapter 10 a technical overview of algae cultivation, harvesting and algal biodiesel production procedures has been presented.

Chapter 11 discusses a case study under-taken to observe the potential of algal biodiesel, produced using power plant exhaust to reduce green house gas emissions. The case study results indicate that using 60% of all coal and gas power plants would unable production of significant amount of biofuel, which can be used for reducing green house gas emissions.

Chapter 12 explores the institutional design of biofuel development across world – Brazil, European Union, United States, Africa and Asia. Chapter 13 focuses primarily on violations of the human rights to food, land and water, with the aim of bringing attention to the social and environmental implications of the biofuel industry.

In chapter 14, different types of materials used for main parts of MFCs (microbial fuel cells) such as anode, cathode, membrane and various substrates are explored. This chapter also demonstrates their resulting performance, limitations as well as future potential materials and substrate.

Chapter 15 provides the current status of biofuel production and sustainability in India. The results show that non-grain based bio-ethanol production can potentially reduce CO_2 emissions from the 2007 levels by 11 million tons and 49 million tons in 2015 and 2030, respectively. Energy crops may also protect natural forests by providing an alternative source of wood, which can be grown on farm or pasture land that is no longer suitable for traditional row crops. Bioenergy-driven restoration of degraded ecosystems can also increase terrestrial carbon sequestration due to large biomass production and root residues as well as slowing decomposition of soil organic materials under no-till conditions.

Table of Contents

1

An Overview to Biofuels

Poonam Bakhshi[1*], Abhay Kulkarni[2] and Jogeshwar Gaurkhede[3]

ABSTRACT

Currently we are majorly dependent on fossil fuels or biomass for generating energy. In this context we rely on sun for providing us energy directly or indirectly. Owing to exhaustion of fossil fuels in coming decades, there is serious lookout for greener sources of energy. Also, there is an increased demand for cleaner and greener energy globally. In terms of finding future resources of energy, biomass can be considered as promising source of energy. Biomass is non-exhaustible resource, which is being produced continuously over earth due to the ongoing photosynthetic activity of plants. Biofuel are primarily derived from biomass. At present, solid, liquid and gaseous fuels can be generated from different biomass resources. Among liquid biofuels the most sought after biofuel, which can be successfully used as transport fuel are ethanol and biodiesel. This paper evaluates the different resources for biofuel generation, specifically ethanol and biodiesel.

***Key words*:** Biofuels, Biodiesel, Biomass, Jatropha, Ethanol, Feedstock

[1]Department of Mechanical Engineering, University of Central London, United Kingdom

[2]Department of Mechanical Engineering, MGM's Jawaharlal Nehru College of Engineering, CIDCO, Aurangabad, Maharashtra

[3]Department of Applied Mechanics, IIT Delhi. New Delhi

**Corresponding author*: E-mail: poonamkaushik@gmail.com

INTRODUCTION

Biofuels are fuels derived from biomass. Biomass is organic matter derived from or produced by plants and animals. It comprises mainly wood, agricultural crops and products, aquatic plants, forestry products, wastes and residues, and animal wastes. In its most general meaning, biofuels are all types of solid, gaseous and liquid fuels that can be derived from biomass. Examples of solid biofuels include wood, charcoal and bagasse. Examples of gaseous biofuels include methane gas and producer gas. However, the focus of this book is on liquid biofuels that are used primarily as transport fuel therefore, this chapter provides an overview of the two major types of biofuels that are commercially available, namely, ethanol and biodiesel (Jain and Sharma, 2010).

FEEDSTOCK FOR BIOFUELS

The various feedstocks that can be used for the production of biofuels may be classified into three groups: cellulosic biomass, sugar and starchy crops, and oil-containing or oil-producing plants.

The first group of feedstock is cellulosic biomass. It is the type of feedstock that is increasingly becoming important for the production of biofuels. Its attractiveness as a feedstock results from the fact that it is made up of very complex sugar polymers that are not usually used as a source of human food. Cellulosic biomass includes a wide range of heterogeneous solid materials including, but not limited to, the following: (a) agricultural residues comprising leftover material from crops such as rice straw, or coconut coir, or the stalks, leaves and husks of corn plants; (b) forestry wastes such as chips and sawdust from lumber mills, dead trees, and tree branches; (c) municipal solid wastes such as paper products; (d) food processing and other industrial wastes such as slops from alcohol distilleries and black liquor from pulp and paper manufacturing; and (e) energy crops grown for fuel purposes such as fast growing trees and grasses (Demirbas, 2001).

The main components of cellulosic biomass are cellulose, hemicelluloses, and lignin. Depending on the source of biomass, cellulose may account for 40% to 60% by weight of the biomass and is thus the most common form of carbon in the biomass. It is a complex sugar polymer, or a polysaccharide, and is made from the six-carbon sugar called glucose. Because of its crystalline structure, it is resistant to hydrolysis, the chemical reaction that enables the production of simple, fermentable sugars from a polysaccharide. Hemicellulose is also a major source of carbon in biomass and may account for 20% to 40% by weight of the biomass. It is also a complex polysaccharide that is made from a variety of five-carbon and six-carbon sugars. Although it is relatively easier to hydrolyze into simple sugars compared to cellulose, the sugars that are produced, however, are not easily fermented to ethanol. Lignin comprises from 10% to 24% by weight of biomass and provides structural integrity and strength in plants. It remains as the residual material after the sugars in the biomass have been converted to ethanol. As a highly complex carbon-containing polymer, it contains a lot of energy and can be burned to produce steam and electricity for use in the biomass-to-ethanol manufacturing process. The second group of feedstock is sugar and starchy crops. These are plants such as sugar cane and sugar beets that can store through photosynthesis the energy from the sun by converting it into simple sugars. In a similar fashion, there are plants such as corn, cassava and sweet potato that store the energy as complex sugars or starches. Although the name sugar is most often used to refer to sucrose or table sugar, in general, sugars are water-soluble carbohydrates that have relatively low molecular weight and usually characterized with having a sweet taste. Carbohydrates, on the other hand, are a group of organic compounds that include sugars, starches, celluloses and gums. They provide a major source of energy in the diet of humans and animals. These compounds are produced through photosynthesis by plants and contain only molecules of carbon, hydrogen and oxygen, usually in the ratio 1:2:1. Simple sugars are called monosaccharides. More complex sugars comprise between two and ten monosaccharides that are linked together. Thus disaccharides are those that contain two monosaccharides, trisaccharides are those that contain three and so on. These biomass products are mainly used as human or animal food. However, because of the need to find alternative sources of energy other than fossil fuels, these products are increasingly being used for the production of biofuels, particularly ethanol as gasoline substitute or blend. The third group of feedstock is composed of oil-containing or oil-producing plants. There are a large number of plants that produce oils, in particular fixed oils, which can be processed to produce biofuels that can be used as diesel substitute or blend (Demirbas, 2003). The common characteristic of these oils is that they may be decomposed into glycerin and one or more acids of the type called fatty acids. The physical properties commonly found in these plant oils are the following: (a) they float on water but are not soluble in it; (b) they are greasy to the touch and have lubricating properties; (c) they are not readily volatile; and (d) they may be burned without leaving any significant residue. Most of these oils such as soybean oil, coconut oil and palm oil have been used mainly for human or animal food or the production of various types of cosmetics and pharmaceuticals but increasing amounts are now being processed for the production of biodiesel. The recent interest in the production of energy from biomass has focused primarily on technologies and applications that produce liquid fuels for the use of the transportation industry. Since ethanol in its anhydrous form can be blended with gasoline, it has become one of the most widely used biofuel. In the same manner, since biodiesel (the methyl ester of plant oil) can be easily blended with or used as a substitute for diesel fuel, its production and use have also become widespread. Their use has the additional advantage of reducing emission of air pollutants, buildup of greenhouse gas and dependence on imported oil. At the same time, the cultivation and harvesting of the feedstocks have positive impacts on the agricultural and rural economies (Faaij, 2006).

Box 1.1 – Biofuels milestones

B.C.E. (Before Common Era)

4000	Sumerians discover the process of fermentation.
10th century	Assyrians use biogas for healing bathing water.

C.E (Common Era)

17th century	Helmont observes that organic matter emits flammable gases.1808 Davy discovers methane as the end product of anaerobic digestion.
Mid-1800s	Transesterification of plant oils is used to distill glycerin during soap production
1858-1864	French biologist Antoine Bechamp experiments with fermentation and concludes that ferments are living organisms.
1864	French chemist Louis Pasteur describes the process of fermentation scientifically.
1875	First successful internal combustion engine is produced.
1880s	First successful internal combustion engine using producer gas is produced.
1892-1893	Rudolf diesel files a patent for a "Working Method and Design for Combustion Engines... a new efficient, thermal engine."
1895	Biogas is used to fuel street lamps in Exeter, Great Britain.

1897	First diesel engine suitable for practical use operates at an efficiency of 75 percent.
1908	Henry Ford's Model T is designed to run on ethanol.
1920-1930s	Attempts to promote ethanol as motor fuel are made. Anaerobic bacteria responsible for methane production are identified.

Source: Asean foundation, 2010

Production of Fuel Ethanol

Ethanol is one type of alcohol that has many properties quite similar to those of gasoline. These similarities make ethanol a highly attractive fuel for use as a gasoline substitute or as an alternative fuel for blending. The densities of ethanol and gasoline are almost identical although the energy content of ethanol is about 30% lowers. On the other hand, since ethanol contains molecular oxygen that promotes more complete combustion, the difference in energy content does not become a major concern at low level of alcohol blends in gasoline.

Ethanol can be produced by the fermentation of carbohydrates from various feedstocks. The feedstocks fall under three main categories: (a) sugar bearing feedstocks such as sugar cane; (b) starchy feedstocks such as cassava or corn; and (c) cellulosic feedstocks such as wood and agricultural residues such as bagasse. The production technologies for the first two types of feedstocks are fully mature and well-developed and are commercially available from various suppliers. However, the processing of cellulosic materials into ethanol using either acid or enzymatic hydrolysis is not yet fully developed commercially but is rapidly becoming technically and economically viable. Fig. 1.1 shows schematically a simplified flow diagram for the production of ethanol from these three different feedstocks.

Ethanol production from feedstock's

Depending on the type and nature of feedstock, the pre-processing steps or operations may differ but there are basically three processes that are common for all three types of feedstocks: (a) fermentation of the sugars into ethanol; (b) distillation to separate the aqueous ethanol (95%) from the fermented mash; and (c) dehydration to produce anhydrous ethanol (>99.5%) suitable for blending with gasoline. In the

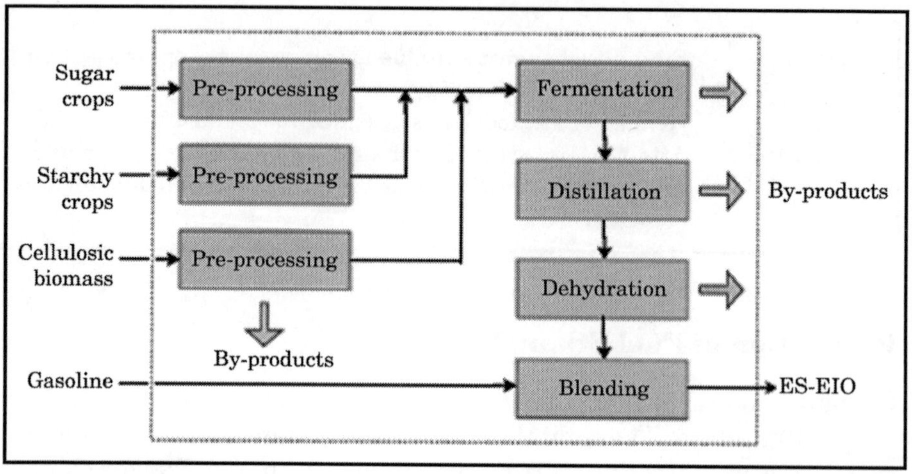

Fig. 1.1: *Source*: www.aboutbioscience.org/pdfs/biofuels.pdf 2006 North Carolina Ass for Bioved research

case of starchy and cellulosic feedstocks, pretreatment through saccharification or hydrolysis is required in order to convert them to sugars that can be fermented by yeast into ethanol (Giampietro *et al.*, 1997).

1940s	First U.S. ethanol plant opens.
1939–1945	Extensive use of blogas to replace gasoline occurs.
1979	Commercial alcohol-blended fuels are marketed.
1984	Number of ethanol plants peaks at 163 in the United States, Producing over 2.2 billion litres of ethanol during the year.
1988	Ethanol is used for first time as an oxygenate to lower pollution caused by burning gasoline.
1990	Ethanol plants begin to switch from coal to natural gas and to adopt other cost-reducing technologies. Ethanol plants are subsidized by the U.S. government to support farmers. Gasohol becomes commonly available in the U.S. Midwest.
1997–2002	Three million U.S. cars and light trucks that could run on E85, a blend of 85 percent ethanol and 15 percent gasoline, are produced but few gas stations sell the fuel. Concerns about climate change cause leading alternative energies such as biofuel, solar and wind to expand by 20 to 30 percent yearly.
2003	California becomes the first state to start replacing the oxygenate MTBE with ethanol. Several other states start switching soon afterward. California consumes 3.4 billion litres of ethanol a year- about a third of all ethanol produced in the United States.

2004	Crude oil prices rise by 80 percent, gasoline prices rise by 30 percent in the United States and diesel fuel prices rise by almost 50 percent. The U.S. ethanol industry makes 225,000 barrels per day in August, an all-time record. Oil companies invest heavily in alcohol fuel.
2005	E85 sells for less than gasoline on average in the United States. More than four million flexible-fuel vehicles (vehicles that run on E85 and gasoline) exist in the United States, mostly in the Midwest. Gasoline Prices rise as ethanol prices stay the same, due to rapidly growing ethanol supply and federal tax subsidies for ethanol.
2006	Indy Racing League switches to a 10 percent ethanol, 90 percent methanol fuel mixture.

Source: Issue Brief: Biofuels, North Caroline Association for Biomedical Research, October 2006. <*www.aboutbioscience. org/pdfs/Biofuels.pdf*>

Overview of History of Biofuels

Fermentation is a series of chemical reactions that convert sugars to ethanol in the presence of suitable strain of yeast, which feed on the sugars. Ethanol and carbon dioxide are produced as the sugar (glucose) is consumed. The simplified fermentation reaction for a 6-carbon sugar, glucose, is as follows:

$$C_6H_{12}O_6 = 2\ CH_3CH_2OH + 2\ CO_2$$

The ethanol concentration in the fermented mash can attain a level of only 4% to 10% depending on the operating conditions and the strain of yeast being used. Ethanol is therefore recovered through distillation but only hydrous ethanol of about 95-96% can be produced through steam distillation of the fermented mash due to the formation of water-ethanol azeotrope. The product ethanol is withdrawn from the top of the distillation column while the spent fermented mash called distillery slops is withdrawn from the bottom and sent for further treatment before final disposal or reuse. To make ethanol fully miscible with gasoline, it is necessary to further remove the residual water to produce anhydrous ethanol with a concentration of at least 99.5%. To attain this concentration, the hydrous ethanol has to undergo a suitable dehydration process or operation.

There are at least three widely-used dehydration processes to further remove water from the azeotropic ethanol-water solution. The first process that was used in many of the earlier ethanol plants is the so-called azeotropic distillation or ternary distillation process (as opposed to a binary or two component distillation process). It consists of introducing a third component, benzene or cyclohexane, to the azeotropic solution which forms a heterogeneous azeotropic mixture in vapor-liquid-liquid equilibrium. When this mixture is distilled, anhydrous ethanol is produced at the bottom of the distillation column. Another early method is called extractive distillation, which consists of adding a ternary component that increases the relative volatility of ethanol. In this case, anhydrous alcohol is produced and withdrawn at the top of the distillation column.

Because distillation processes are energy intensive, a third method has been developed and accepted in majority of modern ethanol plants. This process uses molecular sieves to remove water from ethanol. In this process, ethanol vapor under pressure passes through a bed of molecular sieve beads. The pores of the beads are such that they allow the absorption of water but not ethanol. Two beds are normally used in parallel to allow the regeneration of one bed while the other is in use. This dehydration technology can save significant amounts of energy (upto 840 kJ/l) compared to conventional azeotropic or extractive distillation processes.

For the production of ethanol from corn, there are two main processes in commercial use: dry milling and wet milling. In the dry milling process the entire corn kernel is ground into flour, which is referred to as "meal." The meal is then made into slurry by adding water. Enzymes are added to the mash to convert starch to dextrose, which is a simple sugar. Ammonia is added to control the pH and to provide nutrient for the yeast, which is added later. The mixture is processed at high temperatures to reduce the bacteria levels and transferred and cooled in fermentation tanks. This is where the yeast is added and conversion from sugar to ethanol and carbon dioxide takes place.

The entire process takes between 40 to 50 hours, during which time the fermenting mash is kept cool and agitated in order to facilitate yeast activity. After the process is complete, everything is transferred to distillation columns where the ethanol is removed from the stillage. The ethanol is dehydrated to about 99.5% using a molecular sieve system. A denaturant such as gasoline is added to the anhydrous ethanol to render the product not suitable for drinking. The remaining tillage undergoes a series of processes to produce feed for livestock. The carbon dioxide released from the process is collected and processed to produce industrial or food grade product.

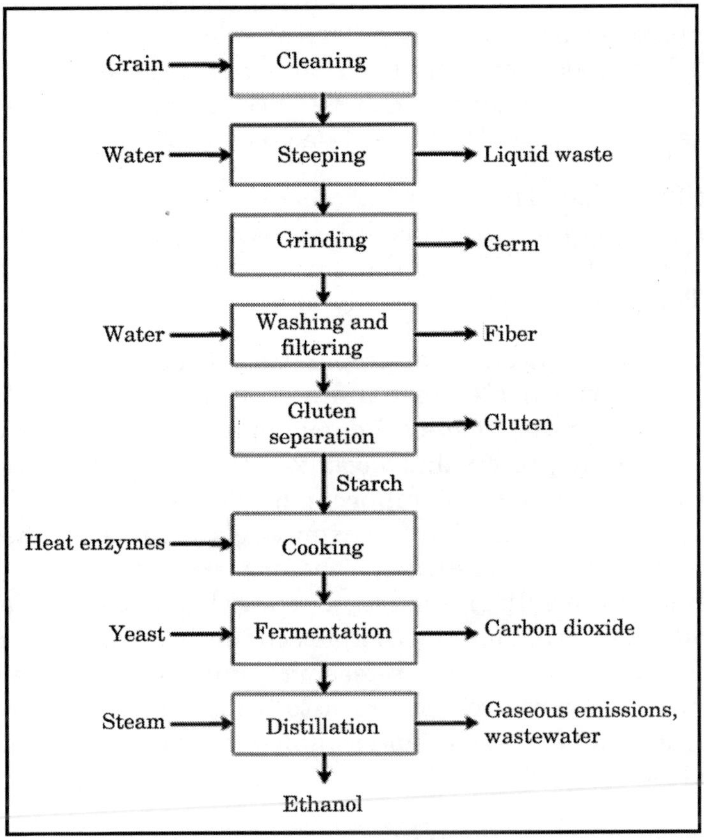

Fig. 1.2: *Source*: Uriarte, 2010

The process of wet milling takes the corn grain and steeps it in a dilute combination of sulfuric acid and water for 24 to 48 hours in order to separate the grain into many components. The slurry mix then goes through a series of grinders to separate out the corn germ. Corn oil is a by-product of this process and is extracted and sold. The remaining components of fiber, gluten and starch are segregated out using screen, hydroclonic and centrifugal separators (Pimentel and Patzek, 2005).

The gluten protein is dried and filtered to make a corn gluten meals coproduct, which is used as a feed ingredient for poultry broilers. The steeping liquor produced is concentrated and dried with the fiber and sold as corn gluten feed to the livestock industry. The heavy steep water is also sold as a feed ingredient. The starch and remaining water can then be processed one of three ways: (a) fermented into ethanol through a similar process as dry milling; (b) dried and sold as modified corn starch; or (c) made into corn syrup. This process is shown schematically in Fig. 1.2. The production of ethanol from corn using the wet mill process has become the technology of choice since it provides more product diversity and flexibility (Hazell and Pachauri, 2006).

To produce ethanol from cellulosic feedstocks, several pre-treatment steps are necessary to convert cellulose into simple sugars that can be converted into alcohol using the conventional yeast fermentation process. This is shown schematically in Fig. 1.3.

The first step is mechanical preparation through size reduction to make the material easier to handle and more susceptible to the subsequent pre-treatment steps. For example, agricultural residues go through a grinding process and wood goes through a chipping process to produce smaller and more uniform particle size. This is followed by acid pre-treatment. In this step, the hemicellulose fraction of the biomass is broken down into simple sugars. A chemical reaction called hydrolysis occurs when dilute sulfuric acid is mixed with the biomass feedstock. In this hydrolysis reaction, the complex chains of sugars that make up the hemicellulose are broken, releasing simple sugars. The complex hemicelluloses sugars are converted to a mix of soluble five-carbon sugars, xylose and arabinose, and soluble six-carbon sugars, mannose and galactose.

The next step is cellulose hydrolysis. In this step, the remaining cellulose is hydrolyzed to glucose. In this enzymatic hydrolysis reaction, cellulase enzymes are used to break the chains of sugars that make up the cellulose, releasing glucose (Carlsson, 2009).

Cellulose hydrolysis is also called cellulose saccharification because it produces sugars. The cellulase enzymes needed for this process may be produced in-house in a separate enzyme production facility or may be purchased from outside suppliers. The glucose is converted to ethanol through yeast fermentation, as discussed earlier. The yeast feeds on the sugars and as the sugars are consumed, ethanol and carbon dioxide are produced. The hemicellulose fraction of biomass is rich in five-carbon sugars, which are also called pentoses. Xylose is the most prevalent pentose released by the hemicellulose hydrolysis reaction. As glucose is converted to ethanol by yeast fermentation similar to those used in the first two types of feedstocks, the pentose (mainly xylose) is subjected to a different type of fermentation. In pentose fermentation, *Zymomonas mobilis* or other genetically engineered bacteria are used instead of yeast. Hydrous ethanol is recovered from the fermented mash through distillation and anhydrous ethanol is produced after dehydration, as in the previous processes described earlier. Lignin and other byproducts of the biomass-to-ethanol process can be used to produce the electricity required for the ethanol production process. Burning lignin actually creates more energy than needed and selling electricity to outside users improves the economic viability of the process (Wyrran, 1999).

Ethanol production from biomass

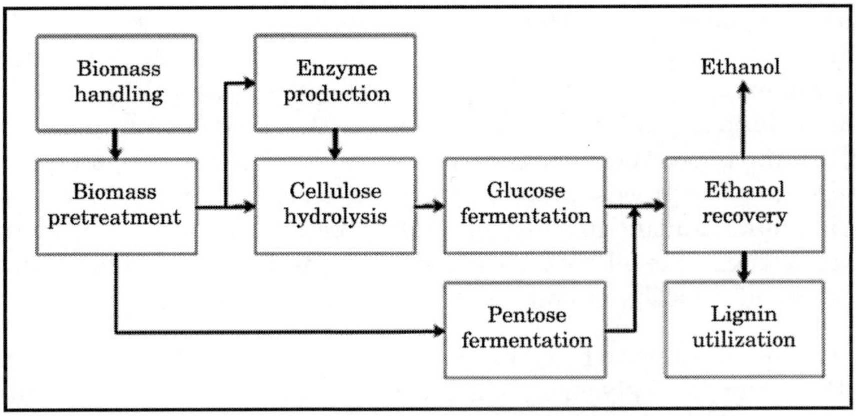

Fig. 1.3: *Souce*: (Uriarte, 2010)
Production of Biodiesel

Biodiesel is produced through a series of physical and chemical processing of the fruits, seeds, or parts of oil-containing plants. The first step usually involves the extraction of the crude oil. This step may involve several different types of operations or processes depending on the kind of feedstock. For example, the process of extracting crude oil from *Jatropha* may involve only one or two simple mechanical operations whereas the process of extracting oil from coconut may involve many. The detailed description of the various unit operations and processes involved in the extraction of crude oil from different types of plant oils such as palm oil, soybean oil, coconut oil and others is outside the scope of this book. However, the two main processes for extracting oil from seed feedstock are mechanical press extraction and solvent extraction. In mechanical press extraction, the oil seed feedstock (*e.g.*, *Jatropha* seed, Sunflower seed, etc.) is first heated to about 110°F. The oil seed is then crushed in a screw press. After most of the oil is removed, the remaining seed meal can be used as an animal feed. The solvent process extracts more of the oil contained in the oil seed feedstock but requires more costly equipment. The process uses a solvent to dissolve the oil. After extraction, a distillation process separates the oil from the solvent. The solvent condenses and can be recycled and reused in the process (Crabbe *et al.*, 2001).

Solvent extraction produces vegetable oil with a higher degree of purity than the mechanical press process. The central problem in using plant oil as diesel fuel is that plant oil is much more viscous (thicker) than conventional diesel fuel. It is 11 to 17 times thicker. Plant oil also has very different chemical properties and combustion characteristics to those of conventional diesel fuel. If the fuel is too thick it will not atomize properly when the fuel injectors spray it into the combustion chamber and it will not combust properly – the injectors may get choked up, leading to poor performance, higher exhaust emissions and reduced engine life. The process of transesterification reduces the high viscosity of plant oil, resulting in a higher quality fuel. In the transesterification process, vegetable oil reacts with alcohol (methanol or ethanol) in the presence of a catalyst (usually sodium hydroxide).

The oil molecules (triglycerides) are broken apart and reformed into methyl esters and glycerin, which are then separated from each other and purified. Biodiesel, which is a mixture of fatty acid alkyl esters, can be produced not only from plant oils, such as coconut oil, palm oil and soybean oil, but also from animal fats or recycled cooking oil and greases. Esters are compounds of alcohol and organic acids. Fatty acid methyl ester is made by bonding methanol to animal fat or plant oil. The process is relatively straightforward, but must consistently achieve prescribed standards to minimize the risk of damaging expensive diesel engines. Biodiesel can be used as a fuel for vehicles in its pure form, but it is usually used as a petroleum diesel blend to reduce levels of particulates, carbon monoxide, hydrocarbons and air toxins from diesel-powered vehicles. Because it is oxygenated, biodiesel dramatically reduces air toxins, carbon monoxide, soot, small particles, and hydrocarbon emissions by 50% or more, reducing the cancer-risk contribution of diesel up to 90% with pure biodiesel. Air quality benefits are roughly proportional for diesel/biodiesel mixtures. Biodiesel's superior lubricity helps reduce engine wear, even as a small percentage additive.

The most common use of biodiesel is as B20 (20% biodiesel, 80% diesel) and B2 (2% biodiesel, 98% diesel) or B1 (1% biodiesel, 99% diesel). The use of these blends requires no engine modifications. But because it gels at higher temperatures than petroleum diesel, however, pure biodiesel requires special management in cold climates. Biodiesel contains slightly less energy than petroleum diesel, but it is denser, so fuel economy tends to fall 7% for every 10% biodiesel in a fuel blend. When applied in its pure form, biodiesel has a higher flash point than fossil based diesels, and its viscosity results in a more intense pulverization, which is another disadvantage. Nevertheless, if properly treated, these shortfalls can be abated – and the advantages of biodiesel greatly make up for its disadvantages. It is nonhazardous, nontoxic and biodegradable, and it reduces air pollutants such as particulates, carbon monoxide, hydrocarbons and air toxins. It burns more efficiently than petroleum diesel, and its higher lubricity can reduce engine wear, prolonging its lifetime (Pimentel and Patzek, 2005).

Key definitions

Biomass –	shall refer to any organic matter, particularly cellulosic or ligno-cellulosic matter, which is available on a renewable or recurring basis, including trees, crops and associated residues, plant fiber, poultry litter and other animal wastes, industrial wastes, and the biodegradable component of solid waste.
Bioethanol –	shall refer to ethanol (C_2H_5OH) produced from feedstock and other biomass.
Biodiesel –	shall refer to Fatty Acid Methyl Ester (FAME) or mono-alkyl esters derived from vegetable oils or animal fats and other biomass-derived oils that shall be technically proven and approved by the Department of Energy (DOE) for use in diesel engines with quality specifications in accordance with the Philippine National Standards (PNS).
Bioethanol – fuel	shall refer to hydrous or anhydrous bioethanol suitably denatured for use as motor fuel, with quality specifications in accordance with the PNS.Biofuel – shall refer to bioethanol and biodiesel and other fuels made from biomass and primarily used for motive, thermal and power generation with quality specifications in accordance with the PNS.
Biofuel – blends	shall refer to gasoline or diesel that has been blended with biofuels such as, but not limited to, bioethanol and biodiesel.

Asean foundation, 2010

Environmental and Social Impacts

The combustion of fossil fuels such as coal, oil and natural gas has been identified as a major cause of the increase in the concentration of carbon dioxide in the earth's atmosphere. Carbon dioxide and other so-called greenhouse gases (*e.g.*, methane) allow solar energy to enter the earth's atmosphere but reduce the amount of energy that can be radiated back into space, thus trapping energy and causing global warming.

One environmental benefit of replacing fossil fuels with biomass-based fuels is that the energy obtained from biomass does not add to the level of carbon dioxide in the atmosphere that causes global warming. Although the combustion of all types of fuel, including fuels produced from biomass, releases carbon dioxide into the atmosphere, the burning of biomass does not cause any net increase in carbon dioxide concentration. The reason for this is that plants use carbon dioxide from the atmosphere to grow (photosynthesis) and the carbon dioxide formed during combustion is balanced by that absorbed during the annual growth of the plants used as the biomass feedstock. In the case of fossil fuels, the carbon content of these fuels has been fixed and contained in the earth's crust for millions of years. By burning these fossil fuels the formerly harmless carbon buried in the earth's crust is released into the atmosphere as carbon dioxide resulting in a net increase in the carbon concentration that leads to global warming. Although some fossil fuel-based energy may also be used to grow, produce and process the biomass feedstock, the general net result, based on a number of product life cycle studies, is still substantially reduced net greenhouse gas emissions. For example, although modern, high-yield corn production is relatively energy intensive, the net greenhouse gas emission reduction from making ethanol from corn grain can still be as much as 20%. If ethanol is produced from sugar cane where waste bagasse is used for steam and power production, the net reduction in greenhouse gas emission is even bigger. And making biodiesel from soybeans can reduce net emissions by nearly 80%. Finally, producing ethanol from cellulosic materials and using the non-fermentable lignin and other biomass for the production of steam, energy and electricity can reduce net carbon dioxide emission by close to 100% (Hazell and Pachauri, 2006).

Since petroleum diesel and gasoline consist of blends of hundreds of different chemicals of varying hydrocarbon chains, many of these are hazardous and toxic. They include volatile compounds such as benzene, toluene and xylene that are identified as responsible for the health hazards and pollution associated with the combustion of petroleum-based fuels. Carbon monoxide (produced when combustion is inefficient or incomplete), nitrogen oxides (produced when combustion occurs at very high temperatures), sulfur oxides (produced when elemental sulfur is present in the fuel), and particulates that are generally produced during combustion are other specific emissions of concern. An important environmental and health benefit of using biofuels as an additive to or replacement of petroleum-based transportation fuels is a reduction in these harmful emissions, in particular sulfur oxides and particulates. Since both ethanol and biodiesel contain oxygen in their chemical structure they can be used as fuel oxygenates to improve combustion characteristics. The presence of chemically-bound oxygen results in more complete combustion, which reduces carbon monoxide emissions. This is significant environmental and health benefit that can result from replacing petroleum fuels with biofuels. Ethanol is typically blended with gasoline to form an E10 blend (5%-10% ethanol and 90%-95% gasoline), but it can be used in higher concentrations such as E85 or in its pure form. Biodiesel is usually blended with petroleum diesel to form B1 blend (1% biodiesel, 99% petroleum diesel) or B20 blend (20% biodiesel and 80% petroleum diesel), although other blend levels can be used up to B100 (pure biodiesel) (Msangi *et al.*, 2006).

The global public debate about the use of ethanol as a world energy source has been going on for more than fifty years. On one hand, ethanol advocates have described ethanol as the "fuel of the future". But critics question the feasibility and advisability of large-scale ethanol production in terms ranging from "net petroleum energy balance" to more general environmental and agricultural impacts. The opponents of widespread use of ethanol as fuel question the use of limited edible resources to meet the world's energy needs. In the same manner, these opponents say that the need for cleaner-burning fuel sources for environmental and health reasons does not fully justify the use of edible plant oils such as soybean oil, palm oil and coconut oil. Many critics also argue that it is wrong to use grain for energy production instead of for food. In the final analysis, it is society that must decide whether the virtues and benefits of using ethanol or biodiesel as an energy source outweigh the potential negative consequences on available food supply of using grain or edible plant oils for biofuel. Society, in general, and the scientific community, in particular, must also resolve the debate about the energy balance of ethanol production, with different studies showing either a net gain or loss when the energy inputs to ethanol production are compared to the energy yield. The future of agricultural and land use policies is complicated by this emerging potential for large-scale bio-energy production. The growing demand for cleaner-burning fuels, such as ethanol, is likely to generate changes in agricultural cropping patterns and land management practices. This will have serious impacts on the environment, possibly threatening the natural resource base. Both biofuels proponents and critics question how production decisions and policy formulation and development surrounding the rapid expansion of the biofuel industry will affect the environment, especially water quality and soil quality. In the same manner, they ask whether the long-touted climate benefits arising from the displacement of petroleum oil use will actually be realized. The answers to these questions are not yet certain. What is certain is that while energy policy will have a significant impact on how the markets for biofuels will develop, agricultural and land use policies will also significantly influence how those markets will impact the environment and the people's health.

CONCLUSIONS

Traditionally production of biofuels had been sourced as from cellulosic biomass, sugary/starchy crops and oil containing plants. Cellulosic biomass is attractive source as it utilizes waste from various industrially important crops which are not rendered as food crops. So basically they do not competes with crops being used as source of food. Owing to their complex polymeric long chain structure, they store lot of chemical energy that can be harnessed as potential source of energy. Sugars are considered as source of food and are grown for same purposes but recently they are also being looked as source of energy and are being identified as excellent candidates. Further, oil producing plants can be easily produced for producing diesel substitutes or blends. As they are very viscous so can't be directly used for energy production. However, they could be easily blended with other fuels to produce energy.

Ethanol has found wide applications as an alternative fuel for blending with other fuels. It can be easily derived from cellulosic or sugary/starchy source of biomass and has found tremendous applications as potential source of energy in future. Further, Biodiesel can be generated from, oil generating plants and owing to its successive advantages, is finding applications in energy generation. Ethanol and biodiesel both are promising as biofuels for generating energy.

REFERENCES

Carlsson, A.S. (2009). *Plant oils as feedstock alternatives* to petroleum–a short survey of potential oil crop platforms. *Biochimie*, 91: 665–670.

Crabbe, E., Nolasco Hipolito, C., Kobayashi, G., Sonomoto, K. and Ishizaki, A. (2001). Biodiesel production from crude palm oil and evaluation of butanol extraction and fuel properties. *Process Biochemistry*, 37: 65–71.

Demirbas, A. (2001). Biomass resource facilities and biomass conversion processing for fuels and chemicals. *Energy Conversion and Management*, 42: 1357–1378.

Demirbas, A. (2003). Biodiesel fuels from vegetable oils via catalytic and non-catalytic supercritical alcohol transesterifications and other methods: A survey. *Energy Convers Manage*, 44: 2093–109.

Faaij, A. (2006). Modern biomass conversion technologies. *Mitigation and Adaptation Strategies for Global Change*, 11: 335–367.

Giampietro, M., Ulgiati, S. and Pimentel, D. (1997). "Feasibility of large-scale biofuel production". *BioScience*, 47(9): 587–600.

Hazell, P. and Pachauri, R.K. (2006). Bioenergy and Agriculture: Promises and Challenges. International Food Policy Research Institute (IFPRI), 2020 focus No 14.

Jain, S. and Sharma, M.P. (2010). Prospects of biodiesel from *Jatropha* in India: A review. *Renewable and Sustainable Energy Reviews*, 14: 763–771.

Msangi, S., Sulser, T., Rosegrant, M., Santos, R.W. and Ringleret, C. (2006). Global scenarios for biofuels: Impacts and implications, International food policy research institute (FPRI).

Pimentel, D. and Patzek, T. (2005). Ethanol production using corn, switchgrass and wood; biodiesel production using soybean and sunflower. *Natural Resources Research*, 14(1): 65–76.

Uriarte, F.A. (2010). Biofuels from Plant Oils, ASEAN Foundation, Jakarta, Indonesia.

www.asean foundation.org/documents/book/biofuel.pdf

Wyman, C.E. (1999). Biomass ethanol: Technical progress, opportunities, and commercial challenges. *Annual Review of Energy and the Environment*, 24(1): 189–226.

2

Biofuels: Perspective of Environment, Economy and Society

Anirban Biswas[1], Abhijit Sarkar[2], Saroni Biswas[3], Pooja Singh[4], Sanjeeb Mohapatra[5] and Rajeev Pratap Singh[4,6]*

ABSTRACT

With the initiation of civilization, mankind's primary fuel was certainly biofuels. The energy production from the biomass is as deep as civilization itself. Wood, dung and charcoal have been utilized as solid biofuel from the time when man discovered fire, and are still being used today around the world. Beside these solid biofuels, liquid biofuels such as - olive oil, whale oil, etc., have also been utilized ever since the ancient times like 1700s. However, till the early 19th century 'fossil fuel' for instance coal, oil, and natural gas which are produced from the residuals of plants

[1]School of Environmental Studies, Jadavpur University, Kolkata-700032, West Bengal, India

[2]Laboratory of Applied Stress Biology, Department of Botany, University of Gour Banga, Malda-732 103, West Bengal, India.

[3]Development Research and Communication Services Center, Kasba, Kolkata-700042, West Bengal, India

[4]Institute of Environment and Sustainable Development, Banaras Hindu University, Varanasi-221 005, India

[5]Centre for Environmental Science and Engineering, Indian Institute of Technology Bombay, Powai, Mumbai-400076, India

[6]Department of Civil Engineering, University of Nebraska Lincoln, Lincoln, United States of America

**Corresponding author*: E-mail: rajeevprataps@gmail.com

and animals fossilized in the earth for millions of years, have been referred as the main fuel sources. Most extensively recognized biofuel is corn or sugar based ethanol, generated in the course of fermentation of plant products. Worldwide the most widely used biofuel is bioethanol whose production augmented from 4.4–46.2 bbl between year 1980 and 2005. USA, Brazil and China are the major bioethanol producers (almost 75% global biofuels) worldwide. The present chapter mainly aims to discuss the environmental, economical and societal perspective of this major alternative energy resource.

***Key words*:** Biofuel, Environment, Society, Economy

1. INTRODUCTION

The term 'biofuels' refers to a range of fuels which are in some way derived from biomass or organic matter (Chhetri and Islam, 2008; Berktay and Nas, 2008; Konwer *et al.,* 2007; Najafpour *et al.,* 2006; Demirbas, 2007a,b,c; Demirbas, 2008; Ben-Iwo *et al.,* 2016). In the present world 'biofuels' are gaining economic and scientific interest owing to rising prices of gas and oil, the conventional fuels (Huang *et al.,* 2012). Ever since the early 19th century term 'fossil fuel' has been commonly used to refer the fuels for example coal, oil, and natural gas which are produced from the residuals of plants and animals that were buried deep in the earth for millions of years. In the 1970s, a new word, 'biofuel' started to be used to explain a different kind of fuel, one derived from more advanced organic matter. Most extensively documented biofuel is corn or sugar based ethanol, manufactured all the way through the fermentation of plant products (NREL, 2012). The sources of biofuel are organic and are reflected in the prefix, *bio*-meaning 'life' or 'living' organisms. The prefix *bio-* was borrowed from the Greek *bios,* denoting 'mode of life'. First known uses of biofuels were in 1970. The International Energy Agency's prediction: biofuels have the possibility to meet more than a quarter of world's requirement for transportation.

Mankind's primary fuels were biofuels. Energy production from the biomass is as old as the civilization itself. Wood, dung and charcoal have been utilizedas solid biofuel from the time when man discovered fire, and are still being used today around the world. Besides liquid biofuels (olive oil, whale oil, etc.) have been used since ancient times of 1700s (Bonner, 1989). In the 1900s, people started depending on the uses of fossil fuels, replacing the woods and in this way it helped to reduce the unsustainable use of timber resources at the time, which had drastically increased in price during the century (Wrigley, 1990).

Primarily transportation fuels were biofuels. The first (1826) patented internal combustion engine was designed to run on ethanol based fuel. In 1925, Henry Ford forecasted, '*The fuel of the future is going to come from fruit like that of sumac, or from apples, weeds, sawdust–almost anything. There is fuel in every bit of vegetable matter that can be fermented*' (Ford Predicts Fuel from Vegetation, N.Y. TIMES, 1925, p. 24). From the 1800s itself, the biofuels were commercially available on a large scale. Simultaneously coal became available, kerosene (liquid hydrocarbon fuel) was invented and commercial oil exploration was started. As a result of the availability and efficiency of the fossil fuels, appetite for biofuel diminished.

During World War-I, the fossil fuels shortage began, and so ethanol was in demand again and idea for blending the biofuels with gasoline became known (Fox, 1924; Wik, 1963). At the same time there were some issues related to oil exploration and export as, in 1973: embargo of OAPEC (Organization of Arab Petroleum Exporting Countries) in 1979. Revolution took place in Iran while in 1990 the Gulf war resultanting high oil prices (Barsky and Kilian, 2001). This led many countries to rethink about the production and uses of biofuels, especially in the countries like U.S. and Brazil. And for the last few years we are concerned about biofuels to save the money, and environment in terms of controlling the climate change phenomenon (FAO, 2008). In this way biofuel came into action rather it is better to say we rediscovered in our own interests.

2. BIOFUELS AND THE ENVIRONMENT

The decline in world's oil reserves and subsequent high prices as well as environmental impacts has resulted in the concerns for the worldwide production and uses of biofuels in recent years. Here lies the debate among nations, international organizations and political leaders to converse about the impacts of using biofuel and in continuous intensification. Till date biofuel production remains a small fraction of the total energy demand, also the potential environmental and social issues related to this should be recognized. The biofuels have been recognized as a cleaner energy source for transport sector. According to Puppa´n (2002), the environmental benefits of biofuels are seen during the combustion, their emissions of CO_2 correspond to the amount that was sequestrated from the atmosphere during the growth of the source plants, and so it creates a closed carbon cycle. Conventional fuel diesel is a dominant black smoke particulate emitter with SO_2, but there was an average decrease in concentrations of (14%) CO_2, (17.1%) CO and (22.5%) smoke after use of the biodiesel and many other studies support the trend of using biodiesel (Laforgia and Ardito, 1994; Cardone *et al.*, 1998; Nas and Berktay, 2007; Utlu, 2007). Vegetable oils also are now a days in discussion for their environmental benefits and as renewable energy resources; *e.g.*, olive oil has been used as an alternative source (Dorado *et al.*, 2003). The use of biodiesel reduces N_2O, and is attractive. Biodiesel and biodiesel blends reduce particulate matter (PM), hydrocarbons (HC) and carbon monoxide (CO) emissions and increase nitrogen oxides (NOx) emissions compared to diesel fuel even if used in an unmodified diesel engine (Tan *et al.*, 2004). Biofuel usage results in greater reduction of overall greenhouse gas emission (US-EPA, 2002; Farrell, 2006).

Biofuels are plant-derived energy sources, to be burned directly or be converted to a kind of liquid fuel such as ethanol or biodiesel. In North and Latin American countries the usage of biodiesel is expanding in recent past (Fernandes, 2007). Scientific world is in search to augment the production and efficiency of biofuels (US Department of Energy, 2006; Ragauskas *et al.*, 2006). These kinds of researches aims to (i) have better energy efficiency; (ii) have positive consequences on nutrient cycling in crop ecosystems; and (iii) necessitate minimal land conversion (US Department of Energy, 2006; Ragauskas *et al.*, 2006; Fargione *et al.*, 2008; Gurgel *et al.*, 2007). The inter-relationship of biofuel production, ecological impacts to land alteration, introduction of new crop species for biofuel production and even genetic modification of source plants are vital determinates of the sustainability of biofuels (Ragauskas *et al.*, 2006). To get the ultimate benefit from biofuel sources we need to consider the environmental cost of producing biofuels, sustainability relative to fossil fuels, holistic assessment of fuel production systems.

Adverse impacts on land, water, biodiversity and overall ecosystem are the side-effects of agricultural production of bio-energy crops. The impacts depend on the amount and process of biofuel production and most importantly the trade of the produce. Despite the benefits of using biofuels, their production and end use may have environmental impacts for instance the utilization of huge amounts of water, forests destruction, the decline in food production and the enhancement in soil degradation. Pimentel and Patzek (2005) reported unfavorable results for the manufacture of biodiesel out of sunflower and soybeans due to low agricultural yield and high energy consumption. The studies showed that depending on the crop type, degree of growth and harvest, there might be positive and or negative effects on the use of the soil and water quality (Wesseler, 2006). A study in Europe concluded that more than 20% land areas can be used for biodiesel producing crops (Hellmann and Verburg, 2008). For more production of biofuel crops, the impact on biodiversity has to be considered because there always remains chance of high losses of biodiversity (OFID/IIASA, 2009). Competition of land use for the production of food and fuel has been recently discussed by Rathmann *et al.* (2010). The swing from soybeans to sugarcane in Brazil and from wheat to corn in the USA in the earlier period could be noticed, resulting a rise in prices and decline in stock of these agricultural commodities (FAO, 2008; Eggen, 2006; Service, 2007; Leemans *et al.,* 1996; Turpin *et al.,* 2009; Sumathi *et al.,* 2008; Hazell and Pachauri, 2006; Mark and Pilgrim, 2011).

The increasing demand for biodiesel increases the food prices; like it accounted for 30% enhancement of global cereal prices between 2000-2007 and for nearly 40% increase in the price of maize as was estimated by International Food Policy Research Institute (IFPRI). Simultaneously IFPRI projects that in 2020, calorie availability will decline, child malnutrition will increase in the countries of Sub-Saharan Africa (Rosegrant, 2008). Increased food prices (Rathmann *et al.,* 2010) are likely to result in calorie deficits and malnutrition due to dietary shift or low consumption and that would be a crime against humanity (Ferrett, 2007).

On the other hand more energy crop production generates work and wealth for small farmers. But foremost crisis is distraction of food and feed crops to biofuel production, as income from biofuel production are often larger than the returns a farmer might get where the same crops sold for food, or for non-biofuel crops (Trostle, 2008). Furthermore, a FAO analysis found worldwide expenses on imported foodstuffs in 2007 increased by about 29% above the record of the preceding year. The mass of the increase was accounted for by growing prices of imported cereals and vegetable oils – commodity groups that feature greatly in biofuel production (FAO, 2008a). Decreased overall food consumption in terms of calories, as well as of other essential nutrients including protein, fat and micronutrients, can result in weight loss; impaired developmental, mental and physical growth in children; and either sub-clinical or clinical micronutrient deficiency in all age groups (FAO, 2008b).

Food versus fuel is broadly discussed (Hellegers *et al.,* 2008; Catic and Rujnic-Sokele, 2008). Environmental impact of biofuels has been estimated using indicators, like net energy values (Brehmer *et al.,* 2008; Su *et al.,* 2009), energy analysis (Talens *et al.,* 2007), carbon footprint (Holzman, 2008). Numerous studies relate the problem oriented approach for estimating the impact of biofuels on categories like climate change, eutrophication, acidification (Kim and Dale, 2005; Renouf *et al.,* 2008). Comparing the outcomes of these studies the question of sustainability of biofuels remains still unanswered.

Increasing biofuel production may also affect water quality as a result of the use of agrochemicals. Report says, bioethanol production in the USA consumes huge quantities of water use, usually from 3.5 to 6.0 gallons of water per gallon of ethanol produced (Dennis and Mark, 2006). Rathmann *et al.* (2010) considered biofuels as main driver of increasing feedstock prices. In future, the switch over to the 2nd and 3rd generation biofuels could at least relieve the side by problems of biofuel production and uses, but still these are in developing stage and need more effort to make these commercially available for application in next 10-20 years (OFID/IIASA, 2009).

3. ECONOMY AND THE BIOFUEL PRODUCTION, SCENARIO OF DEVELOPED AND DEVELOPING NATIONS

Worldwide the most common biofuel is bioethanol on account of which between 1980 and 2005, globally bioethanol production increased from 4.4 bbl to 46.2 bbl (Murray, 2005; Lichts, 2007) US (16.1 bbl), Brazil (16 bbl), and China (3.8 bbl) are the largest (75%) bioethanol producers in the world. In Brazil, large-scale biofuel production was pioneered as a consequence of oil crises in the 1970s (Goldemberg, 2008; Brehmer and Sanders, 2009) and to reduce the dependence on the foreign oil supplies, the military and government introduced mandatory ethanol-gasoline blending but on the other hand this initiative engaged billions of dollars in the activities of infrastructure of distilleries and promotion of biofuel based vehicles. Germany, France, USA and Italy are the leading producers of biodiesel (Balat, 2007; Grassi, 1999; Wiesenthal *et al.*, 2009). In Asia-Pacific, numerous biodiesel approaches are being developed to capitalize on the region's immense palm oil production capacity. The governments of Thailand, the Philippines, Singapore, Japan, Korea, China, and Australia are also aggressively pursuing energy policies intended at increasing the production or use of biodiesel.

The maximum amount of fossil fuels that can be replaced by the uses of biofuels, depends directly on the efficiency with which they can be produced (Malcaa and Freire, 2006). Worldwide bioethanol production basically come from the fermentation of corn (*Zea mays*), sugarcane (*Saccharum* spp.), or other starch-or sugar-rich crops, cellulosic materials (*e.g.*, switchgrass, willow trees, agricultural residues) (*e.g.*, wheat straws), or municipal solid wastes (*e.g.*, paper). Currently the rate of the bio based product is high enough in almost all the developed countries, considering this economy the cost of biofuels is almost three times greater than the conventional petro fuels although it is the costing based on pure economy without considering the environmental aspect, but the situation is completely different in the developing countries and prices of the biofuel are very near to the world market price of petroleum fuel (UN United Nations, 2006). It has been reported that plant size has a foremost effect on cost (Whims, 2002). The plant size can lessen operating costs by 15–20%, saving an additional $0.02–$0.03 per litre. Consequently, a large plant with production costs of $0.29 per litre may be saving $0.05–$0.06 per litre over a smaller plant (Whims, 2002). In the tropical warm countries, ethanol is derived mainly from sugarcane, which is cheaper than to produce from sugar beet, so in these countries (like India, Brazil) sugarcane based bioethanol is being produced. Other competent sugar producing countries such as Pakistan, Swaziland and Zimbabwe have production costs alike to that of Brazil (Dufey, 2006). Anhydrous ethanol, blendable with gasoline, is still fairly more expensive. Prices in India have turned down and are approaching the price of gasoline.

In the present world, production of biofuels offers the vision of new economy for rural people in oil importer and developing countries, because the production to uses life cycle of the biofuels creates jobs with special concern to protect environment (Demirbas, 2008). The biofuel economy, will be governed by the same forces that shaped the development of the petroleum economy over the past century and because of its environmental merits, the share of biofuel in petro-market will be growing in the next decades. To fix the cost of biofuels, the cost of raw crop is a major component and with this co-product credit, chemicals, labor, maintenance, insurance and taxes would be the main driving forces for pricing.

Biodiesel production can be controlled and may turn into a large-scale production centre in a cost-effective manner, with safe handling and storage. In the last 2 years with the implementation of 5–20% biodiesel petroleum diesel blend in developed and developing countries; researches are being conducted globally to improve the method of biofuel production in more cost effective sustainable ways such as uses of enzyme based technology (Bloom *et al.*, 2005), fermentation etc. Which (Yazdani *et al.,* 2007) will favor the reduction of production cost.

4. SOCIETAL PERSPECTIVE OF BIOFUEL PRODUCTION AND UTILIZATION

There are basic environmental and 'good' society assumptions that through photosynthesis, the biofuel plants can remove greenhouse gases and can reduce fossil fuel consumption. But in reality cutting of plants for biofuel production is a problem of mass deforestation *e.g.*, one ton of palm oil generates 33 tons of CO_2 emissions-10 times higher than petroleum. Tropical forests cleared for sugarcane ethanol emit 50% more greenhouse gases than the production and use of the same amount of gasoline. Our general assumption has been that biofuels will bring rural development. But the reality is different in the tropics, 100 hectares of the land used in family farming generates 35 jobs. Whereas plantation and uses of oil-palm and sugarcane generates 10 jobs, specifically the eucalyptus generates two, and soybeans can generate a half job, if all these are planted in the same 100 hectares of lands.

Biofuels are available in the regional and sub-regional market, but their utilization is less, *e.g.*, in the United States also only few farmers have access to the biofuels, as produced in their farms. But things are changing rapidly and a big market is waiting worldwide. But for this large scale business there are some drawbacks also, the producers will be dependent on the high yielding seeds provided by the companies and will have to sell the produce to the company. Small producers will be kicked off from the market, the example can be seen in case of some countries like Brazil, Argentina, Bolivia and Paraguay. Hundreds to thousands of farmers have already been displaced by the soybean plantations in the 'Republic of Soy' a 50-million hectare area in southern Brazil, northern Argentina, Paraguay and eastern Bolivia.

Biofuels also meet an important social issue like hunger and poverty. Hunger results not from scarcity, but poverty. People suffer for want of food when the food prices become higher and higher as a family already spends 50 to 80 percent of household income on food. Now, because food and fuel crops compete for land and resources, both increase the price of land and water. There is need of limit, standard based on limiting land, and market regulation power for the biofuels industry. The North should not shift the burden of overutilization to the South as the tropics have more sunlight, rain and arable land. If biofuels are to be forest- and food-friendly, the biofuel industries need to be regulated, and not disorganized. Sustainable benefits to the poor will be accrued when biofuels will be a complement to the sustainable rural development. Globally the use and expansion of biofuels is needed to develop through regulatory structures and foster conservation and development alternatives. We need the time to make a better transition to food and fuel.

5. BIOFUEL: A MYTH OR A NEED

The biofuels are renewable, clean, green, sustainable and technologically superior so industries, politicians, the World Bank, the United Nations and even the International Panel on Climate Change (IPCC) present these fuels as the next step to fossil fuels and for better economy. The production and uses of biofuels play with the political-economic relationships between land, people, resources and food, but fail to understand the consequences of the industrial revolution of our food and fuel systems. Industrialized countries started the biofuels boom by challenging renewable-fuel targets. But to meet these, a country needs to change a huge percent of its farmland with fuel crops like, the entire corn and soy harvest of the United States would need to be processed as ethanol and biodiesel. Converting most arable land to fuel crops would destroy the food systems and to cope up with the situation one country need to be dependent on the others for most wanted foods and here may began the North and the South debate. Over the past few years, capital investment in biofuels has increased. Private stock is flooding the public research institutions. Behind the scenes, giant oil, grain, auto and genetic engineering corporations are forming partnerships by combining the research, production, processing and distribution chains of food and fuel systems under one industry, under the close watch of most national antitrust laws. Biofuel title holders assure that as the fuel crops are renewable, they are environment-friendly, reduce global warming and will nurture rural development. But the incredible market power of biofuel corporations, coupled with the poor political will of governments to control their activities, make this doubtful. It is assumed and supported that the fuel crops planted on ecologically degraded lands will advance the ecological health and environment. Considering this idea, the government of Brazil opted for agro fuel crop production in some 200 million hectares of dry-tropical forests, grassland and marshes as those were declared as degraded. In reality, these are the bio diverse ecosystems of the Atlantic Forest, the Cerrado and the Pantanal, with the presence of indigenous people, subsistence farmers and extensive cattle ranches. The beginning of agro fuel plantations will push these communities to the agricultural front line of the Amazon where the devastating patterns of deforestation are well known.

There is a myth that ethanol takes more energy to produce than it provides in the tank, but according to scientists at Argonne National Laboratory, today's ethanol production provides roughly 50 percent more useful energy per unit of fossil energy input than gasoline. Cellulosic ethanol is expected to provide a 10:1 return on fossil fuel investment.

Biofuels production is not responsible for the global food crisis, because globally the primary food crops are rice and wheat, very little of which are used for biofuels and the shortage of these tow cereal crops are due to increasing demand not for substitute cultivation (web link-1). Sustainable biofuel production is needed for global energy security and reducing dependence on fossil fuels, to cut short the greenhouse gas emissions, and for economic growth, so biofuels may be considered as the global future.

Biofuels are not causing a rise in food prices. According to the White House Council of Economic Advisors, production of ethanol was responsible for just 3% of the total 43% rise in world food prices in 2007 (web link-2). Increase of energy prices are the largest driver for higher food prices, but also growth and demand in civil life, drought, population growth, growing demand for animal feed in the countries like India and China, and finally political instability are the primary causes of recent food price increases worldwide (web link-3; web link-4)· Again the energy and labor are the dominant food costs to consumers (web link-5). Biotechnology is playing role to increase the fuel crop productions, higher yields mean more grain for food and fuel.

The myth says that there's not enough land to produce food, feed and biofuels, but the demand of biofuels can be meet up without reducing production of food and feed. Today's biofuels production represents less than 5 percent of the transportation fuel supply (web link-6)· Very less amount of arable land (<1%) is used for the production of feedstocks for biofuels (web link-7). Moreover the bio-energy crops can provide food/ feed, fuels, and other high-value co-products from the same crop, and thus highest possible use of the land can be ensured. Agricultural biotechnology also plays role in increasing the biofuel yields, production and efficient uses of the lands (web link-8). Many energy crops nurture well in poorer soils, and can be planted on less productive land, building soil and sequestering carbon in the process.

One of the important debates is that biofuels production is destroying the rainforest. But in reality, one percent of arable land in Brazil has been used for sugarcane plantation. Importantly sugarcane for ethanol is not grown in the Amazon rainforest and moreover the Brazilian government has laws in place to protect deforestation of the Amazon rainforest (web link-9). Some people critics that production of biofuels – by driving up grain prices – is making way to grow soybeans, and therefore indirectly causing rainforest destruction and making way for creating the agricultural land. It has also been reported that Brazilian rainforest has been cleared for complex socio-economic reasons, and that is no way related to ethanol production. Researchers suggest that sustainable biofuels production must go hand-in-hand with sustainable land use policy.

Biofuels does not increase greenhouse gas emissions, the vast majority of research from academia, NGOs, and federal labs have suggested that biofuels have a positive and increasingly beneficial impact on climate (web link-10; web link-11; web link-12; web link-13; web link-14). Corn starch ethanol reduces GHG emissions up to 30 percent compared to gasoline, according to Argonne National Lab. Cellulosic ethanol is expected to reduce emissions by 85 percent or more many non-food feedstock crops continuously sequester carbon even as above-ground biomass is harvested (IPCC, 1996).

The myth is that biofuels consume more water than regular crops. Biotechnology researches are developing drought-resistant fuel crop varieties, which reduce water requirement. Some non-food feedstock crops are highly drought tolerant; require little or no water inputs. One gallon of ethanol requires approximately 3 gallons of water to produce today so from scientific point of view till date it is worth production (web link-15). One interesting aspect about biofuel crop cultivation is that it does not require massive fertilizer input compared to other regular crops. Particularly in case of corn, an important biofuel crop, biotechnology has helped the farmers to become efficient in their fertilizer, pesticide and insecticide applications. Nitrogen and phosphate applications have declined (30%) in last 30 years. Proper agronomic practices like conservation tillage have substantially reduced runoff of all inputs (Source: National Corn Growers Association) (web link-16). On the other hand many non-food feedstock crop varieties does not need of any fertilizer or pesticide inputs.

There are some controversies about ethanol subsidies, as these are waste of taxpayer dollars, and as the biofuel can supply only up to 10% of the transportation fuel, so there is little need to be concerned about this. But the ethanol industry generated an estimated $4.6 billion in tax revenue for the federal government in 2007, or $1.35 for every dollar of federal investment (Source: Renewable Fuels Association) (web link-17). Merrill Lynch estimates that U.S. gasoline prices would be 15 percent–roughly 60 cents a gallon – higher without current ethanol production (web link-18). According to the Natural Resources Defense Council, the United States can produce enough sustainable biofuels to supply half of U.S. transportation fuel needs - 100 percent if vehicle fuel economy doubles (web link-19).

The growth of biofuels around the globe is spurred largely by energy security and environmental concerns and a wide range of market mechanisms, incentives and subsidies have been put in place to facilitate their growth. Developing countries, apart from these considerations, also view biofuels as a potential means to stimulate rural development and create employment opportunities. The Indian approach to biofuels, in particular, is somewhat different to the current international approaches which could lead to conflict with food security. It is based solely on non-food feed stocks to be raised on degraded or wastelands that are not suited to agriculture, thus avoiding a possible conflict of fuel *vs.* food security (National Policy on Biofuel, India). India has been spending on research on biofuel technologies for more than a decade through Council of Industrial & Scientific Research (CSIR), Ministry of New & Renewable Energy (MNRE), Department of Science &

Technology (DST) and Department of Biotechnology (DBT). It was however DBT that took a leap in 2007. Shortly after the USA set up its three famous bio-energy research centres Joint Bioenergy Institute, (JBEI), Bioenergy Science Centre (BESC), Great Lakes Bioenergy Research Centre (GLBRC)), India through DBT set about establishing its own centres for dedicated bioenergy and biofuel related research and development. Thereafter, MNRE and DST also set up a Centre each at Kapurthala and Thiruvananthapuram respectively (Lali, 2016).

6. CONCLUSIONS

The energy strategy of a country aims at efficiency and security and to provide access which being environment friendly and ensures achievement of an optimum mix of primary resources for energy generation. The World Bank, the United Nations, International Panel on Climate Change (IPCC), national industries and countries' policies project biofuel as the next step to fossil fuels towards a better economy and sustainable future as the modern day biofuel is renewable, clean, green, sustainable technological products. However, till date the production and uses of biofuels mainly depends on the political-economic relationships between land, people, resources and food, but fail to understand the consequences of the industrial conversion of our food and fuel. Industrialized countries initiated the biofuels boom by demanding renewable-fuel targets. But to meet this target, any country needs to change a huge percent of its farmland with fuel crops. Most of the countries need to change their fuel policy for ethanol or biodiesel production. Converting most arable land to fuel crops would destroy the food systems and to cope up with the situation one country need to be dependent on the others for foods. The concept of biofuel and the reality of biofuel are still facing a great disparity worldwide. Despite all the problems, worldwide biofuels has become a today's need for better human life, better environment.

7. ACKNOWLEDGEMENTS

AB acknowledges the support of DST in the form of research grant for N-PDF. The authors would like to thank the HOD, Department of Botany, University of Gour Banga, and the Head and Dean, Department of Environment and Sustainable Development and Director, Institute of Environment and Sustainable Development, Banaras Hindu University for providing necessary facilities.

8. REFERENCES

Balat, M. 2007. An overview of biofuels and policies in the European Union. *Energy Sources Part,* B2: 167–81.

Barsky, R.B. and Kilian, L. (2001). Do We Really Know that Oil Caused the Great Stagflation? A Monetary Alternative', *In*: Bernanke, B.S. and Rogoff, K. (*eds.*), NBER Macroeconomics Annual 2001. Cambridge, MA: MIT Press.

Ben-Iwo, J., Manovic, V. and Longhurst, P. (2016). Biomass resources and biofuels potential for the production of transportation fuels in Nigeria. *Renewable and Sustainable Energy Reviews,* 63: 172–192.

Berktay, A. and Nas, B. (2008). Biogas production and utilization potential of wastewater treatment sludge. *Energy Sources Part A,* 30: 179–88.

Bonner, N. (1989). Whales of the world, Facts on File, NY, p. 191.

Brehmer, B. and Sanders, J. (2009). Assessing the current Brazilian sugarcane industry and directing developments for maximum fossil fuel mitigation for the international petrochemical market. *Biofuels, Bioproducts and Biorefining* 3: 347–360.

Brehmer, B., Struik, P.C. and Sanders, J. (2008). Using an energetic and exergetic life cycle analysis to assess the best application of legumes within a biobased economy. *Biomass and Bioenergy,* 32: 1175–1186.

Cardone, M., Prati, M.V., Rocco, V. and Senatore, A. (1998). Experimental analysis of performances and emissions of diesel engines fuelled with biodiesel and diesel oil blends. *In*: Proceedings of MIS–MAC V, Roma; pp. 211–225.

Catic, I. and Rujnic-Sokele, M. (2008). Agriculture products – food for living beings or for machinery. *Gummi Fasern Kunststoffe,* 61(11): 701–708.

Chhetri, A.B. and Islam, M.R. (2008). Towards producing a truly green biodiesel. *Energy Sources Part A,* 30: 754–64.

Demirbas, A. (2007a). Production of biofuels from macroalgae and microalgae. *Energy Edu. Sci. Technol.,* 18: 59–65.

Demirbas, A. (2007b). Converting biomass derived synthetic gas to fuels *via* Fisher–Tropsch synthesis. *Energy Sources Part A,* 29: 507–12.

Demirbas, A. (2007c). Producing and using bioethanol as an automotive fuel. *Energy Sources Part B*, 2: 391–401.

Demirbas, A. (2008). Biofuels sources, biofuel policy, biofuel economy and global biofuel projections. *Energy Conserv Manage*, 49: 2106–16.

Dennis, K. and Mark, M. (2006). Water use by ethanol plants: Potential challenges. Institute for Agriculture and Trade Policy, *http://www.iatp.org/iatp/publications.cfm?accountID¼258&refID¼89449*.

Dorado, M.P., Ballesteros, E.A., Arnal, J.M., Gomez, J. and Lopez, F.J. (2003). Exhaust emissions from a diesel engine fueled with transesterified waste olive oil. *Fuel,* 82: 1311–5.

Dufey, A. (2006). Biofuels production, trade and sustainable development: Emerging issues environmental economics programme, sustainable markets discussion paper no. 2. International Institute for Environment and Development (IIED), London.

Eggen, D. (2006). Corn farmers smile as ethanol prices rise, but experts on food supplies worry. The New York Times.

EPA (US Environmental Protection Agency) (2002). A comprehensive analysis of biodiesel impacts on exhaust emissions. Draft technical report, EPA420-P-02-001.

FAO (Food and Agriculture Organization) (2008). Second FAO technical consultation on bioenergy and food security. Summary proceedings, Roma.

Fargione, J. *et al.* (2008). Land clearing and the biofuel carbon debt. *Science*, 319: 1235–1238.

Farrell, A.E. (2006). Ethanol can contribute to energy and environmental goals. *Science,* 311: 506–508.

Fernandes, S.D. (2007). Global biofuel use, 1850–2000. *Global Biogeochem Cycles 21:* GB2019 DOI: 10.1029/2006GB002836

Ferrett, G. (2007). Biofuels 'crime against humanity. BBC News.

Food and Agriculture Organization of the United Nations (2008a). The state of food and agriculture. Rome: FAO. *<ftp://ftp.fao.org/docrep/fao/011/i0100e/i0100e.pdf>* Available online 14.12.09.

Food and Agriculture Organization of the United Nations (2008b).The state of food insecurity in the world. Rome: FAO. *<ftp://ftp.fao.org/docrep/fao/011/i0291e/i0291e00.pdf>* Available online 14.12.09.

Fox, H.S. (1924). Alcohol Motor Fuels. *In*: Supplementary Report to World Trade in Gasoline. Bureau of Domestic and Foreign Commerce, U.S. Dept. of Commerce Monograph, Trade Promotion Series No. 20., Government Printing Office, Washington, D.C.

Goldemberg, J. (2008). The Brazilian biofuels industry. *Biotechnology for Biofuels,* 1(6): 1–7.

Grassi, G. (1999). Modern bioenergy in the European Union. *Renew Energy,* 16: 985–90.

Gurgel, A. *et al.* (2007). Potential land use implications of a global biofuel industry. *J. Agricult. Food Ind. Org.*, Article 9 5(2). (available at: *http://www.bepress.com/jafio/vol5/iss2/art9*).

Hazell, P. and Pachauri, R.K. (2006). Bioenergy and agriculture: Promises and challenges: 2020 Vision for food, agriculture and the environment. International Food Policy Research Institute.

Hellegers, P., Zilbermann, D., Stedto, P. and McCornick, P. (2008). Interactions between water, energy, food and environment: Evolving perspectives and policy issues. *Water Policy,* 10: 1–10.

Hellmann, F. and Verburg, P.H. (2008). Spatially explicit modelling of biofuel crops in Europe. *Biomass and Bioenergy,* doi:10.1016/j.biombioe.2008.09.003.

Holzman, D.C. (2008). The carbon footprint of biofuels can we shrink it down to size in time? *Environmental Health Perspectives,* 6: 116.

Huang, D., Zhou, H. and Lin, L. (2012). Biodiesel: An Alternative to Conventional Fuel. *Energy Procedia,* 16: 1874–1885.

IPCC (1997). Greenhouse Gas Inventory Reference Manual: Revised 1996 IPCC Guidelines for National Greenhouse Gas Inventories, Report vol. 3, Intergovernmental Panel on Climate Change (IPCC), Paris, France, pp. 1-53. *<www.ipcc.ch/pub/guide.htm>*.

Kim, S. and Dale, B.E. (2005). Life cycle assessment of various cropping systems utilized for producing biofuels: Bioethanol and biodiesel. *Biomass and Bioenergy,* 29: 426–439.

Konwer, D., Kataki, R. and Saikia, P. (2007). Pyrolysis of some indigenous tree species of northeast India: Effect of pyrolysis temperature and heating rate on the products yield and char quality. *Energy Sources Part A,* 29: 1433–42.

Laforgia, D. and Ardito, V. (1994). Biodiesel fuelled IDI engines: Performances, emissions and heat release investigation. *Bioresource Technology,* 51: 53–9.

Lali, A. (2016). Biofuels for India: What, when and how. *Current Science,* 110(4): 552–555.

Leemans, R., Amstel, A., Battjes, C. and Kreileman, E. (1996). The land cover and carbon cycle consequences of large-scale utilizations of biomass as an energy source. *Global Environmental Change* 4.

Lichts, F.O. (2007, 2008). World Ethanol and Biofuel Market Report.

Malcaa, J. and Freire, F. (2006). Renewability and life-cycle energy efficiency of bioethanol and bio-ethyl tertiary butyl ether (bioETBE): Assessing the implications of allocation. *Energy,* 31: 3362–80.

Mark, H. and Pilgrim, S. (2011). The new competition for land: Food, energy, and climate change. *Food Policy,* 36: S40–S51.

Murray, D. (2005). Rising oil prices will affect food supplies. People and Planet.

Najafpour, G., Younesi, H. and Mohamed, A.R. (2006). A survey on various carbon sources for biological hydrogen production *via* the water–gas reaction using a photosynthetic bacterium (*Rhodospirillum rubrum*). *Energy Sources Part A,* 28: 1013–26.

Nas, B. and Berktay, A. (2007). Energy potential of biodiesel generated from waste cooking oil: An environmental approach. *Energy Sources Part B,* 2: 63–71.

National Policy on Biofuel, India, Government of India Ministry of New & Renewable Energy, Block No. 14, C.G.O. Complex Lodhi Road New Delhi-110003, pp. 1–18, *http://mnre.gov.in/file-manager/UserFiles/biofuel_policy.pdf.*

National Renewable Energy Laboratory (NREL) (2012). Biofuels Basics

OFID/IIASA.Biofuels and food security; 2009.

Pimentel, D. and Patzek, T.W. (2005). Ethanol production using corn, switch-Grass, and wood; biodiesel production using soybean, and sunflower. *Natural Resources Research,* 14(1): 65–76.

Puppa´n, D. (2002). Environmental evaluation of biofuels. *Periodica Polytechnica Series, Society Man Science,* 10: 95–116.

Ragauskas, A.J. *et al.* (2006). The path forward for biofuels and biomaterials. *Science,* 311: 484–489

Rathmann, R., Szklo, A. and Schaeffer, R. (2010). Land use competition for production of food and liquid biofuels: An analysis of the arguments in the current debate. *Renewable Energy,* 35: 14–22.

Renouf, M.A., Wegener, M.K. and Nielsen, L.K. (2008). An environmental life cycle assessment comparing Australian sugarcane with US corn and UK sugar beet as producers of sugars for fermentation. *Biomass and Bioenergy,* 32: 1144–1155.

Rosegrant, M.W. (2008). Biofuels and grain prices: Impacts and policy responses. Statement prepared for presentation to the committee on homeland security and governmental affairs, US Senate. <*http://www.ifpri.org/blog/biofuels-and-grain-prices-impacts-and-policy-responses*> Available online 14.12.09.

Service, R. (2007). Biofuel researchers prepare to reap a new harvest. *Science,* 31.

Su Ch, L. and Lee, Y.M. (2009). Development status and life cycle inventory analysis of biofuels in Taiwan. *Energy Policy,* 37: 754–758.

Sumathi, S., Chai, S.P. and Mohamed, A.R. (2008). Utilization of oil palm as a source of renewable energy in Malaysia. *Renewable & Sustainable Energy Reviews,* 12: 2404–21.

Talens, L., Villablba, G. and Gabarrell, X. (2007). Exergy analysis applied to biodiesel production. *Resources Conversion and Recycling,* 51: 397–407.

Tan, R.R., Culaba, A.B. and Purvis, M.R.I. (2004). Carbon balance implications of coconut biodiesel utilization in the Philippine automotive transport sector. *Biomass Bioenergy,* 26: 579–85.

Trostle, R. (2008). Global agricultural supply and demand: Factors contributing to the recent increase in food commodity prices, WRS 0801. Washington, DC: US Department of Agriculture, Economic Research Service. <*http://www.ers.usda.gov/publications/wrs0801/wrs0801.pdf*> Available online 14.12.09.

Turpin, N., Dupraz, P., Thenail, C., Joannon, A., Baudry, J. and Herviou, S. (2009). Shaping the landscape: agricultural policies and local biodiversity schemes. *Land Use Policy,* 26: 273e83.

UN (United Nations) (2006). The emerging biofuels market: Regulatory, trade and development implications. *In*: United Nations conference on trade and development, New York and Geneva.

US Department of Energy (2006). Breaking the biological barriers to cellulosic ethanol: A joint research agenda. DOE/SC-0095, US Department of Energy Office of Science and Office of Energy Efficiency and Renewable Energy (*http://www.doegenomestolife.org/biofuels/*).

Utlu, Z. (2007). Evaluation of biodiesel fuel obtained from waste cooking oil. *Energy Sources Part A,* 29: 1295–304.

Web link-1 *http://www.transportation.anl.gov/pdfs/TA/271.pdf*

Web link-10 *www.unica.com.br/download.asp?mmdCode=407ED204-2194-4AF3-8652-538632DF16B8.*

Web link-11 *http://www.nrdc.org/air/energy/biofuels/contents.asp.*

Web link-12 *http://www.pnas.org/cgi/content/abstract/105/2/464.*

Web link-13 *http://www.esajournals.org/perlserv/?request=get-abstract&doi=10.1890%2F05-2018&ct=1*

Web link-14 *http://www.sciencemag.org/cgi/content/short/311/5760/506.*

Web link-15 *http://www.ncga.com/growersresources/PDFs/TopTenMythsCorn_Ethanol.pdf.*

Web link-16 *http://ct.water.usgs.gov/education/trivia.htm*

Web link-17 *http://www.ncga.com/conservation/pdf/2007/UsingBiotechnologyEconomicEnviromentalSustainability.pdf*

Web link-18 *http://www.ncga.com/ethanol/pdfs/031506Urbanchuk.pdf.*

Web link-19 *http://www.nrdc.org/air/energy/biofuels/contents.asp*

Web link-3 *www.ers.usda.gov/AmberWaves/February08/Features/CornPrices.htm.*

Web link-4 *http://www.kc.frb.org/regionalaffairs/mainstreet/MSE_0108.pdf*

Web link-5 *www.ers.usda.gov/Briefing/CPIFoodAndExpenditures/howchangesininputcostsaffectfoodprices.htm*

Web link-6 *http://online.wsj.com/article/SB120631198956758087.html*

Web link-7 *http://www1.eere.energy.gov/biomass/pdfs/final_billionton_vision_report2.pdf*

Web link-8 *http://www.iea.org/Textbase/techno/essentials2.pdf*

Web link-9 *http://www.liebertonline.com/doi/pdfplus/10.1089/ind.2008.030*

Web link-2 *www.ncga.com/news/publications/PDF/Using%20Technology%20to%20Feed%20%20Fuel%20the%20World*

Wesseler, J. (2006). Communication: opportunities' (costs) matter: A comment on Pimentel and Patzek "Ethanol production using corn, switchgrass, and wood; biodiesel production using soybean and sunflower". *Energy Police,* 35: 1414–16.

Whims, J. (2002). Corn based ethanol costs and margins. Attachment 1, AGMRC, Kansas State University.

Wiesenthal, T., Leduc, G., Christidis, P., Schade, B., Pelkmans, L. and Govaerts, L. (2009). Biofuel support policies in Europe: Lessons learnt for the long way ahead. *Renew. Sust. Energy Rev.,* 13: 789–800.

Wik, R.M. (1963). Henry ford's science and technology for rural America. Technology and Culture, Summer. pp. 247–258.

Wrigley, E.A. (1990). *Continuity, Chance and Change*. Cambridge University Press, Cambridge, U.K.

WWF.WWF–position on biofuels in the EU; 2007. Available in: *http://www.wwf.org.* Acessed: 15/05/2007.

Yazdani, S.S. and Gonzalez, R. (2007). Anaerobic fermentation of glycerol: A path to economic viability for the biofuels industry. *Curr. Opin. Biotechnol.,* 18: 213–219.

3

Biofuels and Their Impact Assessment

Arti Kumari[1], Razi Ahmad[1], Amrik Bhattacharya[1], Shivani Chaturvedi[1] and Sunil Kumar Khare[1]

ABSTRACT

In India, most of GHG emissions are accounted by transport sectors and these figures are growing faster. Now, it is clearly substantiate that emission of GHG, such as carbon dioxide, nitrous oxide and methane rising from fossil fuel combustion. Transportation is the main driving force which accelerated globally demand of oil as it totally relay on oil. One attractive alternative for this fossil based oil is biofuels. It is fuel obtained from microorganism and plant materials. In this chapter we are focusing on the various feedstocks that can be used for the production of biofuels, and the key policy and issues addressing the development of biofuels.

***Key words*:** Biofuel, Greenhouse gas, Feed stock, Biomass, Photobiological Conversion and Biofuel Policy

[1]Enzyme and Microbial Biochemistry Lab, Department of Chemistry, Indian Institute of Technology, New Delhi-110016

**Corresponding author*: E-mail: shivani.d123@gmail.com

1. INTRODUCTION

Current time has been declared as anthropogenic epoch by International Geological Congress in August 2016. This term is given due to increased pollution by human activities and this is very likely due to greenhouse gas (GHG) emissions. In India, most of GHG emissions are accounted by transport sectors and these figures are growing faster. Now, it is clearly substantiate that emissions of GHG, such as carbon dioxide (CO_2), nitrous oxide (N_2O) and methane (CH_4) rising from fossil fuel combustion.

It is well known fact that oil is the main source of fuel energy for transport vehicles in the world. At present, oil requirement is approximately 84 million barrels per day globally and is expected that it will increase up to 116 million barrels per day after almost three decades that is in year 2030 (IEA, 2007). At the same time as the transport sector progressively expanding in Europe and US, growth in economically developing countries such as India and China is envisaged to be considerably greater with a growth rate of at least 3% per year (IEA, 2006). However, access to energy strengthens our existing lifestyle and commuting is supposed to be a central part of this kind of aspirations. Transportation is the main driving force which accelerated globally demand of oil as it totally relay on oil. Unluckily, oil is forecasted to become not only costly but also scarce in the coming time and it's difficult to meet up this demand. One attractive alternative for these fossil based oil is biofuels. It is fuel obtained from microorganism and plant materials. It has all the potential to overtake oil for transportation. It has very bright future and may replace chemical based fossil oil. However, the whole scenario is bit complicated as various types of biofuels have different environmental and socio-economic repercussions. In market biofuels have already established its place. There are many new opportunities available for the efficient production of biofuel. In this chapter we are focusing on the various feedstocks that can be used for the production of biofuels, biological and chemical conversion processes used to convert feedstocks into biofuels and the key policy and issues addressing the development of biofuels.

2. TYPE OF BIOFUELS

Biofuels are broadly classified as liquid and gaseous biofuels (Nigam and Pandey, 2009).

2.1. Gaseous Biofuel

Biogas and biohydrogen are the major gaseous biofuels. Biogas is produced by the process of anaerobic digestion. Biohydrogen, on the other hand, is a relatively new type of gaseous biofuel. It is generally produced by anaerobic fermentation of agro-industrial wastes and residues by using consortium of methanogenic, acidogenic, and hydrogenic bacteria. Sugars present in the biomass are converted into organic acids and hydrogen by the process of thermophilic heterotrophic fermentation (Pandey, 2009; Abreu *et al.*, 2012; Ge *et al.*, 2016). The conventional source of biogas is animal dung, but use of some forest bio-resources (along with cow-dung) are also reported by many researchers. For instance, *Jatropha curcas* DOC (Ali *et al.*, 2010; Prateek *et al.*, 2009; Raheman and Mondal, 2012) and *Jatropha* DOC mixed with *Pongamia pinnata* DOC (Chandra *et al.*, 2012; Gupta *et al.*, 2009) have been reported for biogas production using anaerobic digestion. Cui *et al.* (2010) have presented the potential of pretreated Poplar leaves-a forest-residue for biohydrogen production using anaerobic mixed bacteria. Similarly, *Albizia* biomass has been used for production of biogas through anaerobic digestion by Ge *et al.* (2014).

2.2. Liquid Biofuels

Bioethanol, biodiesel and biobutanol are the main liquid biofuels.

2.2.1. *Bioethanol*

Bioethanol is produced by simple fermentation of sugars. The process of conversion of fermentable sugars to ethanol is mediated by two enzymes–invertase and zymase. The sugars required for the process can be produced by using cheap and renewable carbohydrate rich biomasses. Low cost lignocellulosic forest residues are now identified as one of the major source of fermentative production of bioethanol after recovery of cellulose using various pretreatment strategies. Enzymatic hydrolysis followed by fermentation is one of the most promising routes for the alteration of cellulosic/lignocellulosic materials into ethanol (Delabona *et al.*, 2012). However, high cost of enzyme (cellulase) used for hydrolysis limits its use for broader industrial application (Delabona *et al.*, 2012). This problem can be minimized by cost-effective production of enzymes by using forest based biomasses or its by-products as low cost raw material in the production media.

2.2.2. *Biodiesel*

Biodiesel is obtained by reaction of vegetable oil (triglycerides) with alcohol in presence of biocatalyst or alkali/acid using transesterification reaction. The mono alkyl ester (fatty acid alkyl ester), produced due to above transestrification reaction is popularly known as biodiesel (Nigam and Pandey, 2009). The non-edible oil obtained from forest species are abundantly available and are of cheaper price compared to edible oils like rape seed oil and soybean oil.

2.2.3. *Butanol*

Biobutanol is superior fuel to ethanol due to its less hygroscopic and volatility (Durre, 2007; Qureshi *et al.*, 2008; Zheng *et al.*, 2009). Most of the butanol production is reported by using *Clostridia* sp. under anaerobic fermentation conditions. During the fermentation, sugars are converted first to organic acids (butyrate and acetate) which are then converted to solvents namely acetone, butanol and ethanol in the ratio of 3:6:1 (Qureshi *et al.*, 2008). This above fermentation process for production of butanol is commonly known as ABE fermentation (Qureshi *et al.*, 2008; Zheng *et al.*, 2009; Jang *et al.*, 2012). Most of the reports on butanol production show use of food crops (starch based) or agricultural residues/lignocellulosic biomass (Qureshi *et al.*, 2008; Shorgani *et al.*, 2012; Qureshi *et al.*, 2007; Qureshi *et al.*, 2010a,b). But there are few reports on use of forest biomass or its by-products as raw material for butanol production showing potential of these substrates also. Switch grass has been reported for butanol production using ABE fermentation by Quereshi (2010b).

Except for biodiesel, forest residues are not been exploited very much for biofuels generation. Majorly biofuels productions are reported from agro-industrial wastes or agricultural crops. By seeing the current interest and increase in demand of biofuels, the forest biomass will not remain untouched, as these also posses the potential to be used as raw material for generation of biofuels.

3. BIOMASS AND BIOFUEL PRODUCTION

3.1. Sustainable Feedstocks Used for Biofuel Production

There are wide ranges of sustainable biomass feedstock available for biofuel production. Most of the agricultural crops grown globally and optimized for biomass feedstock for energy production. Plants can produce sugars and starches, either permanent or temporary energy stores and currently these energy sources have been used extensively for biofuel production. One such example is sugar from sugar cane (*Saccharum officinarum*) that has been used for bioethanol production in Brazil. Brazil is manufacturing worlds 40% of bioethanol from sugar, therefore considered as largest producer of bioethanol. On the other hand USA is targeting starch from *Zea mays* for biofuel production, while China is using it for the production of bioethanol. Attention is also given to sorghums (*Sorghum* spp.) for the production of bioethanol (Draye *et al.*, 2001). It is supposed to replace *Zea mays* and sugar cane for the production of bioethanol as it has the property of drought tolerance along with less water is required for good yields of sugar and starch feedstocks (Cherubini *et al.*, 2009). Besides this wheat and sugar beet, are also used for ethanol and biofuel production. Wheat and sugar beet are optimized extensively for food as well as animal feed. The production (domestic) of these feedstocks is totally depends on policies and global commodity market. The former will also include the sustainability criteria of feedstocks in biofuels production.

Other than this people are switching to complex feed stock also, such as lignocelluloses. Lignocelluloses consist of different phenolic compounds, polymers and proteins (Taylor, 2008). This is present in plant cell wall and could be recovered from agricultural side waste products like bagasse and straw of cereal. However, cell wall of most of the plant is rich in cellulose, which is a large reservoir of energy. It is a glucan polysaccharide that provides potential to convert it into biofuels. The accessibility of cellulose and sugars by the cell wall is very limited due to presence of phenolic polymers such as lignin. The nature of cell wall differs with the plant species and so was the lignocelluloses, therefore accessibility of cellulose and sugars vary, which decide the potential of feedstock to be converted into biofuel. Hence, accessibility of this component is very crucial and need some economic and easy way to recover sugars or cellulose. Saccharification is one such process where cellulose and hemicelluloses is converted into hexose and pentose, respectively. Biofuel production from these simplified sugars hexose and pentose has considerable potential. There are numerous research efforts have been put forth all around world to develop technologies for lignocellulosic biofuel production.

3.2. Conversion of Biomass into Biofuels

Rapid outgrowth of population needs continuously increasing amount of energy and its resources. Present sources of energy are not sufficient to overcome the increasing needs. The major energy demand is fulfilled from the conventional energy resources like coal, petroleum and natural gas. These sources are in the edge of getting wiped out. It was estimated that the oil sources might be depleted till 2050. The process of obtaining energy from these sources causes atmospheric pollution, resulting in problems like global warming, acid rain, etc. Viewing the increase in the energy demands and keeping in view of the pollution a shift over to non-conventional sources like wind, sunlight, water, biomass, etc., is inevitable. Biomass is being used from the ancient times as a combustion fuel for cooking, keeping warmth in houses, etc. Biomass is available in abundance and is cheap and its better utilization is to convert it to energy rich products using suitable processes. Biomass is the fourth largest source of energy in the world after coal, petroleum and natural gas, providing about 15% of the world's primary energy consumption (Chen *et al.*, 2003). Renewable biomass is being considered as an important energy resource all over the world. Biomass is used to meet a variety of energy needs, including generating electricity, fuelling vehicles and providing process heat for industries (Bridgwater *et al.*, 1999). Among all the renewable sources of energy, biomass is unique as it effectively stores solar energy. It is the only renewable source of carbon that can be converted into convenient solid, liquid and gaseous fuels through different conversion processes (Ozbay *et al.*, 2001). Biomass comprises all the living matter present on earth. It is derived from growing plants including algae, trees and crops or from animal manure (Garcia-Perez *et al.*, 2002). The biomass resources are the organic matters in which the solar energy is stored in chemical bonds. It generally consists of carbon, hydrogen, oxygen and nitrogen. Sulfur is also present in minor proportions. Some biomass also consists significant amounts of inorganic species. Plants, *via* photosynthesis, produce carbohydrates which form the building blocks of biomass (Demirbas, 2001). Biomass resources can be divided into two broad categories; natural and derived materials. These categories can be further subdivided as wastes, forest products and energy crops (Narvaez *et al.*, 1996). The direct use of biomass to produce heat by combustion is an inefficient way of utilizing its energy and it contributes to enhance environmental pollution. There is a vital need of a process, which can convert biomass into useful energy products like oil, gases, etc. Biomass can be converted into three main products: power/heat generation, transportation fuels and chemical feedstock (White and Plaskett, 1981; Anderson and Tillman, 1977; McKendry, 2002).

The conversion of biomass/feedstocks into biofuels is a complex process. Biomass can be converted into useful forms of energy products using a number of different processes. Biochemical, thermochemical and photobiological conversion processes are common used to convert biomass into biofuels and its derivatives (Frings *et al.*, 1992; Saxena *et al.*, 2009; Goyal *et al.*, 2008; Brennan and Owende, 2010).

3.2.1. *Biochemical conversion*

In biochemical conversion processes, enzymes and microorganisms are used as biocatalysts to convert biomass/feedstocks into desirable products. Bio-gasification and bio-liquefaction are two developed technology for the bioconversion of biomass into energy forms. The anaerobic fermentation of organic matter by a mixed culture of organisms to produce a gaseous mixture of methane and carbon dioxide is known as bio-gasification. Bio-liquefaction is the process to use a pure culture of organisms (generally yeasts) to anaerobically ferment sugars into products such as acetone, butanol and ethanol. Bio-gasification also called anaerobic digestion, can utilise a wide range of feedstocks while bio-liquefaction can only utilise a feedstock that can be converted into fermentable sugars *e.g.*, sugar cane, sugar beet, molasses, starch or cellulose, feedstocks. Biotechnology advances are expected to lead to dramatic biochemical conversion improvements (Frings *et al.*, 1992; Saxena *et al.*, 2009).

3.2.2. *Thermo-chemical conversion*

The thermo-chemical conversion processes have two basic approaches. The first is the gasification of biomass and its conversion to hydrocarbons. In gasification, biomass is heated in an oxygen-starved environment to produce a gas composed primarily of hydrogen and carbon monoxide. The second approach is to liquefy biomass directly by high-temperature pyrolysis, high-pressure liquefaction and supercritical extraction. These processes convert the waste biomass into energy rich useful products (Goyal *et al.*, 2008). In pyrolysis, biomass is exposed to high temperatures in the absence of air, causing it to decompose. Solvents, acids, and bases can be used to fractionate biomass into an array of products including sugars, cellulosic fibers, and lignin (Hamelinck and Faaij, 2002).

3.2.3. *Photobiological conversion*

Photobiological conversion processes use the natural photosynthetic activity of organisms to produce biofuels directly from sunlight. For example, the photosynthetic activities of bacteria and green algae have been used to produce hydrogen from water and sunlight (Brennan and Owende, 2010). Microalgae offer novel aquatic biomass systems with potential source of renewable energy, and they allow the direct generation of desired end products such as biofuel oil and gas. The production of fuel from photosynthetic organisms is a renewable process using solar energy; therefore, biomass is a highly promising resource (Sawayama *et al.*, 1999; Tsukahara *et al.*, 2001). Increase of atmospheric CO_2 by the biological fixation and energy production can be solved the future energy crises (Hirano *et al.*, 1997). Microalgae can assimilate CO_2 gas as the carbon source for growth. Reduction of carbon dioxide concentration in the atmosphere is an important task, and many physical or chemical approaches have been proposed. By using microalgae, beside their capability of fixing CO_2, microalgal oil can be produced through either biological conversion to lipids or hydrocarbons or thermochemical liquefaction of algal cells (Yang *et al.*, 2004). Microalgae have the highest oil yield among various plants such as palm oil coconut, castor bean and sunflower seeds. The oil content of microalgae is between 15% and 40% of dry weight, but some microalgae have up to 80%. Oil content can be estimated to be 64.4% of the total lipid component. By using a transesterification process, algae oil can be converted to biodiesel. The biodiesel product has its main characteristics quite similar to those of conventional Diesel or compatible with conventional petroleum Diesel, and it can also be blended in any proportion with petroleum Diesel (Amin *et al.*, 2009). Research and development activities at present include fundamentals of algae strain improvement, reactor design and process integration, with auxiliary power requirements and specific reactor cost being addressed as most critical issues.

4. BIOFUELS AND THEIR IMPACT ASSESSMENT

There are many supply chain options of biofuels. The long term viability and cost effectiveness with environmental factor should be checked fully.

One has to consider greenhouse gas emission, air quality, land utilization, biodiversity, water and energy consumption. Some key tools for the assessment are sustainability, environmental impact and life cycle. These assessments should also be closely monitored for their economic and social aspects. In economic impact assessment, taxes, subsidies, application of financial incentives, developing and new products should be included in calculations. Biofuels socially has both direct and indirect impacts like social equity, social responsibility and job creations.

The evaluation of the natural resource requirements and its impact on environment from the whole life cycle of a product or service is called life cycle assessment technique (ISO, 2006). In life cycle assessment of biofuels, we try to focus on greenhouse gas emissions, land use and primary energy with respect to conventional oil based transport fuels. With some favourable combinations of options of biofuels one can achieve zero total greenhouse gas emissions or also termed as 'carbon neutral' or 'carbon negative' (Woods and Bauen, 2003). In life cycle assessment of biofuels the geographical scope and technical angle of feedstock should also be important. Life cycle assessment data should include the strategic environmental assessment and environmental impact assessment data. This can provide more holistic impact on environment. In life cycle assessment studies, some of the results mislead not only the policy makes but also the majority of public. It is because there is lack of transparency in studies. The assumptions, source of data, method of calculation should be transparent and well documented. The G-8 countries launched global bioenergy partnership (GBEP) in year 2005. The GBEP is putting efforts to harmonies the methodologies for life cycle assessment of biofuels.

Biofuels and feedstock can also be imported from other countries to fulfill their requirement. Besides this, new feedstock and newer processing technologies might come out. These future probable options should be included in life cycle assessment studies for more refined and realistic policy formations. The reference system in life cycle assessment should be properly established like greenhouse gas emission of previous land use with reference to the present land use in biofuel helps in calculating the earned 'credit'. It is the way to calculate 'maintained set-aside' for life cycle assessment calculations. When food crops are displaced and cultivated in other areas, its life cycle assessment calculation should also be taken into account. The 'comparative reference system' includes all types of resource inputs and outputs before the production of biofuels and after the production has been stored. This before and after conditions greenhouse gas emissions can also be compared. The displaced food crops must be planted elsewhere with better conditions and lower greenhouse gas emissions. Comparative reference systems also consider side products of biofuels like bagasse, straw, corn stover as agricultural waste and some processed products like sugar beet pulp, glycerine, rape meal etc (Mortimer *et al.*, 2006). There are many missing data like emission of N_2O from soil, carbon content changes in soil, impact on water use, biodiversity and quality of air. This data should be accurately quantified and added in realistic data for greenhouse gas emissions or life cycle assessment system.

Due to some agronomic practices, harvesting processes, use of fertilizers and distribution of crop products greenhouse gases get emitted. These gases are nitrous oxide (N_2O), methane, carbon dioxide etc. Plants also emit various types of volatile organic compounds like isoprene. This isoprene in presence of NO_x can form greenhouse gas ozone (Arneth *et al.*, 2007). Large carbon sinks emit carbon dioxide which cannot be quantified because of its uncertain quality. Generally peat lands contain 528 Gt of carbon out of which 42000 Mt is contributed by forest area of tropical peat lands of South East Asia (Hooijer *et al.*, 2006). This area of peat lands in South East Asia drained out and causes up to 100 t/ha/yr or 3 Kg/m^2/day of CO_2 emissions. If this gets burnt then this figure is enhanced to double or treble. This problem should also be addressed for assessment of biofuels. One of the greenhouse gas nitrous oxide (N_2O) has a global warming potential 296 times higher than the CO_2 emission (IPCC, 2007a). This nitrous oxide is produced from nitrogenous fertilizers and also from natural mineralization of nitrogen, by nitrification and denitrification with the help of bacteria. The industrial fertilizers are the main source of atmospheric N_2O. The cause of release of N_2O should also be understood more clearly as far as biofuel crop is concerned. There are also very strong recommendations to use less productive land for biofuel feedstock cultivation. Simultaneously the co-products of feedstock should also be used efficiently like, in wheat grain the bioethanol is produced but its wheat straw can be utilized in heat or electricity production. These policies to encourage the use of biofuel should be incentivised properly. The greenhouse gas savings should also be encouraged and publicised to make the common citizen aware.

In entire supply chain, biofuel have wide range of impact on environment like greenhouse gas emissions, water consumption, air pollution, biodiversity, production of feedstock, conversion of feedstock. These impacts if quantified in a proper way will show overall benefits or even drawbacks by comparing different biofuels to the policymaker and stakeholders. With the lifecycle assessment and other qualitative data the stakeholder makes more balanced decisions for biofuel to be opted in sustainable development. Therefore, it became important to know about the biofuel policy.

5. BIOFUEL POLICY

It is an important aspect of biofuel to discuss policies related to research, innovations, design, development for climate change and reduction in greenhouse gas emission. There is already existing high oil demand from developed countries with the new rising demand from developing countries like India and China, it may be possible in coming time to face high oil prices of fossil fuel. Despite its environmental benefits the production cost of biofuels are higher than the fossil fuel with the exception of ethanol from sugar cane. To reduce the pricing cost of biofuel government has to take some strong initiative in regulatory policies, taxes and lowering its basic production cost by addressing land uses.

The post-tax prices on fossil fuel (petrol and diesel) in Europe and USA increases the prices of fossil fuel making biofuel commercially more attractive. The research and development innovations associated with lignocellulose biofuels have reduced their cost and made them comparable even with the biofuel of other food stuff. The cost effectiveness can also be attained by involved yield and conversion methods of feed stock. Land management practices will not only help to utilize waste or degraded land but also be useful for environmental issues like flooding risks, poor soil quality and water resource management.

There are many issues which have to be considered for biofuel including greenhouse gas emission reduction and climate change benefits. It is dependent upon; the crop selected, the cropping systems, the conversion methods adopted, estimation locations, fertilizer used, the effect of microclimates on N_2O emissions, energy of fuel consumed during processing, processing technology fuel, transportation distances and by product utilization. To develop efficient biofuel greenhouse gas emission reduction per hectare of land is very important. From prior researches it is found that electricity generation from wood is much environmental friendly and found to be more carbon dioxide saving per hectare as compared to food crops or lignocellulose feedstock. Biomass produced electricity is better than coal as far as greenhouse gas mitigation is concern (Havlik *et al.*, 2011).

Some countries have abundant fossil fuel resources like tar sands, oil shale, coal or gas but its conversion to liquid fuels is very costly. These synthetic fuels are more carbon intensive than fossil fuels and their storage is also costly. Many more options are there for environmental and climate change concern in alternative energy and electricity generation like wind, solar, offshore wave, tidal stream, nuclear power energy and hydrogen with fuel cells. They mostly have limited options for transport, requires better techniques to develop, need more infrastructure and safety measures during processing. These conditions not only discourage the popularisation of these favours lignocellulose derived fuels for transport and seem appropriate for commercial feasibility. It also seems logical to utilize degraded agricultural lands, wood lands and watersheds in production of feedstock of lignocellulose materials.

There is a very relevant study done by Stern (2006) about policy instruments for biofuels (Stern *et al.*, 2006). Stern suggested three points for the mitigation of climate change. One is pricing of carbon in the carbon taxes form or the attributed prices of reduction on emissions of carbon. Second is direct support for research and development projects to develop new technologies. Third is environmental standards and regulations to attain wider ecological benefits. In Stern's review policies to improve efficiency in transport is of prime concern. There should be international policies on climate change so that the environmental problem of countries would not be exported to others and every country should be focused on 'collective action' for long term solutions. To promote these policies government should also form long term incentives rather than short duration ones so that its effect could be assessed with in considerable time frame. By addressing these shortcomings in policy making the effectiveness of policies will improve in future.

It not only encourages focused private research and demonstrative effect to reduce greenhouse gas emissions from biofuels but also helps in sustainable development. There is a need to categories various fuel sources and cropping practices for proper pricing system. These pricing systems must consider incentives for greenhouse gas mitigation in biofuel categories. There is a different but small contributory category of fuel which uses organic and municipal wastes. It reduces land requirement and is useful for transport fuels.

Transport fuel policy need improvement in fuel efficiency of vehicles with hybrid and another kind of plug-in hybrids vehicles. By developing these plug-in –hybrid the possibility of charging them by rechargeable batteries through low carbon energy form like renewable and nuclear power can be done conveniently. This step will defiantly be very important in 'carbon foot print' reduction. Vehicle fuel efficiency also improves by reducing congestion and using vehicles which uses alternative fuel such as vehicles runs on hydrogen fuel cells or fully electric. There small level demonstrations are visible but commercially its cost effectiveness must be explored by taking some important policy decisions.

CONCLUSIONS

Several new opportunities for the efficient production of biofuel are available. Focuses of researchers are on the various feedstocks that can be used for the production of biofuels, like biological and chemical conversion processes used to convert feed stocks into biofuels. The key policy and issues addressing the development of biofuels are also evolved with time.

Rising environmental issues, climate risk and decreasing fossil reserves are the key driving force to form supportive policies and implement newer sustainable alternatives to solve this task. Media and public awareness is also picking up day by day. Key stakeholders and public discussion towards policy forming is an improvement milestone. Economists all over the world are also working hard to overcome the shortcoming of biofuel policies in the ray of sustainability as far as cost effectiveness and policy balance is concern.

On international forum combine action, collective initiatives and global unity will definitely be successful in forming better green planate for future generations to live in.

ACKNOWLEDGEMENTS

Dr. Shivani Chaturvedi is grateful to the UGC-PDFWM for providing the Research Fellowship and Dr. Arti Kumari, Dr. Razi Ahmad, Dr. Amrik Bhattacharya are thankful to DST-SERB for providing national postdoctoral fellowship.

REFERENCES

Abreu, A.A., Karakashev, D., Angelidaki, I., Sousa, D.Z. and Alves, M.M. (2012). Biohydrogen production from arabinose and glucose using extreme thermophilic anaerobic mixed cultures. *Biotechnology for Biofuels*, 5: 1.

Ali, N., Kurchania, A.K. and Babel, S. (2010). Bio-methanisation of *Jatropha curcas* defatted waste. *Journal of Engineering and Technology Research*, 2: 038–043.

Al-Shorgani, N.K.N., Kalil, M.S. and Yusoff, W.M.W. (2012). Biobutanol production from rice bran and de-oiled rice bran by *Clostridium saccharoperbutylacetonicum* N1-4. *Bioprocess and Biosystems Engineering*, 35: 817–826.

Amin, S. (2009). Review on biofuel oil and gas production processes from microalgae. *Energy Conversion and Management*, 50: 1834–1840.

Anderson, L.L. and Tillman, D.A. (1977). Fuels from waste. *Fuels from Waste*, 1.

Arneth, A., Schurgers, G., Hickler, T. and Miller, P.A. (2007). Effects of species composition, land surface cover, CO_2 concentration and climate on isoprene emissions from European forests. *Plant Biology*, 9: e87–e98.

Brennan, L. and Owende, P. (2010). Biofuels from microalgae-a review of technologies for production, processing, and extractions of biofuels and co-products. *Renewable and Sustainable Energy Reviews*, 14: 557–577.

Bridgwater, A.V., Meier, D. and Radlein, D. (1999). An overview of fast pyrolysis of biomass. *Organic Geochemistry*, 30: 1479–1493.

Chandra, R., Vijay, V.K., Subbarao, P.M.V. and Khura, T.K. (2012). Production of methane from anaerobic digestion of *Jatropha* and *Pongamia* oil cakes. *Applied Energy*, 93: 148–159.

Chen, G., Andries, J. and Spliethoff, H. (2003). Catalytic pyrolysis of biomass for hydrogen rich fuel gas production. *Energy Conversion and Management*, 44: 2289–2296.

Cherubini, F., Bird, N.D., Cowie, A., Jungmeier, G., Schlamadinger, B. and Woess-Gallasch, S. (2009). Energy-and greenhouse gas-based LCA of biofuel and bioenergy systems: Key issues, ranges and recommendations. *Resources, Conservation and Recycling*, 53: 434–447.

Cui, M., Yuan, Z., Zhi, X., Wei, L., and Shen, J. (2010). Biohydrogen production from poplar leaves pretreated by different methods using anaerobic mixed bacteria. *International Journal of Hydrogen Energy*, 35: 4041–4047.

Delabona, da Silva P., Pirota, R.D.P.B., Codima, C.A., Tremacoldi, C.R., Rodrigues, A. and Farinas, C.S. (2012). Using Amazon forest fungi and agricultural residues as a strategy to produce cellulolytic enzymes. *Biomass and Bioenergy*, 37: 243–250.

Demirbas, A. (2001). Biomass resource facilities and biomass conversion processing for fuels and chemicals. *Energy Conversion and Management*, 42: 1357–1378.

Draye, X., Lin, Y.R., X.-Y., Qian, J.E., Bowers, G.B., Burow, Morrell, P.L., Peterson, D.G., and Durre, P. (2007). Biobutanol: An attractive biofuel. *Biotechnology Journal,* 2: 1525–1534.

Frings, R.M., Hunter, I.R., and Mackie, K.L. (1992). Environmental requirements in thermochemical and biochemical conversion of biomass. *Biomass and Bioenergy*, 2: 263–278.

Garcia-Perez, M., Chaala, A. and Roy, C. (2002). Vacuum pyrolysis of sugarcane bagasse. *Journal of Analytical and Applied Pyrolysis*, 65: 111–136.

Ge, X., Matsumoto, T., Keith, L. and Li, Y. (2014). Biogas energy production from tropical biomass wastes by anaerobic digestion. Bioresource Technology, 169: 38–44.

Ge, X., Xu, F. and Li, Y. (2016). Solid-state anaerobic digestion of lignocellulosic biomass: Recent progress and perspectives. *Bioresource Technology*, 205: 239–249.

Goyal, H.B., Seal, D. and Saxena, R.C. (2008). Bio-fuels from thermochemical conversion of renewable resources: A review. *Renewable and Sustainable Energy Reviews*, 12: 504–517.

Gupta, A., Chandra, R., Subbarao, P.M.V. and Vijay, V.K. (2009). Kinetics of batch biomethanation process of *Jatropha* and *Pongamia* oil cakes and their co-digested substrates. *Journal of Scientific and Industrial Research*, 68: 624.

Hamelinck, C.N. and Faaij. A.P.C. (2002). Future prospects for production of methanol and hydrogen from biomass. *Journal of Power Sources*, 111: 1–22.

Havlik, P., Schneider, U.A., Schmid, E., Bottcher, H., Fritz, S., Skalsky, R., Aoki, K., De Cara, S., Kindermann, G. and Kraxner, F. (2011). Global land-use implications of first and second generation biofuel targets. *Energy Policy*, 39: 5690–5702.

Hirano, A., Ueda, R., Hirayama, S. and Ogushi, Y. (1997). CO_2 fixation and ethanol production with microalgal photosynthesis and intracellular anaerobic fermentation. *Energy*, 22: 137–142.

Hooijer, A., Silvius, M., Wosten, H. and Page, S. (2006). PEAT-CO_2, Assessment of CO_2 emissions from drained peatlands in SE Asia, Delft Hydraulics report Q3943/2006, p. 36.

Jang, Y.S., Malaviya, A., Cho, C., Lee, J. and Lee, S.Y. (2012). Butanol production from renewable biomass by clostridia. *Bioresource Technology*, 123: 653–663.

McKendry, P. (2002). Energy production from biomass (part 2): Conversion technologies. *Bioresource Technology*, 83: 47–54.

Mortimer, N.D., Chambers, W. and Square, C. (2006). The role of life cycle assessment in policy and commercial development: The case of liquid biofuels in the United Kingdom, *In*: 5th Australian Life Cycle Assessment Conference.

Narvaez, I., Orio, A., Aznar, M.P., and Corella, J. (1996). Biomass gasification with air in an atmospheric bubbling fluidized bed. Effect of six operational variables on the quality of the produced raw gas. *Industrial & Engineering Chemistry Research*, 35: 2110–2120.

Nigam, P.S.-N., Pandey, A. (2009). Biotechnology for agro-industrial residues utilisation: Utilisation of agro-residues, Springer Science & Business Media.

Ozbay, N., Putun, A.E., Uzun, B.B., and Putun, E. (2001). Biocrude from biomass: Pyrolysis of cottonseed cake. *Renewable Energy*, 24: 615–625.

Prateek, S., Gopal, R., Mayur, S., and Shilpkar, D. (2009). Biomethanation potential of *Jatropha* (*Jatropha curcas*) cake along with buffalo dung. *African Journal of Agricultural Research*, 4: 991–995.

Presting, S.-X. and Ren, R.A. (2001). Wing, Toward integration of comparative genetic, physical, diversity, and cytomolecular maps for grasses and grains, using the sorghum genome as a foundation. *Plant Physiology*, 125: 1325–1341.

Qureshi, N., Ezeji, T.C., Ebener, J., Dien, B.S., Cotta, M.A. and Blaschek, H.P. (2008). Butanol production by *Clostridium beijerinckii*. Part I: use of acid and enzyme hydrolyzed corn fiber. Bioresource Technology, 99: 5915–5922.

Qureshi, N., Saha, B.C. and Cotta, M.A. (2007). Butanol production from wheat straw hydrolysate using *Clostridium beijerinckii*. *Bioprocess and Biosystems Engineering*, 30: 419–427.

Qureshi, N., Saha, B.C., Dien, B., Hector, R.E., and Cotta, M.A. (2010a). Production of butanol (biofuel) from agricultural residues: Part I-Use of barley straw hydrolysate. *Biomass and Bioenergy*, 34: 559–565.

Qureshi, N., Saha, B.C., Hector, R.E., Dien, B., Hughes, S., Liu, S., Iten, L., Bowman, M.J., Sarath, G. and Cotta, M.A. (2010b). Production of butanol (biofuel) from agricultural residues: Part II-Use of corn stover and switchgrass hydrolysates. *Biomass and Bioenergy*, 34: 566–571.

Raheman, H. and Mondal, S. (2012). Biogas production potential of *Jatropha* seed cake. *Biomass and Bioenergy*, 37: 25–30.

Sawayama, S., Minowa, T. and Yokoyama, S.-Y. (1999). Possibility of renewable energy production and CO_2 mitigation by thermochemical liquefaction of microalgae. *Biomass and Bioenergy*, 17: 33–39.

Saxena, R.C., Adhikari, D.K., and Goyal, H.B. (2009). Biomass-based energy fuel through biochemical routes: A review. *Renewable and Sustainable Energy Reviews*, 13: 167–178.

Stern, N.H., Peters, S., Bakhshi, V., Bowen, A., Cameron, C., Catovsky, S., Crane, D., Cruickshank, S., Dietz, S., and Edmonson, N. (2006). Stern Review: The economics of climate change, Cambridge University Press Cambridge.

Taylor, G. (2008). Biofuels and the biorefinery concept. *Energy Policy*, 36: 4406–4409.

Tsukahara, K., Kimura, T., Minowa, T., Sawayama, S., Yagishita, T., Inoue, S., Hanaoka, T., Usui, Y. and Ogi, T. (2001). Microalgal cultivation in a solution recovered from the low-temperature catalytic gasification of the microalga. *Journal of Bioscience and Bioengineering*, 91: 311–313.

White, L.P. and Plaskett, L.G. (1981). Biomass as fuel, Academic Press Ltd.

Woods, J. and Bauen, A. (2003). Technology status review and carbon abatement potential of renewable transport fuels in the UK, DTI.

Yang, Y.F., Feng, C.P., Inamori, Y. and Maekawa, T. (2004). Analysis of energy conversion characteristics in liquefaction of algae. *Resources, Conservation and Recycling*, 43: 21–33.

Zheng, Y.N., Li, L.-Z., Xian, M., Ma, Y.-J., Yang, J.-M., Xu, X. and He, D.-Z. (2009). Problems with the microbial production of butanol. *Journal of Industrial Microbiology & Biotechnology*, 36: 1127–1138.

4

Biofuels: Potential Role in Mitigation of Air Pollution in India

Sohana Debbarma[1*], Pradeep Goyal[2], Jitendra Nagar[3] and Narendra Singh[4]

ABSTRACT

Worldwide, air pollution has emerged as one of the grave threats to public health with almost 3.7 million deaths in 2012 attributed to outdoor air pollution. 98% of cities in low-and middle income countries do not meet World Health Organization (WHO) air quality guidelines. Delhi being the capital has remained in limelight due to its alarmingly polluted air in the last few years. The exposure to hazardous particulate matter pollution ($PM_{2.5}$ and PM_{10}) is alarmingly high in most Indian cities, with levels exceeding the National Ambient Air Quality Standards for most parts of the year. An assessment of air quality in 168 cities across 24 States and Union Territories pinpointed fossil fuels as one of the main culprits for the deteriorating air quality across the country. Recently biofuels have received global attention as the world seeks to address concerns of fuel efficiency, air quality and energy security. Biofuels are

[1]Centre for Environmental Monitoring, Ministry of Natural Resources and Environment, Vietnam
[2]Govt. Engineering College Ajmer
[3]Dr. B.R. Ambedkar College, Delhi University
[4]Shyam Lal College,Delhi University
**Corresponding author*: E-mail: sohanadebbarma07@gmail.com

produced from biological sources and are a vital alternative to fossil-fuels. In 2003 Indian launched its Ethanol blended programme for 5% ethanol blended petrol. In 2009 Government of India introduced National Policy on Biofuels which aimed to encourage biofuel usage with reduction in fossil fuel dependence along with addressing environmental and fuel efficiency concerns. It is evident that for India a gradual shift from fossil fuels to biofuels would reduce harmful emissions of pollutants and consequently improve the air quality in our cities.

***Key words*:** Biofuels, Mitigation, Air pollution, India

INTRODUCTION

Air Pollution: Present Status

Worldwide, Air pollution has emerged as one of the grave threats to public health with almost 3.7 million deaths in 2012 attributed to outdoor air pollution (WHO, 2014). 98% of cities in low- and middle income countries do not meet World Health Organization (WHO) air quality guidelines (WHO, 2016). Delhi being the capital has remained in limelight due to its alarmingly polluted air in the last few years, while there are several other regions which are equally or polluted in the country (Kaushik *et al.*, 2016). The exposure to hazardous particulate matter pollution ($PM_{2.5}$ and PM_{10}) is alarmingly high in most Indian cities (Apte *et al.*, 2011), with levels exceeding the National Ambient Air Quality Standards for most parts of the year (Kaushik and Borah, 2016). A recently released report revealed that with the exception of few south Indian cities none of the other Indian cities complied with the National Ambient Air Quality Standards. Air quality assessment of 168 cities spanning across 24 States and the Union Territories clearly highlighted the role of fossil fuels as the major culprits responsible for the deteriorating air quality within the country. Year 2015 data reveals that for the 20 most polluted Indian cities the PM 10 levels ranged between 268 $\mu g/m^3$ and 168 $\mu g/m^3$. While, the National Capital Delhi topped the list with a high concentration (268 $\mu g/m^3$), it was closely followed by satellite city in NCR Ghaziabad while cities Allahabad and Bareli in Uttar Pradesh came next. The other cities in the list included Faridabad (Haryana); Jharia (Jharkhand), Alwar (Rajasthan); Ranchi, Kusunda and Bastacola from Jharkhand; Kanpur (Uttar Pradesh) and Patna from Bihar; the PM_{10} levels for these cities ranged between 258 $\mu g/m^3$ and 200 $\mu g/m^3$ (Greenpeace Report, 2017).

Pollutants and their Adverse Impacts

Air pollution is now globally perceived as an important issue. Emissions of various air pollutant species (particulate matter, oxides of sulphur, oxides of nitrogen, carbon monoxide, and volatile organic compounds) released from multiple sources mix, disperse and react in atmosphere to form pollutant concentrations (Table 1). Particulate Matter (PM) refers to tiny particles which remain suspended in air, in the form of either solid or liquid droplets which originate from various sources that pollute the ambient air. Particulate matter comprises of various organic and inorganic components; the major components include acids, ammonia, sodium chloride, black carbon, water and mineral dust. The respirable particulates having aerodynamic diameter de 10 μm (PM_{10}) are an important part of the atmosphere. Six main sources of ambient particulate matter include: traffic, industry, domestic fuel burning, natural sources including soil dust (re-suspended) and sea salt and unspecified sources of pollution of anthropogenic origin (Amato *et al.*, 2009; Belis *et al.*, 2013).

PM is widespread and affects more people than any other ambient air pollutant. These particles have a high probability of deposition deeper into the respiratory tract and are likely to trigger respiratory diseases such as asthma, bronchitis, cardio-pulmonary infections (Green Stone *et al.*, 2015; WHO, 2014). Epidemiological evidence has even attributed PM_{10} in cancer and in some cases even premature death. The relative strength of association of air pollutants with mortality were reported as follows: $PM_{2.5} \geq PM_{10} \geq SO_2 \geq H^+ \geq O_3 \geq NO_x$ (Dockery *et al.*, 1992; Das *et al.*, 2006). Further it has been found that for each 10 μg/m^3 increase in PM_{10} concentration there is an estimated increase in mortality by almost 1 percent (Dockery and Pope, 1994; Ostro, 1996; Das *et al.*, 2006). Each 10 μg/m^3 elevations in fine particulate air pollution has been associated with approximately with 4, 6 and 8% increased risk of all cause, cardio pulmonary and lung cancer mortality respectively (Pope, 2000).

High pollution levels combined with dense population in urban areas results into high mortality and health costs. Conditions in developing Asian cities such as low average incomes, poor health facilities along with inadequate awareness about the sources and treatment of health problems, further contributes to the loss of many lives every year. The Global Burden of Disease recognized that in 2013 approximately 660,000 deaths in India were on account of outdoor air pollution (IHME,

Table 1: Major air pollutants and their adverse effects

Type of vehicles	*Fuel*	*Major primary*	*Secondary pollutants*	*Possible major impacts*	*Severity of problem*
Two-wheelers (two stroke engine, 25)	Gasoline	CO, PM, HC	Secondary organic aerosol (SOA)	Respiratory and cardiovascular health, agriculture, climate	High
Two-wheelers (four stroke engine, 45)	Gasoline	NOx, CO	Ozone	Respiratory health, agriculture	Low
Cars	Gasoline	CO, HC	SOA	Respiratory health	Low
	Diesel	PM (BC), NOx	Nitrates, ozone	Respiratory and cardiovascular health, agriculture, climate	Moderate
Heavy-duty vehicle (HDV)	Diesel	PM (BC), NOx	Nitrates, ozone	Respiratory and cardiovascular health, agriculture, climate	High
	Compressed natural gas (CNG)	NOx, HC	Ozone, nitrates, SOA	Respiratory health, agriculture	Low

Source: Ramanathan *et al.*, 2014

2015). The ground level ozone produced by pollutant gases (such as NOx) released during combustion of fuels leads to over 5 million tons of crop damages (primarily wheat and rice) which could be as high as 25 million tons in India every year. Although not widely recognized, air pollution has significant effects on water both in a direct and indirect manner. Aerosols (comprising mainly of sulphates, nitrates, dust and organics) reduce the sunlight (known as dimming) which reaches the ground level which in turn reduces the rate of evaporation and therefore, precipitation. Aerosols also act as condensation nuclei which nucleate huge quantities of cloud drops, which decrease the precipitation efficiency of low lying clouds. Additionally, aerosols are responsible for causing surface cooling and altering the differential heating pattern of land-ocean, which results in reduced rainfall (Ramanathan *et al.*, 2014).

World Bank's economic estimates also highlight a whooping degradation cost of over Rs. 1.1 trillion on account of outdoor pollution as well as a cost of Rs 0.9 trillion due to indoor air pollution in India. This amount translates to almost 3% of our total GDP. Several air pollutants such as black carbon (it is a component of combustion based PM emissions) and ozone (it is formed at the ground level with the reactions between precursor pollutants like NOx and VOCs in the presence of sunlight) as these also have significant warming potential these, are commonly known as the short lived climate pollutants or the SLCPs (TERI, 2016).

Emissions

Within the past two decades India has treaded over a path of steep economic growth. A close look at the sector wise energy consumption scenario in the country reveals that power plants and the domestic sector together account for the largest share within the overall energy usage in the country (Fig. 1). However the transport sector despite having a lesser share in the energy mix has greater presence within densely populated cities and its share which is rapidly growing emphasizes towards the attention it deserves.

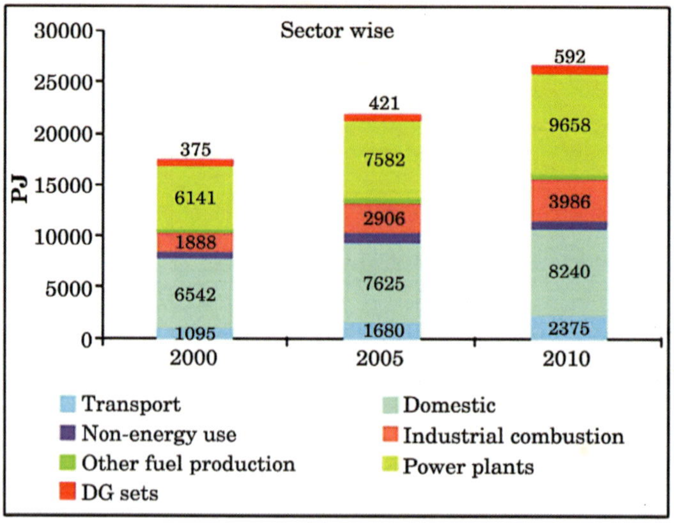

Fig. 1: Sectoral energy consumption in India. *Source*: The Energy and Resource Institute (TERI, 2013a)

Transport Sector

A total of 419 source apportionment records from studies conducted in cities of 51 countries were used to calculate regional averages of sources of ambient particulate matter. Based on the available information, globally 25% of urban ambient air pollution from $PM_{2.5}$ is contributed by traffic, 15% by industrial activities, 20% by domestic fuel burning, 22% from unspecified sources of human origin, and 18% from natural dust and salt (Karagulian *et al.*, 2015). Since independence India's transport sector has grown in an exponential manner. With growing income levels (per capita), highly competitive automobile-markets, absence of an effective public transportation system and network have together contributed to this increased demand for personal transport or vehicles. Further, the growth of vehicles is greatly skewed towards the cities. Of the total number of registered vehicles almost one third are present in 53 cities, with populations greater than a million. It is the second tier cities which are currently showing a spurt within the vehicle population. Almost 28,000 two wheelers and over 4,200 cars find place in India's vehicular fleet on a daily basis (in 2011) (MoRTH, 2012). In the year 2011, India revealed a fleet of 141 million registered vehicles, a drastic increase from less than a million vehicles in 1950.

Present Status: Contribution of Transport Sector towards Air Quality

Urban areas suffer from high concentrations of air pollutants which are contributed by the transportation sector. The transport sector emissions are mainly dependent upon: (1) Estimated numbers of on-road vehicles (2) The technological distribution of plying vehicles (in terms of sizes, engine capacity, power) (3) Fuel-wise classification of vehicles (as gasoline, gas or diesel) (4) Age of vehicles (their compliance with existing emission norms) and (5) Inspection & Maintenance schedules as well as practices. The fuel-wise vehicle distribution is significant as different pollutants are emitted in variable proportions while the vehicles are made to run on different fuels. While the gasoline powered vehicles emit hydrocarbons as well as carbon monoxide (CO), which are released due to incomplete gasoline combustion (Ramanathan *et al.*, 2014). On the contrary, vehicles driven on diesel emit greater concentrations of particulates and oxides of nitrogen on account of combustion taking place at high temperature (Fig. 2 a-b).

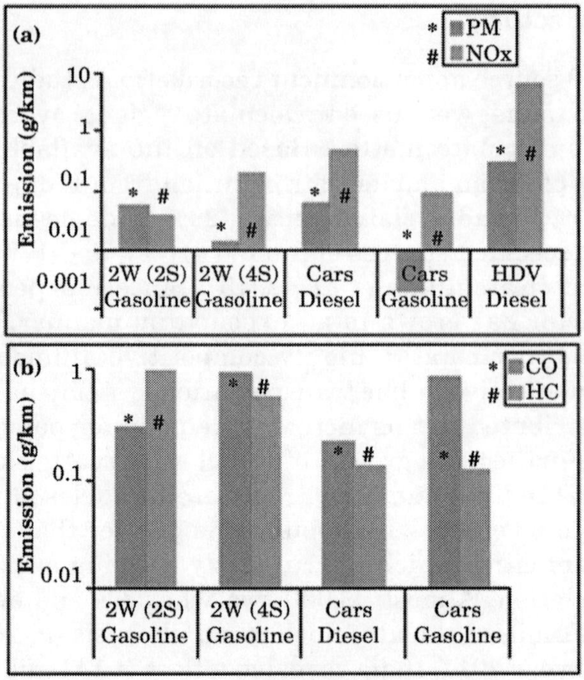

Fig. 2: **(a-b)** Emissions (g/km) from different types of vehicles. *Data source*: Automotive Research Association of India (ARAI), 2010

It is evident from Fig. 2 that the pollutant emissions are dependent on the type of fuel. Heavy duty diesel powered vehicles are the worst polluters in terms of highest PM emissions among all the vehicle classes. Even the (Two-stroke) two-wheelers are notorious for their combustion inefficiencies and concomitant high PM emissions. However the gasoline vehicles are considered to be comparatively less polluting, despite their emission of gases like CO and the HC (ARAI, 2010).

Emission estimates: The share of transport sector

Fig. 3 a-b (below) presents the share of different sectors in current PM_{10} and $PM_{2.5}$ concentrations prevalent in the three major cities. While, road and soil dust were implicated as the major contributor in the PM_{10} concentrations, it was the transport sector which was the dominant contributor towards the $PM_{2.5}$ fractions (CPCB, 2011).

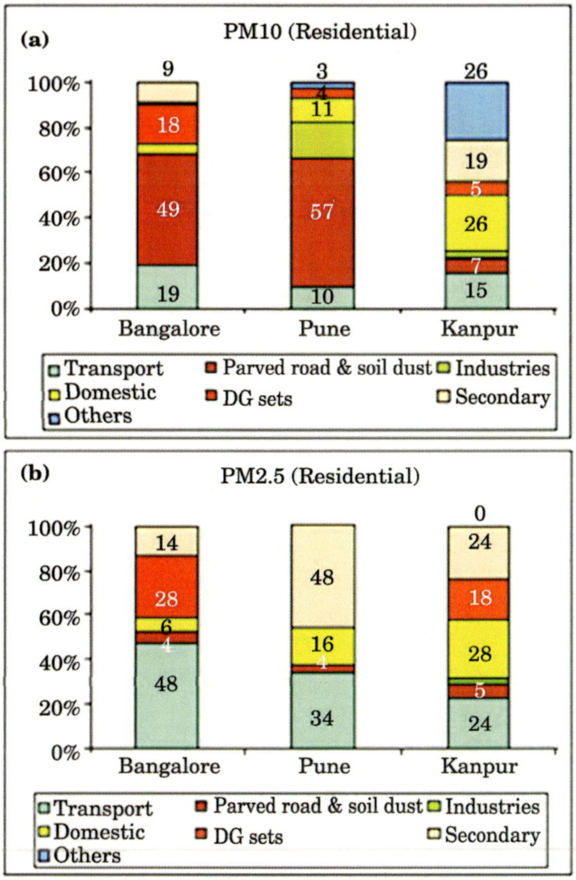

Fig. 3: **(a-b)**Results of source apportionment studies in Indian cities. *Source*: Central Pollution Control Board (CPCB, India) 2011

Rationale for focusing on transport sector

Though the results from source apportionment studies reveal that the emissions from industrial and the domestic sector are dominant at the National level, the current review is focused on the transport sector for the below mentioned reasons:-

a. Transport sector contributes much more significantly at the city level, generally where the resident populations are several times greater.

b. According to WHO, 2012 estimates diesel exhausts have been classified as potent carcinogens.

c. Further as the growth of this sector is much higher as compared to industrial and domestic sectors if the emissions of transport sector are not controlled these are bound to grow much faster.
d. Also since transport is an organized sector, it is much easier for the implementation of strategies as well as gathering quick benefits.
e. The stake of black carbon within the PM is greater from transportation as compared to domestic or industrial emissions which provide another important reason for reduction of air pollution and control of global warming.
f. It also needs to be noted that the various strategies which would be implemented for the mitigation of vehicular emissions would also lead to co-benefits with regard to decreased congestion, fuel savings as well as the impact on climate.

Vehicle-wise contribution of emissions from the transportation sector

Fig. 4 shows the vehicle wise distribution for year 2010 with regard to emissions of various pollutants in India. It is clear that it is the heavy-duty vehicles which have the major proportion with respective shares of 52% and 83% in $PM_{2.5}$ and NOx emissions from transport sector in the country. However, the gasoline combustion within vehicles results in greater CO and HC emissions (TERI, 2010).

Therefore, unless the alternative fuels based on indigenously produced renewable feedstock are developed to substitute or supplement fossil fuels, India's transport sector would continue to be a source of air pollution. Biofuels are emerging as the most promising alternative options to conventional fuels, as they can be produced locally, and can substitute diesel or gasoline to meet the transportation sector's energy requirements. Biofuels could have positive implications for national energy security, local air quality and GHG mitigation, employment generation and rural development (Renewable Energy Policy Network for 21st Century, 2015).

Current Status of Biofuels

Out of 127.7 billion litres of global biofuel production in 2014, bioethanol accounted for 74% and biodiesel for 23% of production (Fig. 5). USA and Brazil have dominated the global ethanol market accounting for

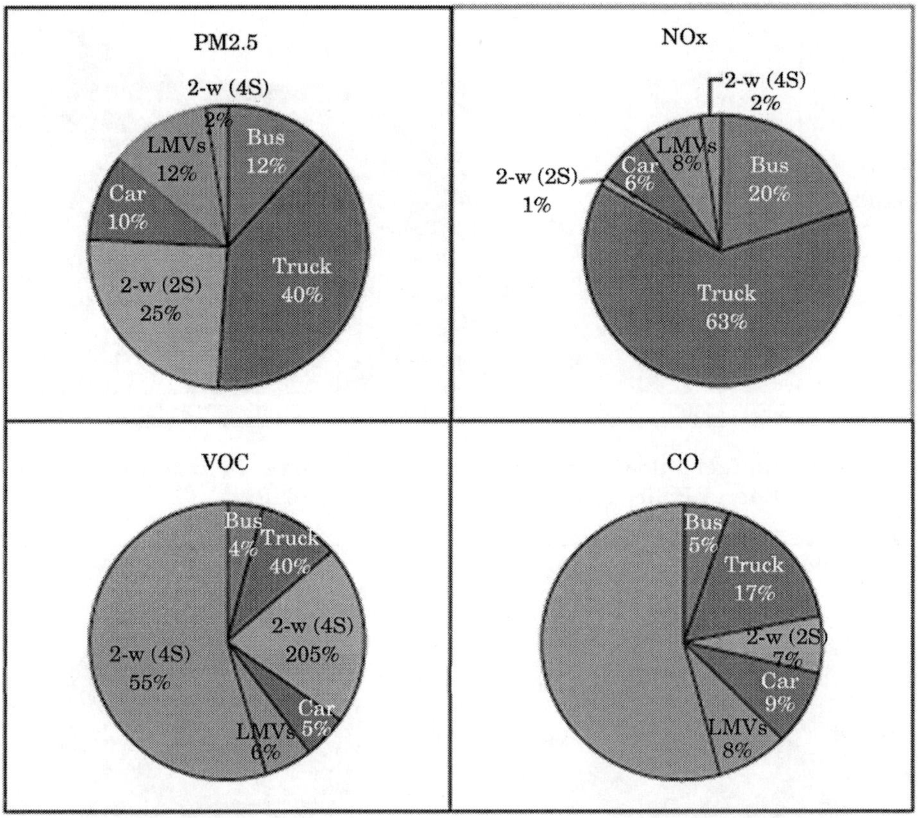

Fig. 4: Vehicular pollutants emissions in India. *Source*: TERI, 2010

over 80% of the ethanol production. The European Union (led by Germany and France), China, Canada and Thailand were other countries with significant bioethanol production (REN 21, 2015).

Nevertheless, biofuels accounted for just 3.5 percent of global road transport fuels (energy content basis) in 2014 (REN 21, 2015). Fig. 6 shows the production of ethanol and biodiesel by country and region in 2014. India's biofuel production currently accounts for nearly 1 percent of global production.

The Government of India (GoI) in 2003 launched its Ethanol Blended Petrol (EBP) Programme for 5% ethanol blended petrol. In 2009 GoI introduced a National Policy on Biofuels which focused on encouraging biofuel usage and reducing the prevailing dependence on fossil fuels while it sought to mitigate environmental and fuel efficiency concerns. Policy also recognized the opportunity offered by biofuels to agricultural

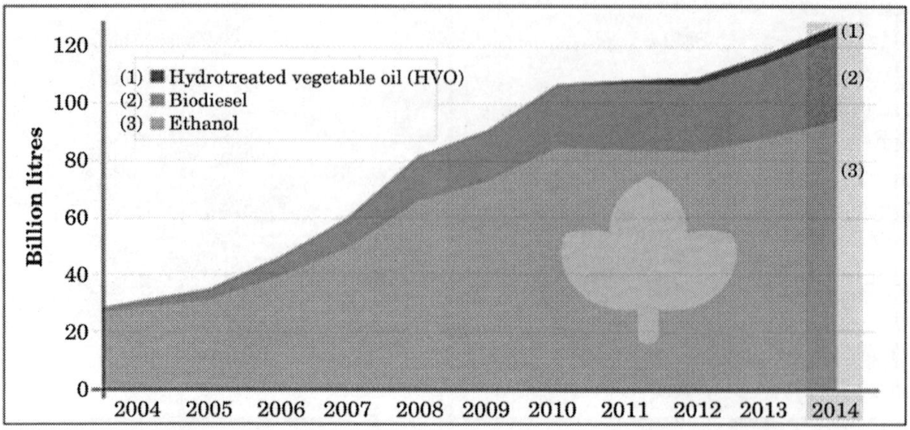

Fig. 5: Biofuels global production trends, 2004-14. *Source*: Renewable Energy Policy Network for 21st Century, 2015

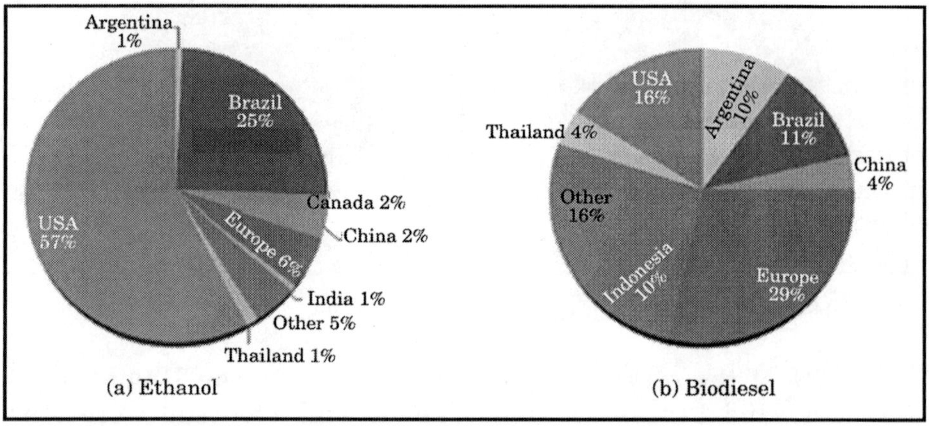

Fig. 6: Biofuel production. *Source*: Purohit and Subash Dhar, 2015

and industrial sectors. The policy has received fresh impetus with recent initiative of government to expand domestic capacity. In 2015–16 significant investments have been made in improving the storage and blending infrastructure. Further public investments (INR 7000 crores) are being made to support second generation biofuels processing and production. This alone is expected to generate an additional 350 million litres of ethanol by 2019. The National Policy on Biofuels has set a target of 20% blending of biofuels, both for biodiesel and bioethanol. However, India has managed to achieve an average blending rate of close to 5% for the first time only in 2016. Biofuels offer significant economic and environmental benefits.

Biofuels and Air Pollution

Numerous research studies have been conducted on comparison of various combinations of natural gas, diesel and diesel or biodiesel blends concerning bus emissions. The results of these studies have been well discussed and published (López *et al.*, 2009; Janulis, 2004; Krahl *et al.*, 2009; Soltic *et al.*, 2009; Coronado *et al.*, 2009; Demirbas, 2009).

Effect of Biodiesel on Engine Emissions

The blending of low levels of biodiesel into diesel significantly reduced the emissions of PM_{10}, VOC, CO as well as SOx during the process of combustion. The biodiesel blends also revealed lower life-cycle emissions for these three pollutants (CO, PM_{10}, and SO_x) relative to diesel (Sheehan *et al.*, 1998).

According to United States Environmental Protection Agency's (EPA) report (2002) as the percentage level of biodiesel blend was gradually increased from B0 (initial 0 percent biodiesel) through all the way to B100 (100 percent biodiesel), there were considerable reductions within emissions of HC, CO as well as PM however, there was a little increase in the emissions of NOx. Over the years it was that slight increase in NOx emissions that had been a significant source of controversy within several studies. Further, in most of the real-world applications it has been observed that biodiesel is generally used at low concentrations which usually range from B5 to B20. Therefore, it is in those biodiesel blend ranges that it is quite significant to examine the fate of NOx emissions. It was concluded that this disparity in the experimental emissions of NOx was dependent on the testing procedure applied, and it also illustrated the reason for complexity faced in the estimation of the "real impacts" of fuel modifications on the emissions from engine when these are applied across the total vehicular fleet. Another factor was that the different studies were undertaken in a wide range of conditions *e.g.*, various forms of biodiesel which were then blended within different base fuels in addition there were several other differences. Also the data scatter for the emissions of HC, CO, and PM emissions was as large as those for NOx emissions (EPA, 2002). Another research found that ethanol blending with gasoline (at lower levels) in form of an oxygenate was able to lower the emissions of volatile organic compounds (VOCs), carbon monoxide (CO) as well as those of harmful particulate matter with an aerodynamic diameter ≤ 10 µm (PM_{10}) upon combustion (Brinkman *et al.*, 2005). The oxygenated fuel blends of diesel possess great potential to significantly reduce the particulate matter (PM) emissions and can serve as an important substitute to diesel fuel. Results obtained have shown that bioethanol supplementation to the diesel fuel may be required for decreasing the PM generation from diesel during the process of combustion (Granda *et al.*, 2007; Ozkurt, 2009). The total number of particles as well as the total mass of the PM from the ethanol-diesel blend fuels was found to be lower by almost 11.7–26.9% (Kim and Choi, 2008). In order to evaluate the impacts of use of ethanol blends with regular diesel fuel (5% and 10% on a volume basis) an experimental investigation was undertaken. The performance as well as the exhaust emissions from a completely instrumented, heavy duty, 6-cylinder, turbocharged and after-cooled, direct injection (DI), Mercedes–Benz engine was evaluated. The fuel consumption, exhaust smoke and the exhaust controlled gas emissions (such as those of NOx, CO and total unburned HC) were measured. The engine's performance and exhaust emissions when operated on neat diesel fuel as well as when operated on the ethanol-diesel fuel blend (use of anhydrous 10% bioethanol blended with typical diesel) were ascertained and compared (Lapuerta *et al.*, 2008). The resultant emissions were then related with those obtained from pure diesel (Graham *et al.*, 2008). From the results of the statistical analysis it was evident that the utilization of E10 resulted in statistically significant reductions within CO emissions (16%); however, (statistically significant) increases were found in the emissions of acetaldehyde (108%), 1,3-butadiene (16%) and benzene (15%). But there were no statistically significant changes within the emissions of NOx, CH_4, CO_2, N_2O or those of formaldehyde emissions. The statistical analysis also suggested that the usage of E_{85} resulted in (statistically significant) declines in emissions of 1,3-butadiene (77%), benzene (76%) and NOx (45%). However, it was found that there were statistically significant increments in emissions of formaldehyde (73%) as well as acetaldehyde (2540%), but no significant changes were observed in the emissions of CO and CO_2 (Corro and Ayala, 2008).

Experimental analysis of tail pipe exhaust of a four-stroke (SI) engine was undertaken. The pollutant emissions from ethanol–gasoline blends (ranging from 0% to 20%) were analyzed. With the introduction of ethanol blends the emissions of CO and HC within the exhaust pipe were observed to be decreased. This observation was explained on the basis of the fact that there is high oxygen percentage in ethanol which brings about cleaner combustion. But in a complete contrast, the concentrations of CO_2 and NOx were found to be increased with the introduction of ethanol (Rakopoulos *et al.*, 2008).

The use of biodiesel within a typical diesel engine is associated with substantial reductions in the emissions of carbon monoxide, polycyclic aromatic hydrocarbons, nitrated polycyclic aromatic hydrocarbons, unburned hydrocarbons, sulphates as well as the particulate matter. It was found that with the increase in percentage of biodiesel blended into diesel there were reductions in these emissions except those of NOx emissions which increased (Demirbas, 2009). This fact that emissions of NOx increase with the increasing concentrations of biodiesel could prove to be a disadvantage in those areas which are suffering with higher concentrations of ozone. In the city of Helsinki over a period of three years, in 300 buses of 17 types Neste trialed its NExBTL renewable diesel. Some buses used a 30% blend while the others used 100% renewable diesel. It was found that the buses which ran on 100% *renewable diesel* did not experience any adverse issues with their engines, even when the ambient temperatures dropped to as low as 25°C. Also the average reduction in the particulate emissions was significantly high as much as by 30%.

Emissions from B20 Fuel

B20 has been of specific focus in research studies with regard to emissions as it is the upper end of the most widespread range of biodiesel application levels. In an important study undertaken by Robbins *et al.* (2011) the results were classified on the basis of engine classes namely medium and heavy duty engine as well as the medium and heavy duty engine on a chassis and the light duty engine. The study found that overall large reductions were observed within the emissions of HC, CO as well as the PM along with a small increase in the NOx emissions; however the results revealed a large scatter in them. It is important to note that the results in this study (HC, –17.4; PM, –17.2; CO, –14.1; NOx, +1.8) actually agreed very well with the results of the previous EPA study (EPA, 2002) (HC, –21.1; CO, –11.0; PM, –10.1; NOx, +2.0). With regard to the emissions of aldehydes, despite limited data availability, the use of biodiesel did not emerge out to impact the emissions in a consistent or significant manner. The impacts of biodiesel on polycyclic aromatic HC emissions have been hard to establish, but the scarce data which exists is suggestive of modest effect, in case of any (Institute of Medicine, 2014).

Literature from U.S.A (Tin, 2016) has shown that since 1995 when reformulated gasoline (with 10% ethanol) was introduced in Illinois there has been a 67% decline in NOx, 51% in Ozone, 31% in carbon monoxide, 25% in Sulphur dioxide and 29% reduction in Particulate matter. There was also a reduction in lung cancer incidences by 19-22%.

CONCLUSIONS

In India worsening air quality is threatening health, agriculture and economy. However, despite various measures taken by the government air quality has not been a priority. For a rapidly developing country like India biofuels offer various benefits in terms of foreign exchange savings, reduction in pollution, rural development, employment generation and better air quality. Literature has shown that the use of biofuels (ethanol/biodiesel) in a conventional diesel engine substantially reduces emissions of unburned hydrocarbons, carbon monoxide, sulfates, polycyclic aromatic hydrocarbons, nitrated polycyclic aromatic hydrocarbons, and particulate matter. These reductions increase as the amount of biodiesel blended into diesel fuel increases. However in order to better quantify the long term effects of biodiesel on engine performance as well as air quality need of the hour is additional research and monitoring. Another important point to be considered is that though numerous policies regarding biofuels exist on paper it is time Indian government undertakes their strict implementation.

REFERENCES

Apte, J.S., Kirchstetter, T.W., Reich, A.H., Deshpande, S.J., Kaushik, G., Chel, A., Marshall, J.D. and Nazaroff, W.W. (2011). Concentrations of fine, ultrafine, and black carbon particles in auto-rickshaws in New Delhi, India. *Atmospheric Environment*, 45(26): 4470–4480.

Brinkman, N., Wang, M., Weber, T. and Darlington, T. (2005). Well-to-Wheels Analysis of Advanced Fuel/Vehicle Systems: A North American Study of Energy Use, Greenhouse Gas Emissions, and Criteria Pollutant Emissions (Argonne Natl. Lab, Argonne, IL).

Coronado, C.R., de Carvalho, Jr. J.A. and Silveira, J.L. (2009). Biodiesel CO_2 emissions: A comparison with the main fuels in the Brazilian market. *Fuel Process Technol.*, 90: 204–11.

Corro, G., Ayala, E. (2008). Bioethanol and diesel/bioethanol blends emissions abatement. *Fuel*, 87: 3537–42.

Demirbas, A.H. (2009). Inexpensive oil and fats feedstocks for production of biodiesel. *Energy Educ. Sci. Technol. Part A*, 23: 1–13.

EPA (U.S. Environmental Protection Agency). A comprehensive analysis of biodiesel impacts on exhaust emissions: A draft technical report. 2002. [October 10, 2013]. (EPA420-P-02-001). Available at *http://www.epa.gov/otaq/models/analysis/biodsl/p02001.pdf*.

Forouzanfar, M.H. *et al.* (2015). Global, regional, and national comparative risk assessment of 79 behavioral, environmental and occupational, and metabolic risks or clusters of risks in 188 countries, 1990-2013: A systematic analysis for the Global Burden of Disease Study 2013. *The Lancet,* doi: 10.1016/ S0140-6736(15)61455-6. Specific numbers for India were obtained through the Institute for Health Metrics and Evaluation "GBD Compare" tool at *http://vizhub.healthdata.org/gbd-compare/#*

Graham, L.A., Belisle, S.L. and Baas, C.L. (2008). Emissions from light duty gasoline vehicles operating on low blend ethanol gasoline and E85. *Atmos Environ.*, 42: 4498–516.

Granda, C.B., Zhu, L. and Holtzapple, M.T. (2007). Sustainable liquid biofuels and their environmental impact. *Environ. Prog.*, 26: 233–50.

http://apps.who.int/iris/bitstream/10665/250141/1/9789241511353-eng.pdf?ua=1

http://wsds.teriin.org/concept-note/Air-Pollution-India-CN.pdf.

http://www.biofuels2050.eu/reducing-emissions/air-quality.

http:*//www.greenpeace.org/india/en/Press/Not-just-Delhi-Air-in-most-Indian-cities-hazardous-Greenpeace-report/*.

http://www.who.int/phe/health_topics/outdoorair/databases/cities/en/

Janulis, P. (2004). Reduction of energy consumption in biodiesel fuel life cycle. *Renewable Energy*, 29: 861–71.

Jason Hill, Erik Nelson, David Tilman, Stephen Polasky, and Douglas Tiffany (2006). Environmental, economic, and energetic costs and benefits of biodiesel and ethanol biofuels PNAS 2006 103 (30) 1206-11210; published ahead of print July 12, 2006,

Kaushik, G. and Borah, P. (2016). Application of Clean Air Scorecard Tool to Aurangabad City. Royal Society of Chemistry-National Environmental Engineering Research Institute symposium on Urban Air Pollution Chemistry, 2 and 3rd December 2016, NEERI Nagpur, India.

Kaushik, G., Chel, A., Shinde, S. and Gadekar, A. (2016). Assessment of PM_{10} in Aurangabad city of Central India. *International Journal of Environment*, 5(2): 61–74.

Kim, H. and Choi, B. (2008). Effect of ethanol–diesel blend fuels on emission and particle size distribution in a common-rail direct injection engine with warm-up catalytic converter. *Renew Energy*, 33: 2222–8.

Krahl, J., Knothe, G., Munack, A., Ruschel, Y., Schröder, O., Hallier, E. *et al.* (2009). Comparison of exhaust emissions and their mutagenicity from the combustion of biodiesel, vegetable oil, gas-to-liquid and petrodiesel fuels. *Fuel*, 88: 1064–9.

Lapuerta, M., Armas, O. and Herreros, J.M. (2008). Emissions from a diesel–bioethanol blend in an automotive diesel engine. *Fuel*, 87: 25–31.

López, J.M., Gómez, Á., Aparicio, F. and Sánchez, F.J. (2009). Comparison of GHG emissions from diesel, biodiesel and natural gas refuse trucks of the City of Madrid. *Appl. Energy*, 86: 610–5.

Ozkurt, I. (2009). Qualifying of safflower and algae for energy. *Energy Educ. Sci. Technol. Part A*, 23: 145.

Political, economic and environmental impacts of biofuels (2009). A review Ayhan Demirbas. *Applied Energy*, 86: S108–S117.

Rakopoulos, D.C., Rakopoulos, C.D., Kakaras, E.C. and Giakoumis, E.G. (2008). Effects of ethanol– diesel blends on the performance and exhaust emissions of heavy duty DI diesel engine. *Energy Convers Manage*, 49: 3155–62.

Ramanathan, V., Sundar, S., Harnish, R., Sharma, S., Seddon, J., Croes, B., Lloyd, A., Tripathi, S.N., Aggarwal, A., Al Delaimy, W., Bahadur, R., Bandivadekar, A., Beig, G., Burney, J., Davis, S., Dutta, A., Gandhi, K.K., Guttikunda, S., Iyer, N., Joshi, T.K., Kirchstetter, T., Kubsh, J., Ramanathan, N., Rehman, I.H., Vijayan, A., Waugh, M. and Yeh, S. (2014). India California Air Pollution Mitigation Program: Options to reduce road transport pollution in India. Published by The Energy and Resources Instiute in collaboration with the University of California at San Diego and the California Air Resources Board.

REN 21 (2015). "Renewables 2015 Global Status Report," REN 21 Secretariat, Paris. Promoting low carbon transport in India. Biofuel Roadmap for India Pallav Purohit and Subash Dhar UNEP DTU Partnership.

Robbins, C., Hoekman, S., Ceniceros, E. and Natarajan, M. (2011). Effects of biodiesel fuels upon criteria emissions. (SAE Technical Paper 2011-01-1943).

Roundtable on Environmental Health Sciences, Research, and Medicine (2014). Board on Population Health and Public Health Practice; Institute of Medicine. Biofuels Impact on Air, Atmosphere, and Health Washington (DC): National Academies Press (US).

Sheehan, J., Camobreco, V., Duffield, J., Graboski, M. and Shapouri H. (1998). Life Cycle Inventory of Biodiesel and Petroleum Diesel for Use in an Urban Bus (Natl. Renewable Energy Lab, Golden, CO) NREL Publ. No. SR-580-24089.

Soltic, P., Edenhauser, D., Thurnheer, T., Schreiber, D. and Sankowski, A. (2009). Experimental investigation of mineral diesel fuel, GTL fuel, RME and neat soybean and rapeseed oil combustion in a heavy duty on-road engine with exhaust gas after treatment. *Fuel*, 88: 1–8.

Tin, A. (2016). Clean air choice improving the air we breathe. US India Business Council Air Quality Seminar in Delhi.

5

Prospects and Constraints of Bioethanol Production in Bangladesh: An Overview

SAMAVI FARNUSH BINT-E-NASER[1], LAILA HOSSAIN[1] AND MOHIDUS SAMAD KHAN[1]*

ABSTRACT

The ongoing development in industrial and agricultural sectors has led to an increasing demand of petroleum based transportation fuel in Bangladesh. At present, along with petrol, octane and diesel, public vehicles use compressed natural gas as transportation fuel. However, because of limited natural gas reserves of the country, the government will inevitably restrict the access of natural gas to transportation vehicles in the coming years. Therefore, there will be a subsequent rise in use of petroleum based transportation fuel in the upcoming years. It is imperative for Bangladesh to look for a renewable alternative fuel source for achieving sustainable economic growth and for reducing the dependence on international oil market. Bioethanol, obtained from the conversion of carbon-based feedstock, is a quasi-renewable energy source. To meet the future energy demand and to save foreign exchange, bioethanol blend petrol and octane could be attractive alternatives for the country. This article outlines the prospects of producing bioethanol locally, possible usage of bioethanol blended petrol and octane as

[1]Department of Chemical Engineering, Bangladesh University of Engineering & Technology, Dhaka-1000, Bangladesh

**Corresponding author*: E-mail: mohid@che.buet.ac.bd

transportation fuel and the associated impacts on food sector, national economy and environment in the light of technical advances and regulatory requirements.

Key words: Bioethanol, Petrol, Octane, Energy, Economy, Environment, Emission

1. INTRODUCTION

Two key challenges of the 21st century are to meet the growing demand of energy for transportation, heating and industrial processes as well as to reduce the increasing levels of greenhouse gases in the environment. In the recent years, to address the growing concerns of energy security and climate change, there has been a greater interest in using automotive fuels derived from biomass (Fulton *et al.*, 2004). Bioethanol derived from biomass, could be used as a potential alternative to be used either as a fuel or as a gasoline enhancer.

Approximately 80% of the ethanol produced in the world is obtained from fermentation; the remainder comes largely by synthesis from the petroleum product ethylene (Lin and Tanaka, 2006). Ethanol fermentation is a biological process in which organic material is converted by microorganisms to simpler compounds such as sugars. These fermentable compounds are then fermented by microorganisms to produce ethanol and CO_2 (Lin and Tanaka, 2006). In Brazil, as of 1984, approximately 7.9 million metric tons (MMT) of ethanol was produced by fermentation of sugarcane sucrose. The US is also substantially increasing its fuel alcohol production, because of the fluctuation in petroleum prices and the subsequent need for developing alternative energy sources (Lin and Tanaka, 2006). Currently, Brazil has more than 300 bioethanol plants, producing 6190 million gallons per year and supplying 3 million cars with pure ethanol. In the US, there are more than 80 plants producing 14300 million gallons of ethanol per year. China produces 635 million gallons of ethanol per year. India's annual production of ethanol is 155 million gallons, and Europe's production figure is 1445 million gallons. In Canada, 510 million gallons of bioethanol per year is produced (Renewable Fuels Association).

In Bangladesh, the demand for fossil fuel is growing each day with the development of industrial and agricultural sectors. The annual demand for petroleum products in Bangladesh is met primarily by refining imported crude oils from overseas. To meet the future energy demand, to save foreign exchange and to reduce environmental pollution, it is essential to look for domestically produced renewable fuel sources. Bioethanol, produced from biomass could be a potential alternative fuel for Bangladesh.

Bangladesh does not commercially produce bioethanol till date. Recently, there are few initiatives within the private sector in this regard and therefore, it is important to understand the prospects and limitations of bioethanol production in Bangladesh. For this purpose, Bangladesh current energy scenario, pros and cons of bioethanol use, the availability of raw materials, possible impact on food security, possible yield of bioethanol, environmental and economic impact of bioethanol production and performance of bioethanol blends as transportation fuel are evaluated in this article.

2. BANGLADESH CURRENT ENERGY SCENARIO

The present annual demand of petroleum products in Bangladesh is about 5.25 million metric tons (MMT). The annual demand is met primarily by refining imported crude oil from overseas, which is processed with a small quantity of oil from Haripur Gas Field at Sylhet, Bangladesh. Fig. 1 shows the annual consumption of petroleum products over the last five fiscal years. Fig. 2 shows the annual consumption of petrol and octane over the last five fiscal years. Reducing the annual consumption of petroleum based transportation fuel by alternative option such as bioethanol will reduce import requirement and save foreign currency. Moreover, domestic production of alternative fuel will also reduce the dependence on the fluctuating oil prices in the international market and will provide economic stability to the country.

3. BIOETHANOL AS AN ALTERNATIVE SOURCE

On account of the finite nature of fossil fuel, fluctuating oil prices in international market and the negative impact of fossil fuel on the environment and human life, the world has been gradually shifting towards developing sustainable energy supply. Biomass energy has been gaining popularity as a renewable energy source due to its fewer adverse impacts on environment and human health. Bioethanol derived from biomass is often advocated as analternative to our needs for a sustainable transportation fuel.

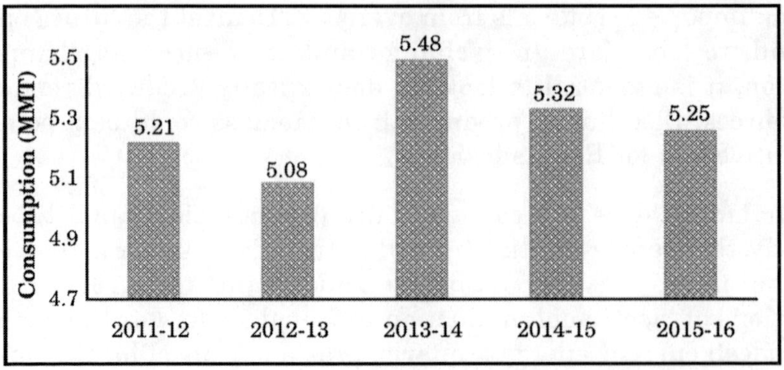

Fig. 1: Annual consumption of petroleum products (last five fiscal years) (Bangladesh Petroleum Corporation)

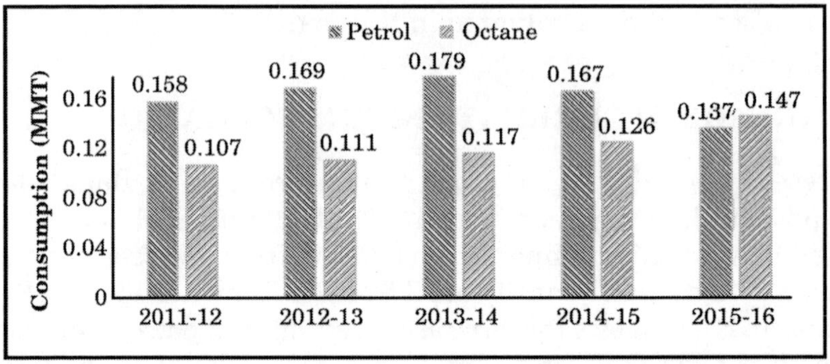

Fig. 2: Annual consumption of petrol and octane (last five fiscal years) (Bangladesh Petroleum Corporation)

3.1. Advantages

Manufacture from renewable sources

Bioethanol is manufactured from crops, such as maize, rice, and Sugarcane. In Bangladesh, the above crops are likely to be grown in high quantities and carry a much lesser threat of running out than conventional fossil fuels. Bioethanol can be produced and utilized on an infinitesimally shorter time scale than fossil fuels, which need millions of years to naturally decompose and form.

Biodegradable

Bioethanol is biodegradable and water soluble. Therefore, spills of bioethanol will not cause any form of pollution like other gasoline spills and can be converted into nontoxic compounds when spilled, 74% of bioethanol is broken down within 5 days (Green The Future).

Higher oxygen content

The oxygen content in bioethanol is around 35% (wt. basis) (Jeuland *et al.*, 2004). The increased percentage of oxygen of bioethanol blend petroleum allows a better oxidation of the gasoline hydrocarbons with the consequent reduction of CO and aromatic compounds (Sanchez and Cardona, 2008).

Higher octane rating

Bioethanol has a higher octane rating than gasoline, enabling combustion engines to run at a higher compression ratio and thus giving a superior net performance (Zaldivar *et al.*, 2001). Due to higher compression ratio, vehicles run smoothly without any change in gears and removes free water, which can plug fuel lines in cold climates.

Safer transportation and storage

The autoignition temperature and flash point of bioethanol is higher than those of gasoline, which makes it relatively safer for transportation and storage than conventional gasoline (Hsieh *et al.*, 2002).

Increased engine efficiency

Mixing bioethanol increases the antiknock value of automotive gasoline. Since the heat of combustion of the fuel-air mixture decreases with an increase of oxygen concentration in the fuel, the maximum temperature of combustion decreases. This makes it possible to increase the degree of compression of the fuel-air mixture and thus increase engine efficiency (Karpov, 2007).

Increased power outputs

The vapor pressure of bioethanol is greater and the heat of vaporization is higher than that of gasoline, which is primarily responsible for the increased power output using bioethanol (Zaldivar *et al.*, 2001).

Eco-friendly

In comparison with gasoline utilization, using bioethanol reduces greenhouse gas emission. The growing plants consume approximately equivalent amount of carbon dioxide, which is produced when the bioethanol is burnt; therefore, bioethanol could be considered carbon dioxide-neutral. Several studies have shown that Sugarcane based ethanol reduces greenhouse gases by 86 to 90% if there is no significant land use change (Ibeto *et al.*, 2011). Due to its higher oxygen content, it reduces particulate and NO xemissions from combustion (Balat and Balat, 2009).

Decrease in ozone formation

The emissions produced by burning bioethanol are less reactive in presence of sunlight than those produced by burning gasoline, which results in lower ozone forming potential.

Positive net energy balance

According to U.S. Department of Energy, the net energy balance of making fuel ethanol from corn grain is 1.34. The report also indicates that using a crop with a higher sugar content than corn, such as sugar beets, would result in production with a much higher positive net energy balance (Russian Biofuels Association).

Usable by-products

Two major by-products of corn-based ethanol are carbon dioxide and distillers dried grains with soluble (DDGS) which can be utilized in other industries. The CO_2 can be used in food and beverage industries, and in algae culture. DDGS can be used for cattle feed or further crushed to extract corn oil, for food or biodiesel usage.

Employment generation

Domestic bioethanol production creates job opportunities. According to the renewable fuel association, ethanol industry jobs have proven to be stable, well paid, fulfilling and safe (Renewable Fuels Association).

3.2. Disadvantages

Lower energy content

Bioethanol has lower energy content than gasoline (about two-third of the energy content of gasoline on a volume base) *i.e.*, it requires more bioethanol than gasoline to travel the same distance.

Higher corrosivity

Bioethanol is completely miscible with water in all proportions, while gasoline and water are immiscible (Furey and Perry, 1991). This may cause bioethanol blended fuel to contain water and further result in the corrosion problems on the mechanical components, especially for components made of copper, brass, iron or aluminum (Coelho *et al.*, 1996).

Lower reid vapor pressure (RVP)

Reid vapor pressure of bioethanol is very low which makes it difficult to start engine at low ambient temperature. Without aids, engines using bioethanol cannot be started at temperature below 20ºC.

Phase separation

When bioethanol absorbs enough moisture to reach its saturation point, phase separation occurs. Reduction of octane rating and severe engine damage can occur due to phase separation.

Pollutant emissions

Bioethanol-gasoline blended fuel may increase the emission of formaldehyde, acetaldehyde and acetone. Although using bioethanol as a fuel will increase the emission of aldehyde, the damage to the environment by the emitted aldehyde is far less than that by the poly-nuclear aromatics emitted from the aromatic gasoline (Chao *et al.*, 2000).

Food for fuel

Bioethanol production from food crops may increase the price of food due to competition with the crops. This is a particular problem for the developing countries where using food crops for fuel could lead to food shortages. Cynics have suggested that the agriculture industry in the western world has used bioethanol as a route to obtain increased subsidies and profits. Regardless of the motivation, many people are uneasy at the idea of having food and fuel production in direct competition (eejits' guides).

Monoculture

Monoculture refers to practice of producing same crops year after year, rather than producing different crops through a farmer's field over time. While, this might be economically attractive for farmers, but growing the same crop every year deprives the soil of nutrients that are put back into the soil through crop rotation (Conserve Energy Future).

Use of fertilizers

Biofuels are produced from crops and these crops, need fertilizers to grow better. The downside of using fertilizers is that they can have harmful effects on surrounding environment. Fertilizers contain nitrogen and phosphorus. They can be washed away from soil to nearby lake, river or pond to cause water pollution (Conserve Energy Future).

High cost of production

The production cost of bioethanol is greater than that of gasoline, which is still preventing the use of bioethanol from being more popular.

3.3. Possible Solutions to Overcome the Drawbacks

The ignition limit of the gasoline-bioethanol blend is higher than that of pure gasoline. Operation on lean blends is possible because of this characteristic. By selecting alcohol-resistant polymeric materials, fuel losses from evaporation can be eliminated (Karpov, 2007).

To reduce the corrosivity problem on fuel delivery system, materials of copper, brass, iron or aluminum should be avoided in preparation the mechanical component of fuel delivery system. Alcohol can react with rubber and block the fuel pipe. Therefore, it is advised to use fluorocarbon rubber if bioethanol blend is used (Naegeli *et al.*, 1997).

Phase separation caused by the presence of small amount of water can be overcome by adding different modifiers and stabilizers to blended fuels. Also it is recommended to prevent water from entering the fuel tank (Karpov, 2007). Different chemicals are proposed as stabilizers for bioethanol blend gasoline: C_3-C_{12} aliphatic alcohols of normal and branched structure, phenols, alkyl acetates, ethers and esters (their organometallic derivatives, ketones, amines, surfactants) and glycols (with their ethers, aldehydes, ketals, acetals, alkyl carbonates, carboxylic acids, and blends of these compounds). Stabilizers prevent separation of alcohol-containing gasoline (Karpov, 2007).

The cost of bioethanol production is higher than petrol and octane price. However, bioethanol production process is constantly improving, which is bringing down the cost of bioethanol production. Government can support bioethanol production in Bangladesh by offering subsidies, exemption from excise and sales tax, and encouraging pilot plants and R&D projects.

4. SCOPE OF BIOETHANOL PRODUCTION IN BANGLADESH

137,000 MT of petrol and 147,000 MT of octane were consumed as transportation fuel in the last fiscal year (2015–16) in Bangladesh. Bioethanol, produced by fermentation of sugars, starch or lingo-cellulosic materials can be blended with petrol or octane in various ratios. These blended fuels can in turn, be used to reduce the annual consumption of petrol and octane. For example, in order to produce 10% bioethanol blend petrol and octane, 57 million litres of bioethanol will be required annually; hence, the annual consumption of petrol and octane will be reduced by 10%. Using such blended fuels may have substantial impact on the food sector, economy, and environment. For further evaluation, petrol-bioethanol and octane-bioethanol blends containing 5, 10, 15 and 20% of bioethanol (referred to, in the text so-forth, as E5, E10, E15 and E20 respectively) are considered in this article.

4.1. Impact on Food Sector

In Bangladesh, potential raw materials for producing bioethanol include corn, wheat, broken rice, sugarcane, molasses, and sugarcane bagasse. Table 1 enlists the amount of potential raw materials required to produce petrol-bioethanol and octane-bioethanol blends (E5, E10, E15, and E20) in Bangladesh. As an example (according to the current estimate) it will require 114 million litres bioethanol per year (347,000 litres bioethanol per day; considering 330 working days per year) to replace 20% of the annual consumption of petrol and octane in Bangladesh (Table 1).

Table 1: Raw material required for bioethanol production

Material	*Unit*	*Annual production (MMT)*	*Required amount*			
			E5	*E10*	*E15*	*E20*
Bioethanol	million L/yr.	—	28.6	57.2	86.0	114.3
Dry corn	MMT/yr.	1.9	0.06	0.12	0.18	0.24
	% of annual production	100	3	6	10	13
Wheat	MMT/yr.	1.3	0.08	0.17	0.25	0.34
	% of annual production	100	6	13	19	26
Broken rice	MMT/yr.	4.2	0.07	0.14	0.21	0.28
	% of annual production	100	2	3	5	6
Sugarcane	MMT/yr.	7.5	0.41	0.82	1.22	1.63
	% of annual production	100	5	11	16	22
Molasses	MMT/yr.	0.1	0.11	0.23	0.34	0.45
	% of annual production	100	113	226	339	452
Bagasse	MMT/yr.	0.8	0.20	0.41	0.61	0.82
	% of annual production	100	26	51	77	102

Annual corn production in Bangladesh was 2.27 MMT (million metric ton) in 2013 (World: Maize production in metric tonnes). Generally, maize/corn contain around 16% moisture which indicates that 2.24 MMT corn would be equivalent to 1.9 MMT dry corn. 1 ton of dry corn grain yields 124.4 gal or 471 litres of bioethanol. To produce 114 million litres of bioethanol annually, 242 kT (kiloton) of dry corn grain will be required per year, which is only 13% of total dry corn produced in Bangladesh.

Bangladesh produced 1.3 MMT of wheat in 2016 (Index Mundi). Bioethanol yield from wheat is around 340–350 litre per ton of wheat (Sanchez and Cardona, 2008). To produce 114 million litres of bioethanol annually, 340 kT of wheat is required per year, which constitutes 26% of our annual wheat production. Wheat is mainly used in Bangladesh for production of flour and therefore, it may not be feasible to use wheat as the sole raw material for bioethanol production in Bangladesh. However, a significant portion of the bioethanol required to reduce the combined annual consumption of petrol and octane in Bangladesh, can be produced using wheat as the raw material.

In 2015, annual paddy production of Bangladesh was 51.8 MMT (International Rice Research Institute). It is reported that upon milling of Asian rice or paddy rice, around 8% to 16% of broken rice is produced (Gbabo and Ndagi; Hisano and Mohacsi, 2003; Kahlon, 2009). Assuming 8% broken rice is produced during milling, annual broken rice production in Bangladesh is around 4.2 MMT. Again, 1 ton of broken rice yields about 415 to 425 litres of bioethanol (Dongare). To produce 114 million litres of bioethanol annually, around 250 kT of broken rice is needed per year. Currently, broken rice is used for making rice flour and poultry feed in our country. If broken rice is used as the principal raw material for the production of required bioethanol, about 6% of the total broken rice produced in Bangladesh will be required. Consuming 6% broken rice to produce bioethanol is expected to have very minimum effect on the food security of the country.

The annual production of sugarcane, molasses and bagasse in Bangladesh was about 7.5 MMT, 100 kT and 800 kT respectively in 2014 (Banglapedia). 1 ton of sugarcane yields 70 litre of bioethanol (Sugarcane in Philipines). Thus, production of 114 million litres of bioethanol annually would require 1.63 MMT of sugarcane per year that is 22% of the total annual production of the country. It would not be possible to produce the required amount of bioethanol only using molasses or bagasse due to their limited availability and production.

A bioethanol production plant can process different types of raw materials. Therefore, depending upon the yearly production and usages of the above raw materials and to avoid any adverse impact of the food security, Bangladesh may consider using a combination of available raw materials to produce required amount of bioethanol to reduce the annual consumption of petrol and octane up to 20%.

4.2. Commercial and Economical Aspect

The international market of crude oil is often very unstable. At the moment, the crude oil price is low; however, it may increase drastically in future. At present, production costs of petrol and octane in Bangladesh are BDT 52.0 per litre of petrol and BDT 58.0 per litre of octane (BPC purchase price) respectively (Bangladesh Gazette, 2016). Using E10 blends will reduce the annual consumption of petrol by 14 kT, which is equivalent to annual saving of about 12.5 million USD. On the other hand, by using E10 blends there would be reduction in annual octane consumption by 15 kT, which is equivalent to 14.25 million USD. Table

2 presents possible reduction in annual consumption of petrol and octane, and corresponding savings of foreign currency by using bioethanol blend as petrol and octane.

At current crude oil price, production cost of bioethanol (BDT 85 per litre) is higher than the processing cost of both petrol and octane. However, partial replacement of petrol and octane consumption in Bangladesh using domestically produced bioethanol will create employment for local skilled, semi-skilled and unskilled labor and contribute towards increase in national GDP. In addition, local production of bioethanol will reduce national expenditure related to import of refined petroleum products required to meet the demand of transportation fuel

Table 2: Possible reduction in annual consumption of petrol and octane and savings in foreign currency (based on consumption in the year 2015-16 and purchase price of 2016)

	Annual consumption (kT) (2015-16)	***Reduction in annual consumption (kT)***				***Savings in currency (million USD)***			
		E5	***E10***	***E15***	***E20***	***E5***	***E10***	***E15***	***E20***
Petrol	137.4	7.0	14.0	20.6	27.4	6.0	11.9	17.9	23.7
Octane	147.6	7.4	15.0	22.1	29.5	7.1	14.3	21.4	28.5

4.3. Environmental Impact

Bangladesh has different raw materials to produce bioethanol, which can replace the annual consumption of petrol and octane upto 20%. Although the cost of production of bioethanol is higher than that of conventional petroleum, but using bioethanol as transportation fuel is more environment friendly.

Carbon dioxide (CO_2) is one of the major contributors to the greenhouse effect. Bioethanol produces less carbon dioxide emissions than petrol or octane. Petrol and octane contains approximately 90% carbon by mass and 1 ton of petrol or octane produces 3.3 tons of CO_2 upon combustion (The Engineering Toolbox). At current level of consumption, petrol and octane produces 878 kT CO_2 per year. Bioethanol on the other hand, contains only 52% carbon by mass and 1 ton of bioethanol produces 1.91 tons of CO_2 upon combustion. However, since 1 gallon of petrol or octane is equivalent to 1.5 gallons of bioethanol, gross annual CO_2 production remains almost unchanged if bioethanol blends are used to replace petrol and octane. Nevertheless, bioethanol is considered carbon neutral, *i.e.,* the amount of CO_2 released in the bioethanol production process would be consumed during photosynthesis of the crops/biomass. Therefore, the net annual CO_2 emission associated with the use of bioethanol blends is less than that associated with use of petroleum fuels. Annual CO_2 emission can be reduced up to 5.88% by using bioethanol blends containing upto 20% bioethanol. Table 3 contains gross and net annual CO_2 production associated with using petrol, octane and bioethanol blends as transportation fuels.

Bioethanol contains an oxygen atom that prompts more complete and cleaner combustion. As a result, more of the carbon in the fuel is converted to carbon dioxide (CO_2) and carbon monoxide (CO) emission is reduced. Thus, from environmental viewpoint, using bioethanol blends in Bangladesh is highly favorable, since bioethanol blends can reduce pollutant emissions.

Table 3: Annual CO_2 production associated with use of petrol and octane and bioethanol blends

Parameters		***Fuel***				
		E0	***E5***	***E10***	***E15***	***E20***
Required amount	Petrol and Octane	142.4	270.6	256.4	242.1	228.0
(kT/yr.)	Bioethanol	-	23.1	46.2	69.2	92.3
Gross CO_2 emission	kT/yr.	878	878	878	878	878
Net CO_2 emission	kT/yr.	878	834	790	746	703
% reduction	%	0	5	5.26	5.56	5.88

5. FUEL PROPERTIES: BIOETHANOL AND BIOETHANOL BLEND FUELS

In this section, the key fuel properties such as calorific value, flashpoint, density, reid vapor pressure, viscosity, octane number and water content of the E5, E10, E15 and E20 blends are evaluated. Table 4 contains physico-chemical properties of petrol, octane and bioethanol used for determining the properties of bioethanol blends for the purpose of evaluation with respect to those of petrol, octane and bioethanol.

The calorific value of bioethanol is lower than that of petrol and octane. In general, 1.5 gallons of bioethanol has the same energy content as 1 gallon of petrol or octane (as evident from the data in Table 4). As

Table 4: Fuel properties of petrol, octane and bioethanol (Gary *et al.*, 2007; Sujan *et al.*, 2015)

Parameter	*Unit*	*Petrol*	*Octane*	*Bioethanol*
Lower heating value	MJ/kg	42.7	43.5	26.8
Flashpoint	°C	–7.2	–8	17.2
Density	kg/L	0.74	0.755	0.79
Reid vapor pressure	psi	7	8	2
Viscosity	cSt	0.64	0.645	1.2
Octane number	–	80	95	110

a result, more bioethanol is needed than either petrol or octane to perform the same function. On the other hand, the higher flash point of bioethanol (Table 4) may increase the flash points of the blends to some extent. This is highly favorable in our country especially in hot weather conditions. Higher flash point of fuel will improve safety during transportation and storage by reducing the possibility of ignition in presence of any source of spark; and contribute towards improved engine performance by enhancing anti-knock properties.

Again, density of bioethanol is slightly higher than that of petrol and octane (Table 4). However, blending of bioethanol with petrol and octane causes insignificant increase (maximum 1.3% increase when E20 blend is used) in density (Table 5). Table 5 represents the density, Reid vapor pressure, viscosity and octane number of E5, E10, E15 and E20 blends.

Table 5: Fuel properties of bioethanol blends (Gary *et al.*, 2007)

Parameter	*Petrol and petrol-bioethanol blends*				*Octane and octane-bioethanol blends*			
	E5	*E10*	*E15*	*E20*	*E5*	*E10*	*E15*	*E20*
Density (kg/L)	0.75	0.76	0.76	0.76	0.75	0.76	0.76	0.76
Reid vapor pressure (psi) @ 37.8°C	6.75	6.8	6.35	6.1	6.95	7.2	6.9	6.85
Viscosity (cSt)	0.655	0.68	0.69	0.71	0.655	0.69	0.7	0.71
Octane number	82	83	85	86	96	97	97	98

Reid vapor pressure (RVP) of bioethanol is lower than petrol and pure octane (Table 4). As a result, RVP of bioethanol-blends are lower than RVP of pure petrol or pure (Table 5). This is advantageous from environmental viewpoint since the higher the RVP of a fuel, the higher are its emissions. According to RFA guidelines, unless other more volatile blending components are used, the addition of ethanol should not create a vapor pressure increase above 1.0 psi in conventional fuel which is complied within all blends (Table 5) (Renewable Fuels Association, 2005).

The viscosity of bioethanol is higher (almost double) than that of petrol and octane (Table 4), however, similar to density, the blending of bioethanol up to 20% volume content causes only a small increase in viscosity. A maximum of 10% increase in viscosity is seen for E20 blends in both the cases (Table 5).

Since spark-ignition (SI) engines rely on ignition of air and fuel compressed together as a mixture, high compressibility of the fuel is significant for SI engines. The higher the octane number of the fuel, the more compression it can withstand before igniting. Using gasoline with lower octane numbers may lead to the problem of engine knocking or premature combustion. Bioethanol has a high octane rating and can be used as octane booster to enhance octane number of petrol and octane (Table 5). As a result, using bioethanol-blends lead to improved engine performance.

Determining the water content of fuels and biofuels is important for quality control, meeting trade specifications, protecting financial value, enhancing process optimization and for taking necessary steps to reduce risks from corrosion, safety problems and infrastructure damage (Bint-E-Naser *et al.*, 2016). Bioethanol is completely miscible with water, which causes phase separation in bioethanol blends when water content of blended fuels exceeds a tolerance limit. For example, E10 blend, containing 10 vol% ethanol, has a water tolerance of ~0.45% at 60°F and if an E10 blend is contaminated with over 0.5% water the ethanol and water mixture will separate from the gasoline, petrol and octane supplied in local market in Bangladesh is usually free of water, and bioethanol production processes generally yield almost pure ethanol (minimum 99.8% alcohol). Thus E10 blends in Bangladesh are expected to have a maximum water content of 0.02 vol% which is well below the tolerance limit. Several studies have examined the effect of E10 on fuel tanks and fuel system components and have concluded that ethanol up to 10% does not increase corrosion in normal, everyday operation (Royal Automobile Association). Vehicles using greater than E10 may experience clogged fuel filters, fuel pumps and carburetors (Road Guardians). As ethanol is more corrosive than gasoline, motor vehicle engine manufacturers advise their warranties would be voided where greater than E10 blends are being used. E5 and E10 blends can be used in conventional engine without any engine modification. It is reported that various developed countries including USA, Australia, and European countries, use corrosion inhibitor chemicals along with bioethanol blend fuels (E10 and higher number, more than 10% bioethanol). If greater than E10 is to be used, it is recommended to avoid the metals containing copper, brass, aluminum or iron in fuel delivery system. However, the bioethanol blends should be stored in a water-free environment to prevent additional water being pulled into the stored fuel and consequent phase separation.

6. CONCLUSIONS

The global interest in alternative energy sources increased because of the inevitable depletion of world's fossil fuels. Bioethanol produced from biomass is a renewable energy source. The availability of feedstocks of bioethanol can vary considerably from season to season and depend on geographic locations. Bangladesh does not commercially produce bioethanol to-date; however, there are few initiatives at the private sector level in this regard. From this critical review, bioethanol is found to produce less pollutant emissions. It was also found that Bangladesh has raw materials to locally produce bioethanol. Using bioethanol blend petrol and octane as transportation fuel will not only save the foreign currency, but also create employment opportunities and reduce pollution load. However, if higher blends of bioethanol are used, the demand and price of raw materials may grow higher and affect the food cycle. In dealing with challenges in bioethanol production, Bangladesh government should prepare its long term bioethanol production plan and supporting policies associated with the plan.

LITERATURE CITED

Balat, M. and Balat, H. (2009). Recent trends in global production and utilization of bio-ethanol fuel. *Applied Energy,* 86: 2273–2282.

Bangladesh Petroleum Corporation, Marketing and Distribution Division: Sale of Petroleum Products During Last Five Years. Bangladesh [online] Available at: *http://www.bpc.gov.bd/contactus.php?id=*22 [Accessed January 2, 2016].

Banglapedia, Sugarcane. [online] Available at: *en.banglapedia.org/index.php?title=Sugarcane* [Accessed November 15, 2016].

Bint-E-Naser, S.F., Hossain, L., Barua, P.P., Debnath, M. and Khan, M.S. (2016). Analyzing Physico-Chemical Properties of Bioethanol and Bioethanol Blended

Fuels. International Conference on Petroleum Engineering 2016 (ICPE-2016), Bangladesh University of Engineering & Technology (BUET), Dhaka-1000, Bangladesh.

Chao, H.R., Lin, T.C., Chao, M.R., Chang, F.H., Huang, C.I. and Chen, C.B. (2000). Effect of methanol-containing additive on the emission of carbonyl compounds from a heavy-duty diesel engine. *Journal of Hazardous Materials,* 73: 39–54.

Coelho, E.P., Moles, C.W., dos Santos, M.A., Barwick, M. and Chiarelli, P.M. (1996). Fuel injection components developed for Brazilian fuels. SAE Technical Paper.

Conserve Energy Future, What are Biofuels? [online] Available at: *http://www.conserve-energy-future.com/advantages-and-disadvantages-of-biofuels.php* [Accessed November 10, 2016].

Dongare, M.K., Fuel Ethanol: An Opportunity for Renewable Energy & Sustainable Development in Bangladesh.

Eejits' guides, E85 Bioethanol-What's All The Fuss? [online] Available at: *http://www.eejitsguides.com/environment/ethanol-fuel.html* [Accessed November 10, 2016].

Fulton, L., Howes, T. and Hardy, J. (2004). Biofuels for Transport. International Energy Agency, Paris, France.

Furey, R.L. and Perry, K.L. (1991). Composition and reactivity of fuel vapor emissions from gasoline-oxygenate blends. SAE Technical Paper.

Gary, J.H., Handwerk, G.E. and Kaiser, M.J. (2007). Petroleum refining: technology and economics. CRC press.

Gbabo, A. and Ndagi, B. Performance Evaluation of a Ricemill Developed in NCRI.

Green The Future, Ethanol Pros and Cons. [online] Available at: *http://greenthefuture.com/ETHANOL_PROSCONS.html* [Accessed November 9, 2016].

Hisano, S. and Moha'csi, G. (2003). The role of rice processing industries in Bangladesh: A case study of the Sherpur district. *Journal of Hokkaido University*, 57: 121–133.

Hsieh, W.D., Chen, R.H., Wu, T.L. and Lin, T.H. (2002). Engine performance and pollutant emission of an SI engine using ethanol–gasoline blended fuels. *Atmospheric Environment,* 36: 403–410.

Ibeto, C., Ofoefule, A. and Agbo, K. (2011). A global overview of biomass potentials for bioethanol production: A renewable alternative fuel. *Trends in Applied Sciences Research,* 6: 410.

Index Mundi, Bangladesh Wheat Production by Year. [online] Available at: *http://www.indexmundi.com/agriculture/?country=bd&commodity=wheat&* [Accessed November 15, 2016].

International Rice Research Institute, World Rice Statistics Online Query Facility. [online] Available at: *http://ricestat.irri.org:8080/wrs2/entrypoint.htm* [Accessed February 16, 2016].

Jeuland, N., Montagne, X. and Gautrot, X. (2004). Potentiality of ethanol as a fuel for dedicated engine. *Oil & Gas Science and Technology,* 59: 559–570.

Kahlon, T.S. (2009). Rice bran: Production, composition, functionality and food applications, physiological benefits. CRC Press, Taylor & Francis Group: Boca

Raton, FL, USA.

Karpov, S. (2007). Ethanol as a high-octane, environmentally clean component of automotive fuels. Current aspects of use. *Chemistry and Technology of Fuels and Oils,* 43: 355–361.

Lin, Y. and Tanaka, S. (2006). Ethanol fermentation from biomass resources: Current state and prospects. *Applied Microbiology and Biotechnology,* 69: 627–642.

Naegeli, D.W., Lacey, P.I., Alger, M.J. and Endicott, D.L. (1997). Surface corrosion in ethanol fuel pumps. SAE Technical Paper.

Renewable Fuels Association, Economy. Washington, DC [online] Available at: *http://www.ethanolrfa.org/issues/economy/* [Accessed November 9, 2016].

Renewable Fuels Association (2005). Fuel Ethanol: Industry Guidelines, Specifications, and Procedures.

Road Guardians, Understanding Ethanol Related Fuel Problems. Wisconsin, USA [online] Available at: *https://roadguardians.org/september-2012-newsletter-biker-chads-corner/* [Accessed November 20, 2016].

Royal Automobile Association, Ethanol and E10. [online] Available at: *http://www.raa.com.au/motoring-and-road-safety/fuel-information/ethanol-and-e10* [Accessed November 20, 2016].

Russian Biofuels Association, What is bioethanol. [online] Available at: *http://www.biofuels.ru/bioethanol/What_bioethanol/* [Accessed November 9, 2016].

Sanchez, O.J. and Cardona, C.A. (2008). Trends in biotechnological production of fuel ethanol from different feedstocks. *Bioresource Technology,* 99: 5270–5295.

Sugarcane in Philipines, Israel [online] Available at: *http://www.netafim.com/story/sugar-cane-in-philippines* [Accessed March 6, 2016].

Sujan, S., Jamal, M., Hossain, M., Khanam, M. and Ismail, M. (2015). Analysis of gas condensate and its different fractions of Bibiyana gas field to produce valuable products. *Bangladesh Journal of Scientific and Industrial Research,* 50: 59–64.

The Engineering Toolbox, Coumbustion of Fuels- Carbon Dioxide Emission. [online] Available at: *http://www.engineeringtoolbox.com/co2-emission-fuels-d_1085.html* [Accessed November 14, 2016].

World: Maize production in metric tonnes, [online] Available at: *http://www.geohive.com/charts/ag_maize.aspx* [Accessed March 20, 2016]

Zaldivar, J., Nielsen, J. and Olsson, L. (2001). Fuel ethanol production from lignocellulose: A challenge for metabolic engineering and process integration. *Applied Microbiology and Biotechnology,* 56: 17–34.

6

Scope, Development and Significance of Biomass Energy in India: A Review

A.K. CHATURVEDI[1]* AND ROMESH SHARMA[1]

ABSTRACT

Living and solid biomass including industrial waste are major sources of energy. In this Chapter, scope and significance of the energy generated using biomass in India and related environmental and social issues have been discussed. Also an assessment has been made, about the biomass resources (agricultural and forest) available within India and their potential for the Biofuel production. The scope of biomass conversion technologies used within the country with their future trajectory of growth has also been reviewed. Despite substantial biomass production in India, only first generation biomass feed stock utilization is taking place, which does not fulfill the requirement of the sustainable biomass energy generation. Therefore, more intense research (second, third and fourth generation) needs to be done in conjunction with scientists and the policy makers. Finally, an assessment has been attempted with respect to social and environmental impact of the Bio-mass exploitation. These two aspects need to be minimized in near future to exploit the abundant biomass wealth of India for energy generation with the consideration of rural biomass energy development programmes.

Key words: Biomass, Bio-fuel, Bio-energy, India

[1]College of Engineering Roorkee Uttarakhand, India
**Corresponding author*: E-mail: ajay_bengal@yahoo.co.in

INTRODUCTION

The primary energy consumption in India is the third largest after China and USA with 5.3% global share as per recent data (2015) (Renjish and Vimala, 2016).

Total primary energy consumption from crude oil (33.6%), natural gas (12.5%), coal (49.5%), nuclear energy (1.1%), hydroelectricity (3.2%) and renewable power (2.21%) (Excluding traditional biomass use) in 2015, (BP., 2015). This resulted in primary energy consumption rising by 5.9% per annum (P.A.). The growth was led by coal and oil, meeting the 49.5% and 33.6% of the growth in demand respectively. Although natural gas consumption also increased during this period but it did so, from a very small base. In 2013, India's net imports were nearly 144.3 million tons of crude oil, 16 Million Tons of Oil Equivalent (Mtoe) of Liquefied Natural gas (LNG) and 95 Mtoe coal totaling to 255.3 Mtoe of primary energy which is equal to 42.9% of total primary energy consumption. About 70% of India's electricity generation capacity is from fossil fuels. India is largely dependent on fossil fuel imports to meet its energy demands. By 2030, India's dependence on energy imports is expected to exceed 53% of the country's total energy consumption. In 2009–10, the country imported 159.26 million tons of crude oil which amounted to 80% of its domestic crude oil consumption at that point in time and 31% of the country's total imports were oil imports. By the end of calendar year 2015, India had become a power surplus country with huge power generation capacity idling for want of electricity demand (Deb and Applaby, 2015).

Due to rapid economic expansion, India has one of the world's fastest growing energy markets and is expected to be the second-largest contributor to the increase in global energy demand by 2035, accounting for 18% increase in worldwide energy consumption. Given India's growing energy demands and limited domestic fossil fuel reserves, the country has ambitious plans to expand its renewable and most worked out nuclear power programme. India has the world's fifth largest wind power market and also plans to add about 100,000 MW of solar power capacity by 2020. India also envisages increasing the contribution of nuclear power to overall electricity generation capacity from 4.2% to 9% within next 25 years. The country has five nuclear reactors under construction (third highest in the world) and plans to construct another 18 additional nuclear reactors (second highest in the world) by 2025. However, India is still quite low on the primary energy consumption index. which is 775 Mtoe (Deb and Applaby, 2015). Another aspect which is relevant in India's case is that one form of energy is incapable of meeting all our energy needs because of regional and seasonal variation. Thus, in India's case the energy basket has to comprise of more than one form of energy. The current energy basket of India is as tabulated below:-

Table 1: Energy basket of India

Sl. no	*Type*	*Contribution to energy basket in percentage (%) terms*
1.	Coal	44
2.	Biomass and waste	23
3.	Petroleum and other Liquids	23
4.	Natural gas	06
5.	Nuclear	01
6.	Hydroelectric	02
7.	Other renewable	01

Source: Renjish and Vimala, 2016

SECOND-GENERATION BIO-FUELS

Gasification of bio- mass, yields wood gas or Syn-gas which can be converted into substitute natural gas by methanation. Nearly 750 million tons of non edible (by cattle) biomass is available annually in India (Dwivedi *et al.*, 2013), which can be put to higher value addition and then it could act as a substitute to imported crude oil, coal, LNG, nuclear fuels etc. It is estimated that renewable and carbon neutral biomass resources of India can replace present consumption of all fossil fuels when used productively. Huge quantity of imported coal is being used in pulverised coal-fired power stations. Raw biomass cannot be used in the pulverised coal mills as they are difficult to grind into fine powder due to caking property of raw biomass. However, biomass can be used after torrefaction in the pulverised coal mills for replacing imported coal. North, West and Southern regions can replace imported coal use with torrefied biomass as surplus agriculture/crop residual biomass is available. Cheaper production cost of algae oil from algae particularly in tropical countries like India would displace the prime position of crude oil in near future (Dwivedi *et al.*, 2013).

Former President of India, Late Dr. A.P.J. Abdul Kalam, was one of the strongest advocates of *Jatropha* cultivation for production of biodiesel. According to him out of a total of 6,00,000 km^2 of waste land that is available within India, over 3,00,000 km^2 is suitable for *Jatropha* cultivation. Once the plant is grown, it has a useful lifespan of several decades. During its entire life cycle the *Jatropha* requires very little water as compared to other cash crops. A plan for supplying incentives to encourage the use of *Jatropha* has been coloured with green stripes (NGOs and cooperatives are raising nurseries for *Jatropha* plantation and supplying saplings to others for further cultivation).

Biomass Energy

Energy generation from Biomass is affected by two elements involved in the generation chain, namely; production practices and the energy conversion technologies. Technological advancements in biomass energy conversion comes from three sources - enhanced efficiency of biomass energy conversion technologies, improved fuel processing technologies and enhanced efficiency of end-use technologies. The main sources of biomass energy are trees, crops and animal waste. Until the middle of the 19th century, biomass dominated the global energy supply with a 70 percent share (Grubler and Nakicenovic, 1988). Presently; the biomass sources contribute 14% of global energy supply as a whole and 38% of energy supply in the developing countries (Woods and Hall, 1994). Among the biomass energy sources, wood fuel is the most important energy source, which is now getting developed around the world, especially to reduce the consumption of fossil fuels. On a global scale, biomass; supplies more than 1% of the electricity demand, *i.e.*, some 257 TWh per year (IEA, 2009). Versatility of modern biomass technologies to use variety of biomass feedstock has enhanced the supply potential. Small economic size and co-firing with other fuels has also opened up additional applications. Biomass integrated gasifier/combined cycle (BIG/CC) technology has potential to be competitive (Reddy *et al*, 1997; Johansson *et al*, 1996), since biomass as a feedstock is more promising than coal for gasification due to its low sulfur content and less reactive character.

Biomass remains the primary energy source in the developing countries in Asia especially in South and South East Asia. Share of biomass in energy basket varies - from a very high (over three quarters) in countries such as Bhutan, Nepal, Laos, Sri Lanka, Cambodia and Myanmar to nearly half in those of Vietnam, Pakistan and Philippines. Its share is nearly a third in India and Indonesia while China and

Malaysia it has a lower proportion (almost 10 percent) (FAO, 1997). In the wake of rapid industrialization and marketization during past last decades, the higher penetration of commercial fossil fuels in most Asian developing nations has caused decline in the share of biomass energy. The absolute consumption of biomass energy has however been rising unabatedly during past two decades, growing at an annual rate of over two percent (FAO, 1997).

The term biomass literally means living matter. However biomass is often used to describe any organic material obtained from plant and animals tissues. This includes agricultural biomass, forest biomass, industrial biomass and other kind of biomass as well as algae. These materials are referred as feed stocks in bio refining and are classified into four generations: first, second, third and fourth generation (Fig. 1). As far as India is concerned, first generation biomass energy is taken into consideration and other three forms have been focused on future generation energy.

Potential of Biomass Energy in India

A total of 94,125 MW of renewable energy based power generation were achieved up to 31st March 2013. Out of which wind power contributed to 52.20%, small hydro power by 20.98%, biomass power contributed 18.63%, cogeneration bagasse provided 5.31% and waste to energy contributed marginally to 2.88%. Punjab has highest potential for bio-energy by 2013, 3517 MW were generated. Ministry of New and Renewable Energy (MNRE) is promoting Biomass Gasifiers, which comprises a process of converting solid biomass fuel into a gaseous combustible gas; called producer gas; through a sequence of thermo-chemical reactions. The gas is a low-heating value fuel, with a calorific value between 1000–1200 kcal/Nm^3 (Kilo Calorie per normal cubic metre) through power plants electricity using locally available biomass resources. Various projects related to biomass power generation have been installed in various states of India for fulfilling energy requirement by biomass gasification. India, being located in the tropical region, is endowed with vast potential of renewable energy. It is at the forefront of harnessing renewable energy and is one of the largest programmes in the world. The goal of the Eleventh Five-year Plan was to achieve at least 10% of grid-interactive power generation capacity from Renewable Energy Systems (RES) by the end of the plan period. As on 31 December 2010, grid-interactive renewable power contributed about 15,691.43

megawatts (MW) of electricity, *i.e.*, about 10% of the total installed capacity in the country at that time. The leading state for biomass power projects are Chhattisgarh, Uttar Pradesh, Maharashtra, Andhra Pradesh and Tamil Nadu. The states which have taken leadership in bagasse cogeneration projects are Andhra Pradesh, Karnataka, Maharashtra, Tamil Nadu and Uttar Pradesh (Sylvain *et al.*, 2009). All the states lay emphasis on first and second generation Bio-fuels. However third and fourth generation of Bio-energy generation needs to be focused on, because these have great potential in India (Fig. 1).

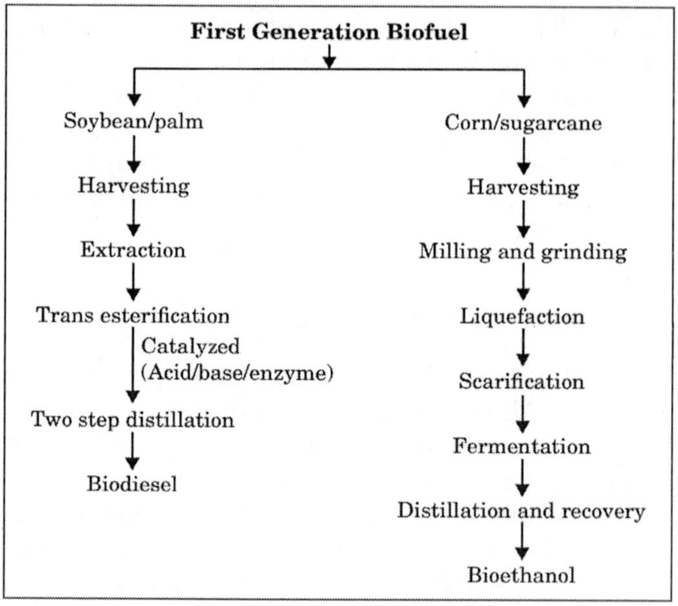

Fig. 1: Four levels of biomass energy generation

Problems Associated with the Use of Biomass as a Source of Energy

Most of the biomass energy in the World in general but particularly in India is derived from the sources owned by private individuals; like farm trees, agriculture or cattle, or that which is collected by the households. The biomass energy consumption is much limited in meeting the cooking needs of households and traditional industries and services in rural areas as well as in urban area as well. In developing countries, due to surplus labor availability, biomass acquires no resource value so long as it is not scarce. In the absence of an energy market, the traditional biomass fails to acquire exchange value in substitution. Absence

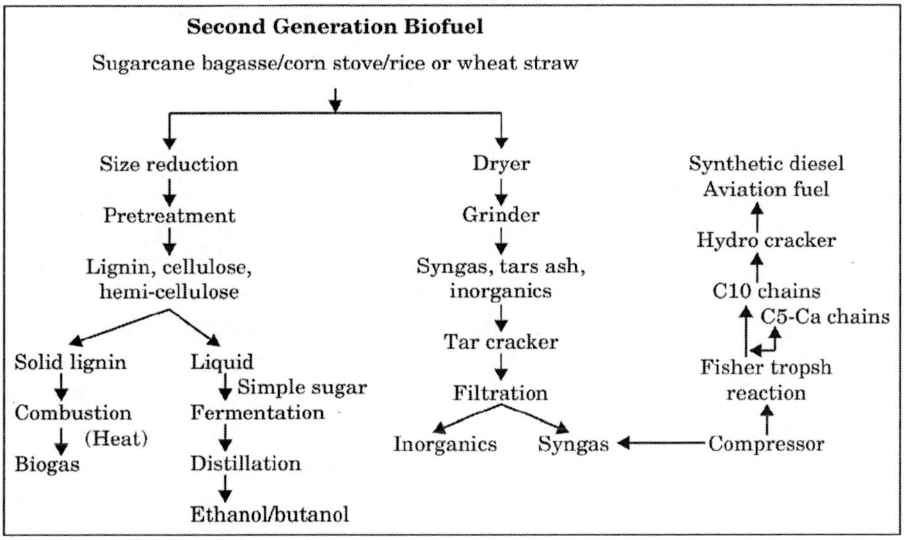
Second Generation Biofuel
Sugarcane bagasse/corn stove/rice or wheat straw
Size reduction
Pretreatment
Lignin, cellulose, hemi-cellulose
Solid lignin
Combustion
(Heat)
Biogas
Liquid
Simple sugar
Fermentation
Distillation
Ethanol/butanol
Dryer
Grinder
Syngas, tars ash, inorganics
Tar cracker
Filtration
Inorganics
Syngas
Compressor
Fisher tropsh reaction
C5-Ca chains
C10 chains
Hydro cracker
Synthetic diesel
Aviation fuel

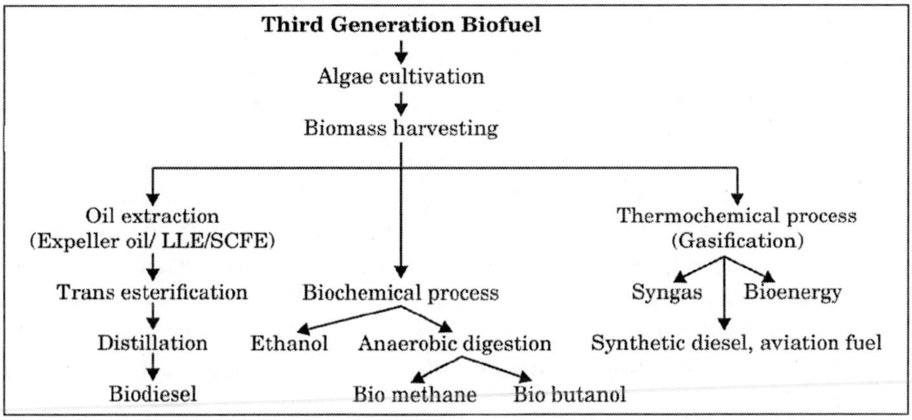
Third Generation Biofuel
Algae cultivation
Biomass harvesting
Oil extraction (Expeller oil/ LLE/SCFE)
Trans esterification
Distillation
Biodiesel
Biochemical process
Ethanol
Anaerobic digestion
Bio methane
Bio butanol
Thermochemical process (Gasification)
Syngas
Bioenergy
Synthetic diesel, aviation fuel

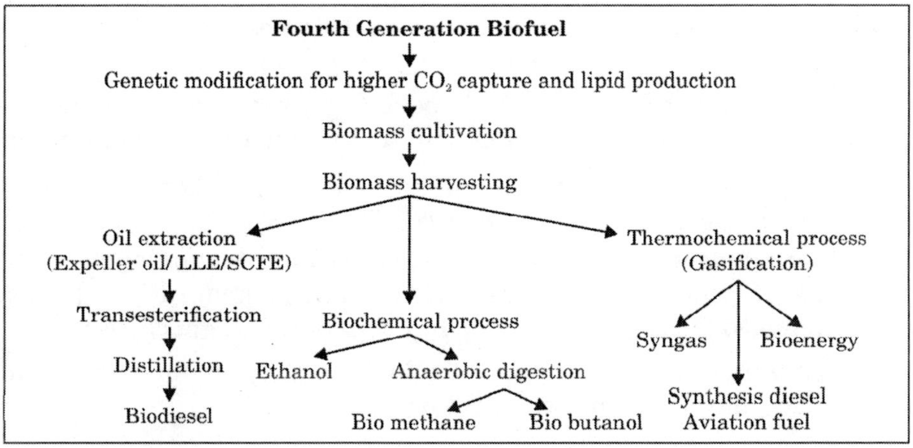
Fourth Generation Biofuel
Genetic modification for higher CO2 capture and lipid production
Biomass cultivation
Biomass harvesting
Oil extraction (Expeller oil/ LLE/SCFE)
Transesterification
Distillation
Biodiesel
Biochemical process
Ethanol
Anaerobic digestion
Bio methane
Bio butanol
Thermochemical process (Gasification)
Syngas
Bioenergy
Synthesis diesel
Aviation fuel

of market thus acts as a barrier to the penetration of efficient and clean energy resources and technologies which puts a check on its growth and further development.

An additional problem with the traditional biomass use is the social costs linked with an excessive environmental pollution. The incomplete combustion of biomass in traditional stoves releases excess pollutants in environment in the form of release of toxic gases like; carbon monoxide, methane, benzo(a)pyrene, benzene, nitrogen oxides formaldehyde, aromatics and respirable particulate matter. These pollutants cause considerable health hazards, especially for women and small children who are exposed to indoor pollution for long duration hours due to poor cooking practices and faulty accommodation design especially in the rural areas (Smith, 1987; Smith, 1993; Patel and Raiyani, 1997). Other problems with the traditional biomass energy conversion are; the energy conversion inefficiency of the technologies entailed due to inefficient combustion and unsustainable harvesting practices.

Biomass Energy Use in India

India has a long history of energy planning and programme interventions. Programmes for promoting biogas and improved cooking-stoves began as early as fourth decade of the twentieth century. Afforestation and rural electrification programmes are being pursued since fifth decade of the twentieth century. A decade before the oil crisis of 1973, India appointed an Energy Survey Committee. Biomass policies followed a multi-pronged strategy (Shukla, 2007). Salient aspects of this strategy are as follows:-

a. Improving efficiency of the traditional biomass use (*e.g.*, improved cooking-stove programme).
b. Improving the supply of biomass (*e.g.*, social and commercial forestry, wasteland development and reclamation programme).
c. Sustainable technologies for improving the quality of biomass use (*e.g.*, biogas, improved cook-stoves).
d. Introduction of biomass based renewable technologies (wood gasifiers for irrigation as fuel, electricity generation through biomass) for services provided by conventional energy sources.
e. Establishing institutional and logistic support for programme formulation and implementation.

The institutional response resulted in establishment of Department of Non-conventional Energy Sources (DNES) in 1982 and state level nodal energy agencies during the early part of the decade of 1980.

The aim of modern biomass programmes in India is to overcome problems mentioned in the earlier section. In this context; some of the programme achievements like; introduction of efficient and clean technologies for household energy use such as improved cooking-stoves (22.5 million), family sized biogas plants of two to four cubic metres per day capacity (2.4 million) and community biogas plants (1623) which have been added (till March 1996) to the technology stock (CMIE, 1996). Although, the biogas and the improved cooking-stove programmes have been only moderately successful, their overall impact on rural energy remains marginal (Ramana *et al.*, 1997). An analysis of the programme revealed two deficiencies in policy perspectives which have contributed to the slow progress in the penetration of biomass technology. Firstly, the biomass was viewed solely as a traditional fuel for meeting rural energy needs and secondly, the policies primarily focused on the supply-side push. Market instruments had little role in biomass policies. Under the circumstances, neither the modern plantation practices for augmenting the biomass supply nor the growing pool of advanced biomass energy conversion technologies could penetrate the Indian energy market.

Potential of Biomass Based Energy Generation in India

Biomass contributes over a third of primary energy in India and is a vital source of energy for the household and industrial energy requirements. It is still the most commonly used domestic fuel, besides; being the energy source for several small-scale industries and fuel for independent power plants. Biomass fuels are predominantly used in rural households for cooking, water heating and as by the traditional and artisan industries. Sector wise usage is as tabulated below in Table 2.

Table 2: Sector wise usage of biomass to meet energy needs

Sl. no	*Sector*	*Percentage of total biomass energy consumption*	*Authority*
1.	Domestic use		
	a. Rural	–90	NCAER-1992
	b. Urban	–40	
2.	Wood fuels	56	Sinha *et al.*, 1994

Note: Consumption of wood grew annually at 2 percent during past two decades (FAO, 1996).

As per an estimate, in 2009 the total financial investment in clean energy in India was at 135 billion (REDA, 2015–16). Apart from this, Indian Renewable Energy Development Agency (IREDA) and other public sector agencies are also actively funding renewable energy projects. The absolute consumption of biomass energy has steadily risen unabatedly during last two decades, growing at an annual rate of over 1–2 percent (FAO, 1997). Various factors like over population and scarcity of commercial fuels in rural areas and traditional sectors have sustained the growing biomass use. However, it needs to be noted that the demand of timber has already led to considerable deforestation and/ or loss of forest cover. The deforestation and the land degradation with changing land use pattern have made tropical Asian forests the net emitters of atmospheric CO_2 (Dixon *et al.*, 1994). Presently the Asian governments are working to increase afforestation (by a ratio of 8.5:1) (Houghton, 1996). The sustainable growth of biomass energy in Asia therefore would require augmenting existing biomass resources with modern plantations, of energy crops and by introducing efficient and sustainable biomass energy conversion technologies.

India's biomass based renewable energy sources have the potential to generate 68,000 MW of green power costing under Rs 6/- (€0.098) per unit, according to a World Bank report. The study, titled Potential of Renewable Energy in India, states: 'Developing indigenous renewable energy sources, which have low marginal costs of generation is more economically viable in the long run ((Bioenergy India, 2011). The current availability of biomass in India is estimated at about 500 million metric tons per years. The MNRE has estimated biomass availability at about 120–150 million metric tons per annum covering agricultural and forestry residues corresponding to a potential of about 18,000 MW. In addition to this, about 5,000 MW of additional power could be generated through bagasse based co-generation in the country's 550 sugar mills (Bioenergy India, 2011).

Current status

In the last six decades, India's energy use has increased 16 times and the installed electricity capacity by 84 times. In 2008, India's energy use was the fifth highest in the world (Bio-energy India, 2011). Still, India as a country suffers from significant electricity deficit and faces 15 percent shortfall during peak hours (IEA, 2008). As of December 2010, the projected total installed capacity of biomass based power in

India was to be 2,559 MW in the near future (MNRE, 2010). In this background with the huge potential for the development of renewable energy, a grant of €200.5 million (approximately 1,200 cores) was sanctioned as a Line of Credit by The KfW Development Bank, Germany for Promotion of Renewable Energy Projects in India (Bioenergy India, 2011).

Future of Biomass Energy in India

As far as future potential is concerned, India is poised to become one of the world leaders in power generation from biomass in the near future. India has a potential to generate an additional 20 GW (Shukla, 1998) of electricity from biomass residues. In order to realize the potential effectively, various fiscal incentives are being provided by the Government. Under key incentives like capital subsidy, renewable energy certificates and Clean Development Mechanism (CDM) which can be utilized effectively to make the project economically attractive.

Table 3: Fuel wood consumption in million tons

Sl. no	*Type of demand*	*Quantity*	*Cumulative total*
1.	Household		
	a. Forested rural	78	78
	b. Non-forested rural	74	152
	c. Urban areas	10	162
2.	Cottage industry	25	187
3.	Rituals	04	191
4.	Hotels	10	201

Source: Rai and Chakrabarti, 1996

In India wood fuels contribute 56 percent of total biomass energy (Sinha *et al.*, 1994). Consumption of wood has grown annually at 2 percent rate over past two decades (FAO, 1996). However, it needs to be noted that in India, estimates of biomass consumption remain highly variable (Ravindranath and Hall, 1995; Joshi *et al.*, 1992), since most biomass is not transacted on the market. On the basis of supply-side estimates (Ravindranath and Hall, 1995), the types and the quantum of the biomass energy reported are; fuel wood for domestic sector- 218.5 million tons (dry), crop residue- 96 million tons (estimate for 1985), and cattle dung cake-37 million tons. A recent study (Rai and Chakrabarti, 1996) estimates that the demand in Indian subcontinent for fuel wood is 201 million tons (Table 3). Supply of biomass primarily has been from the fuels that are home grown or collected by households for own needs. The Government sponsored social forestry programme has added to fuel-wood supply to the tune of 40 million tons on an annual basis (Ravindranath and Hall, 1995).

Modern Biomass Technologies in India

A decade of experience with modern biomass technologies for thermal, motive power and electricity generation applications exists in India. Gasifier technology has penetrated the applications such as village electrification, captive power generation and process heat generation in industries producing biomass waste. Over 1600 gasifier systems, having 16 MW total capacity, have generated 42 million Kilo Watt hour (KWh) of electricity, replacing 8.8 million litres of oil on an annual basis (CMIE, 1996). An important aspect of small gasifier technology in India is the development of local manufacturing base. The large sized gasifier based power technologies are at R&D stage and at best in pilot demonstration stage. The thrust of the biomass power programme is now on the grid connected megawatt scale power generation with multiple biomass materials such as rice straw, rice husk, bagasse, wood waste, wood, wild bushes and paper mill waste. Nearly 55 MW of grid connected biomass power capacity has already been commissioned and another 90 MW capacity is under erection. Such an enhanced scale of operations has improved the economics as well as the technology of biomass power generation. Technology improvement has also derived from the joint ventures of the Indian firms with the leading international manufacturers of the turbines and the electronic governors, (Bioenergy India, 2011).

Status of Renewable Energy in India

India, being located in the tropical region, is endowed with a vast potential of renewable energy. It is at the forefront of harnessing renewable energy and has one of the most ambitious programmes in the world. The goal of the Eleventh Five year Plan was to achieve at least 10% of the grid interactive power generation capacity from Renewable Energy Systems (RES) by the end of the plan period (Ministry of New and Renewable Energy, India). As on 31 December 2010, grid interactive renewable power contributed about 15,691.43 megawatts (MW) of electricity, *i.e.*, about 10% of the total installed capacity in the country at that time. Today, India has a significant potential for the generation of power from renewable energy sources like wind, small hydro, biomass, solar energy. India is at the 5th position among the countries exploiting wind power in the world, in terms of exploitation of wind power, after Denmark, Germany, Spain and USA, with an estimated total wind potential of 45195 MW. The country has an estimated Small Hydro Plants (SHP) potential of about 15000 MW along with other renewable energy technologies, including solar photovoltaic, solar thermal and biomass power which are also spreading. The bio-power (agro residue) is estimated around 16900 MW and co-generation from bagasse and waste to energy is about 5000 MW and 2700 MW respectively (Ministry of New and Renewable Energy, India). After wind and solar energy, biomass energy has the third largest potential for power generation through renewable energy sources with a total potential of 24,600 MW (including agro residue, co-generation from bagasse and waste to energy) however, only 3526.87 MW has been harvested till now. Therefore, there is a wide scope to tap the biomass potential to fulfil the needs of society through renewable energy sources (Table 4). To overcome the dependency on the imported fuel we have to look through other sources of energy like biomass, solar, hydro, geothermal, wind and others (Dwivedi *et al.*, 2014).

Systematic Plan of Biomass Energy Development in India

Table 4: Biomass power in India – now and future

Parameter	***Now (2012)***	***Future (2020)***
Distributed electricity	Biomass power (agro-wastes/ residues) has a minor contribution of 13% of the total grid connected renewables while it contributes to nearly 73% of of off-grid power.	Estimates of potential in the future for power from biomass in India varies from about 18,000–50,000 MW generation
Use in co-firing in power plants	Fewer than 1% of power plants use biomass	A much larger proportion of plants will be powered by biomass
Use of feedstock	Primarily waste biomass and assorted	Dedicated energy crops
Related revenue streams	Some additional revenue streams already present	A more established end user market for co-products
Standalone renewable power source	Primarily stand alone mode	Will be used in conjunction with other renewable electricity sources

Source: Bioenergy India, 2011

India currently produces around 1,70,229 MW of energy from all sources, but the nation's demand for electricity is expected to rise by 7.4 percent a year during the next quarter of century. This calls for generation capacity to increase at least fivefold in India to meet this growing demand. Our government has set a target, which, if met, could see at least 40,000 MW of additional renewable energy capacity added.

Source: http://www.bioenergy-news. com/index.php?/Industry-News?item_ id=3215

Various States have Formulated Programmes for Bioenergy which are Discussed in Subsequent Paragraphs. These are as mentioned in the subsequent paragraphs

Punjab

In controlling the pollution levels, the state of Punjab of India aims to generate 1,000 MW of renewable energy from agricultural waste by 2015. Excessive amount of cow dung is proving problematic for cities such as Ludhiana and is a major source of pollution for the region. In addition to livestock waste, which is the major source for waste, Punjab also produces 21 million tons of rice straw and other biomass crop residues annually. The Punjab government now plans to generate 10 percent of its total energy output from renewable sources in the next five years. Already in operation near Ludhiana is a 1 MW bio-methanation plant – a facility dedicated to enhancing the production of methane through the addition of microbes known as methanogens. The facility produces 18,000 kWh of electricity from 235 tons of cattle dung. It also makes 45 tons of organic manure a day which can be used to fertilize the fields. To date, the Punjab Energy Development Agency has commissioned 318 MW of renewable energy projects. These include 250 MW of biomass co-generation plants, 28 MW of biomass power plants, 37 MW of small hydro projects and 2 MW of solar projects, in addition to a 1 MW bio-methanation facility in Ludhiana. However, according to the agency, there are projects currently under development to add an additional 132 MW of renewable power capacity, including 100 MW of cogeneration, 20 MW of standard biomass and 11 MW of hydro. This means that by 2012 around 700 MW of renewable energy had already been added in the state. According to former deputy chief minister Sukhbir Singh Badal, farmers were now earning around 4,000 (€65) per acre per year by selling agro residues to biomass plants, Ministry of New and Renewable Energy, (Bioenergy India, 2011).

Source: http://www.bioenergy-news. com/index.php?/Industry-News?item_ id=2781

Karnataka

In Bangalore, the capital city of the State of Karnataka, a landfill waste-to-electricity plant has been installed at K.R. Puram. Two decades of research in Scalene Energy Research Institute (SERI) provided the technology for the power plant. Spiral Protium Accelerated Reactor Super Enrichment (SPARSE), SERI claims that its technology is the first of its kind in the world. The technology has not yet been commercialized but according to the advisor to MNRE Mr. A. Shukla the cities would become cleaner and less polluted, should the technology be developed on a large-scale basis. A number of companies have expressed interest in erecting similar plants in the region, including the UB Group, Global Green Energy Parks and Malankara Plantations.

Source: http://www.bioenergy-news. com/index.php?/Industry-News?item_ id=2709

Biomass energy for rural India (BERI), Tumkur, Karnataka

Initiated in 2001 by UNDP and the Government of Karnataka, the project aimed at biomass gasifiers to provide electricity to the 24 project villages and community biogas systems for the provision of clean energy. The Indian Institute of Science (IISc), a premier R&D institution, was engaged to supervise, render advice and train locals on the gasifier plant operation. Extensive community mobilization was actuated through the creation of 26 Village Bioenergy Management Committees (VBEMC), 26 Village Forest Committees (VFC) and 72 Self Help Groups (SHGs) led by women, and the strengthening of 68 old SHGs, 31 Water User Associations and 33 Biogas User Groups. The development of biomass was activated through 'energy' plantations. About 2015 ha of common land was taken under forestation. A nursery with nearly two million seedlings was set up, alongside tree-based farming over 900 hectares of land. To address the immediate need for a cleaner cooking fuel, community biogas plants were built. Irrigation problems were reduced through drip irrigation. As of July 2010, a total 383 MWh of green energy had been generated, leading to reductions of 11,880 tones of CO_2 after taking into account carbon sequestration, (*Project sponsored by UNDP and Government of Karnataka, India in year 2010*).

IFC to invest in SEGL

The International Finance Corporation (IFC), the private investment arm of the World Bank, invested up to $30 million (€21.5 million) in Andhra Pradesh, India based renewable energy company, Shalivahana Green Energy (SGEL). $15 million of the funds were assigned to financing SGEL's renewable power projects, while the other $15 million were provided to the fully owned subsidiary of SGEL – Rake Power – in project debt financing. To date SGEL's five operational biomass power plants generated 47 MW between them. The company would add an additional 200 MW to its production capacity subsequent to completion of its other projects, currently at various stages of implementation. Three of SGEL's biomass plants were expected during Q2 2011, generating approximately 45 MW. Rake Power planned two 23 MW plants (10 MN capacity) for Jharkhand and Orissa, which costed $25 million each. IFC's $15 million loan for Rake Power was for 12 years.

Source: http://www.bioenergy-news. com/index.php?/Industry-News?item_ id=3324

Thermal, biomass power plants to come up in Bihar

Bihar State has a daily requirement of 2,200–2,500 MW of energy. However, it produced only 45–50 MW of power. Electricity-starved Bihar set up five thermal power plants of a total capacity of 7,000 MW and others based on gas and biomass, this sector attracted private investment worth hundreds of crores of rupees. Other than anything else, Bihar attracted private investment in the power sector till March 31, 2011" according to Bihar State Investment Promotion Board (SIPB). The thermal power plants, of 1,320 MW each, were set up at Kajra in Lakhisarai, Piparpainti in Bhagalpur, Areraj in East Champaran and one in Banka district. Besides, a plant of 2,640 MW was set up at Rajauli in Nawada district. SIPB officials said investment proposals to set up gas- and biomass based power plants in Rohtas, Gaya and Kaimur districts were also cleared. In view of the shortage of power, Bihar Chief Minister Mr. Nitish Kumar asked top officials to review progress in procuring power from non-conventional energy sources since residents in Bihar are still living in the lantern age with electricity being a luxury for people in most parts. Capital Patna being an exception but most small towns and district headquarters have been badly hit by the power shortage. While the state has a daily requirement of 2,200–2,500 MW, it is able to generate only 45–50 MW of power.

Source: http://articles.economictimes. indiatimes.com/2011-04-14/ news/ 29417579_1_power-plants-sipbpower-sector

Railways to build biodiesel plants

Indian Railways decided to promote use of alternative fuels such as biodiesel in a big way for powering its vast fleet of over 4000 diesel locomotives in its endeavour to adopt green technologies. Indian Railways is the single largest bulk consumer of diesel in the country and as mentioned in Railway Budget 2014–15, it would start using biodiesel up to 5% of the total fuel consumption in diesel locomotives. According to a report in the Economic Times, two major plants were to be built at Raipur and Chennai during the next two years, the other two units would be built later. Each plant is expected to cost Rs 300 million and would produce up to 30 tons of biodiesel each day. According to the report these plants will use waste oil, fatty acid and nonedible vegetable oil as a feedstock and the biodiesel will be blended with the HSD oil for running the trains the report says. The use of biodiesel is expected to earn the railways Rs 20 million a year in carbon credits. The Railways currently consumes 2.2 billion litres of diesel a year. Indian Railway Organisation for Alternative Fuels has been formed to take up projects to introduce alternative fuels such as CNG and biodiesel.

Source: http://www.thebioenergysite. com/news/8082/railways-to-buildbiodiesel-plants

Modern Biomass Technology in India: Experiences

Understanding research and development on bio fuel crops is a major mandate of the National Bio-fuel policy. It is stated in the policy that research and development efforts would be focused on plantations, production and processing technologies of bio-fuels, as well as maximizing efficiencies of different end used applications and utilization of bye products. The policy envisages to accord high priority on indigenous R&D and technology development based on the local feed stocks and needs, which would be benchmarked with international efforts.

The specific areas identified in the policy framework on R&D are production and development of quality planting materials and high sugar containing varieties of sugarcane, sweet sorghum, sugar beet and cassava advanced conversion technologies for first and second generation bio-fuels including conversion of ligno-cellulosic materials; technologies for end-use applications; utilization of by-products, etc. When the national biodiesel mission was launched in 2003, several above-mentioned R&D efforts were initiated along with its other mandated activities. The NOVOD Board established a "National Network on *Jatropha* and Karanja" in 2004 by involving Indian Council of Agricultural Research (ICAR), State Agricultural Universities (SAUs), Council of Scientific and Industrial Research (CSIR), Indian Council of Forestry Research and Education (ICFRE), Central Food Technological Research Institute (CFTRI), Indian Institute of Technology (IIT, Delhi) and The Energy Research Institute (TERI). Research was mainly focused on issues such as identification of elite planting material, tree improvement to develop HYVs with better quality of reliable seed source, intercropping trials, developing suitable package of practices, post-harvest tools and technology and detoxification of oil meal of important TBOs. The Central Soil Salinity Research Institute (CSSRI) under ICAR is conducting field trials on various cultivars of *Jatropha* and *Pongamia* for developing site-specific genotypes that can tolerate adverse climatic condition. The Central Research Institute for Dry land Agriculture (CRIDA), undertook the studies on genetic diversity, variability and other biotechnological traits on *Jatropha* (Annual report MoPNG, 2009–10).

A decade of experience with modern biomass technologies for thermal, motive power and electricity generation applications exists in India. Gasifier technology has penetrated the applications such as village electrification, captive power generation and process heat generation in industries producing biomass waste. Over 1600 gasifier systems, having 16 MW total capacity generated 42 million Kilo Watt hour (KWh) of electricity, replacing 8.8 million litres of oil annually (CMIE, 1996). An important aspect of small gasifier technology in India is the development of local manufacturing base. The large sized gasifier based power technologies are at R&D and pilot demonstration stage. The thrust of the biomass power programme is now on the grid connected megawatt scale power generation with multiple biomass materials such as rice straw, rice husk, bagasse, wood waste, wood, wild bushes and paper mill waste. Nearly 55 MW of grid connected biomass power capacity was commissioned and another 90 MW capacity has been under construction (Shukla, 1998). Enhanced scale has improved economics as well as the technology of biomass power generation. Technology improvement is also derived from joint ventures of Indian firms with leading international manufacturers of turbines and electronic governors.

Scope and Modernization of Biomass Energy in Asia

Modernization in biomass energy use in Asia happened in the last two decades along three routes:-

a. Improvement of technologies in traditional biomass applications such as for cooking and rural industries.
b. Process development for conversion of raw biomass to superior fuels (such as liquid fuels, gas and briquettes).
c. Penetration of biomass based electricity generation technologies. These developments have opened new avenues for biomass energy in several Asian nations, besides India.

China

During early 1980's, China initiated a nationwide programmes to disseminate improved cook stove and biogas technologies. The programme led to raising energy efficiency of cook stoves to 20 percent, saving nearly a ton of wood fuel per household (Shuhua *et al.*, 1997). In 1995, nearly 6 million biogas digesters produced 1.5 billion m^3 gas annually (Baofen and Xiangjun, 1997). Another 24,000 biogas purification digesters, with a capacity of 1 million m^3, were in use for treating waste water for 2 million of urban population (Keyun, 1995). Two hundred small biogas based power plants, adding to a capacity of 3.5 MW, produced 3 GWh of electricity on an annual basis (Ravindranath and Hall, 1995). China has focused on a process for converting high quality Chinese sorghum breed into liquid fuel, pyrolysis technology and gasification of agriculture residue and wood. Biomass based electricity generation technologies have penetrated the Chinese market lately, with a penetration of 483 MW and 323 MW respectively in sugar industry in two major sugar cane producing provinces Guandong and Guangxi (Baofen and Xiangjun, 1997). The policy support points to a promising future for modern biomass in China.

Philippines

Philippines is a major biomass using nation, where 44 percent of energy is contributed by biomass. Philippines were among the first nations to initiate the modern biomass programme. In 1970's, three quarters of electricity in Philippines was generated from oil and diesel fired power plants, and a third of the imported oil in 1979 was used for electricity generation (Bawagan and Semana, 1980). A biomass combustion based (Dendro thermal) power programme was launched in 1979 it was aimed at reducing the share of imported oil fired electricity plants to 30 percent (Durst, 1986). Programme met with failure in 1980's due to myriad circumstances - like political instability, declining oil prices, inappropriate technology, over reliance on a single tree species, inadequate institutional support and lack of functioning biomass energy market.

Thailand

Thailand is a large user of biomass energy, which contributes a quarter of total energy. A third of biomass energy is consumed in industry. Bagasse is used in sugar mills as a boiler feedstock (Panyatanya, 1997). The policy of purchase of power from Small Power Producers - SSP announced in 1992 by Electricity Generating Authority of Thailand (EGAT) was favorable to biomass electricity producers (Verapong, 1997). However, the response on the SSP policy was slow. A cogeneration potential of 3100 MW biomass based power was identified in chemical, agro processing and textile industries (Verapong, 1997).

Malaysia

Malaysia is gifted with conventional energy resources such as oil and gas as well as renewables like hydro, biomass and solar energy. Malaysia's commercial demand for energy is projected to continue its upward trend, from 1,244 Peta Joule (PJ) in 2000 to an estimated 2,218 PJ in 2010 (Mekhilef, 2011). It also has considerable biomass resource base. Nearly 60 percent land is under forests and 15 percent under cultivation (Mekhilef, 2011). The forest and agriculture industry generate substantial quantities of wastes and residues which are available cheaply. Wood briquettes from saw dust has grown around domestic and export demand. Palm oil industry is a major source of residues. Another vast biomass source is rice husk. In 1995, there were 328 rice mills producing 430 thousand tons of rice husk (Ang, 1997). Several fish meal manufacturers use rice husk for drying. Cogeneration systems (350 KW size) using rice husk have been established in three rice mills. Seven demonstration plants for cogeneration and efficient biomass combustion have also been promoted (Ang, 1997).

Environmental Impact of Biomass

The growing demand for clean energy, the concern about climate change and the need for GHG emission reduction have challenged most countries to source for alternative forms of energy. Environmental sustainability of biofuels is primarily defined in terms of Green House Gases (GHG) (CO_2, Methane and NO_2), emission reduction and other emissions resulting from agricultural practices such as the use of fertilizers and pesticides, irrigation, soil tillage and harvesting. Additionally, land use prior to biofuel conversion is a critical factor in evaluation of the environmental impacts. GHG reduction potential suffers markedly if grassland or forests are used for biofuels.

Climate Change and Agricultural Biomass

Indian agriculture is highly dependent on the onset, retreat and magnitude of monsoon precipitation, particularly in the rain fed areas of East, North-East and South India. Climate modelers and Intergovernmental Panel on Climate Change (IPCC) documents have projected possibilities of increasing irregularities in the onset of monsoon, drought, flood and cyclone in a warmer world. Despite expansion of area under irrigation, droughts caused by inadequate and uneven distribution of rainfall, continue to be the most important climatic aberration, which influences the agricultural production in India. The severity of a drought will be intensified in a warmer world. Intense and frequent flooding due to climate change would be a major problem in the Indian subcontinent.

For the past few decades, the gaseous composition of earth's atmosphere is undergoing a significant change, largely through increased emissions from energy, industry and agriculture sectors; widespread deforestation as well as rapid changes in land use and land management practices. These anthropogenic activities are resulting in an increased emission of gases, *viz.*, carbon dioxide (CO_2), methane (CH_4) and nitrous oxide (NO_2), popularly known as the 'greenhouse gases' (GHGs). These GHGs trap the outgoing infrared radiations from the earth's surface and thus raise the temperature of the atmosphere. The global mean annual temperature at the end of the 20th century, as a result of GHG accumulation in the atmosphere, has increased by 0.4–0.7°C, above that recorded at the end of the 19th century (Pathak *et al.*, 2012). The past 50 years have shown an increasing trend in temperature @ 0.13°C/decade, while the rise in temperature during the past one and half decades has been much higher. Reduction in crop yield rise in the mean temperature above a threshold level will cause a reduction in agricultural yields. A change in the minimum temperature is more crucial than a change in the maximum temperature. Grain yield of rice, for example declined by 10% for each 1°C increase in the growing season's minimum temperature above 32°C (Pathak *et al.*, 2003). The climate change impact on the productivity of rice in Punjab (India) has shown that with all other climatic variables remaining constant, temperature increases of 1°C, 2°C and 3°C, would reduce the grain yield of rice by 5.4%, 7.4% and 25.1%, respectively (Aggarwal *et al.*, 2009b).

Potential Impacts of Climate Change on Different Sectors of Agriculture Biomass

a. ***Impact on crop***

- Increase in ambient CO_2 concentration is beneficial since it leads to increased photosynthesis in several crops, especially those with C3 mechanism of photosynthesis such as wheat and rice, and decreased evaporative losses. Despite this, yields of major cereals crops, especially wheat are likely to be reduced due to decrease in grain filling duration, increased respiration, and/or reduction in rainfall/irrigation supplies.
- Increase in extreme weather events such as floods, droughts, cyclones and heat waves will adversely affect agricultural productivity.
- Reduction in yields in the rain fed areas due to changes in rainfall pattern during monsoon season and increased crop water demand.
- Incidence of cold waves and frost events may decrease in future due to global warming and it would lead to a decreased probability of yield loss associated with frost damage in northern India in crops such as mustard and vegetables.
- Quality of fruits, vegetables, tea, coffee, aromatic, and medicinal plants may be affected.
- Incidences of pest and diseases of crops would be altered because of more enhanced pathogen and vector development, rapid pathogen transmission and increased host susceptibility.
- Agricultural biodiversity is also threatened due to the decrease in rainfall and increase in temperature, sea level rise, and

increased frequency and severity of droughts, cyclones and floods.

b. ***Impact on livestock***

- Climate change will affect fodder production and nutritional security of livestock. Increased temperature would enhance lignification of plant tissues, reducing the digestibility. Increased water scarcity would also decrease production of feed and fodder.
- Major impacts on vector-borne diseases will be through expansion of vector populations in the cooler areas. Changes in rainfall pattern may also influence expansion of vectors during wetter years, leading to large outbreaks of diseases.
- Global warming would increase water, shelter, and energy requirements of livestock for meeting the projected milk demands.
- Climate change is likely to aggravate the heat stress in dairy animals, adversely affecting their reproductive performance.

c. ***Impact on biodiversity***

Loss of species of animals and plants are estimated to disappear at a rate which would be about 100-times faster than the historical record, largely as a result of human activities. A detailed assessment of the 394 species of primates from South America to Indonesia has indicated that 29% are in danger of disappearing due to hunting, habitat loss and climate change.

Future of Biomass Energy in India

Biomass use is growing globally and rapidly. Despite advancements biomass energy technologies, most bioenergy consumption in India still remains confined to traditional uses. Modern technologies offer possibilities to convert biomass into synthetic gaseous or liquid fuels (like ethanol and methanol) and electricity (Johansson *et al.*, 1993). Lack of biomass energy market has been the primary barrier to the penetration of modern biomass technologies. Growing experience with modern biomass technologies in India suggests that technology push policies need to be substituted or augmented by market pull policies.

A primary policy lacuna hampering the growth of modern biomass energy is the implicit environmental subsidy allowed to fossil fuels. Increasing realization among policy makers about positive externalities of biomass has now created conditions for biomass to make inroads into the energy market. Modern biomass has potential to penetrate in four segments, which are as follows:-

a. Process heat applications in industries generating biomass waste.
b. Cooking energy in domestic and commercial sectors (through charcoal and briquettes).
c. Electricity generation.
d. Transportation sector with liquid fuels.

Economic reforms have opened the doors for competition in energy and electricity sectors in India. Future of biomass energy lies in its use with modern technologies. An analysis under competitive dynamics in energy and electric power markets using the Indian-MARKAL model (Shukla, 1996; Loulou *et al.*, 1997) suggests that biomass energy has significant potential to penetrate the Indian energy market under strong global greenhouse gas mitigation scenarios in future.

Future of biomass energy depends on providing reliable energy services at competitive cost. In India, this will happen only if biomass energy services can compete on a fair market. Policy priorities should be to orient biomass energy services towards market and to reform the market towards fair competition by internalizing the externalities of competing energy resources. Most economical option is utilization of waste materials. Potential availability of agro residues and wood processing waste in India can sustain 10,000 MW power. Biomass waste however shall be inadequate to support the growing demands for biomass resources. Sustained supply of biomass shall require production of energy crops (*e.g.*, wood fuel plantations, sugar cane as feedstock for ethanol) and wood plantations for meeting growing non-energy needs. Land supply, enhanced biomass productivity, economic operations of plantations and logistics infrastructure are critical areas which shall determine future of biomass in India. Policy support for a transition towards a biomass based civilization in India should consider the following:

Short-term policies (1 to 5 years): These are recommended to be as follows:-

a. Enhanced utilization of crop residues and wood waste.
b. Information dissemination.

c. Niche applications (*e.g.*, remote and biomass rich locations).
d. Technology transfer (*e.g.*, high pressure boiler), v) co-ordination among institutions.
e. Demonstration projects.
f. Participation of private sector, community and NGOs.
g. Waste land development.
h. Subsidy to biomass technologies to balance the implicit subsidies to fossil fuels.

Medium-term (5 to 20 years): Some of these could be as follows:-

a. R&D of conversion technologies.
b. Species research to match agro-climatic conditions.
c. Biomass plantation.
d. Scale economy based technologies.
e. Local institutional developments.
f. Removal of distortions in fossil energy tariffs.

Long-term (over 20 years): Some of the areas which can be included in the long term policy objectives are as follows:-

a. Infrastructure (logistics, T&D).
b. Multiple biomass energy products (*e.g.*, gas, liquid, electricity).
c. Institutions and policies for competitive biomass energy service market.
d. Land supply for biomass generation.

Experience of operating the modern biomass plantations and energy conversion technologies is growing in India. The learning effects and the shared knowledge from innovations in conventional technologies are rapidly enhancing the efficiency and reliability of biomass production systems and conversion technologies. Although present penetration of modern biomass energy services is little, technological developments and policy reforms which propose to eliminate energy subsidies and internalize externalities from fuel cycle are set to be advantageous to biomass technologies. Realization of biomass potential shall help many developing countries to make a smooth transition from the present inefficient biomass energy use in traditional sectors to a competitive, commercial and efficient biomass energy use in the future. This will reduce their energy import and conserve scarce finances for national development. The government policies in India during the next decade shall play a decisive role in the penetration of biomass energy. Global climate change policies shall also have significant influence on future of biomass. Myriad economic, social, technological and institutional barriers remain to be overcome. Future of biomass technologies depends on will and ability to overcome these barriers. A key issue before Indian policy makers is to develop a fair market for biomass energy services. Significant social and environmental benefits make biomass a deserving alternative for support from governments committed to sustainable development. Governments have in the past promoted new energy technologies like nuclear power in France (Johansson *et al.*, 1996), wind power in Denmark (Johansson *et al.*, 1996) and India (Naidu, 1997), and ethanol from sugarcane in Brazil (Goldemberg *et al.*, 1993). The challenge with the Indian policy makers is to support the energy transformation away from exhaustible and polluting energy resources towards renewable and clean energy resources on the road to sustainable development. Modern biomass technologies provide viable options for such an energy transformation, on the way to a sustainable energy system of the future.

Biomass Stock in India

Larger population of India lives in rural area. According to census 2011, 68.84% population of India lives in rural area. There are 0.638 million villages in India, therefore to plan for electrification in villages; biomass will be vital option as a renewable source of energy. Ministry of New and Renewable Energy Sources (MNRE) has proposed to reach a total of 4324.22 MW of power generation based on biomass power and gasification as well asco-generation (biomass produce by anaerobic Digestion). The MNRE, Govt. of India, has taken initiative like central financial assistance and fiscal incentives for promoting the use of bio-energy from agro residues, plantations and from various waste of urban and industries. MNRE is using the methodology for providing the subsidies based on co-generation and generation by biomass gasification (Kishore *et al.*, 2004, Balachandra, 2011). Overall the consumption of energy in India is distributed into thermal, hydro, nuclear, new renewable sources and biomass resources (Fig. 2).

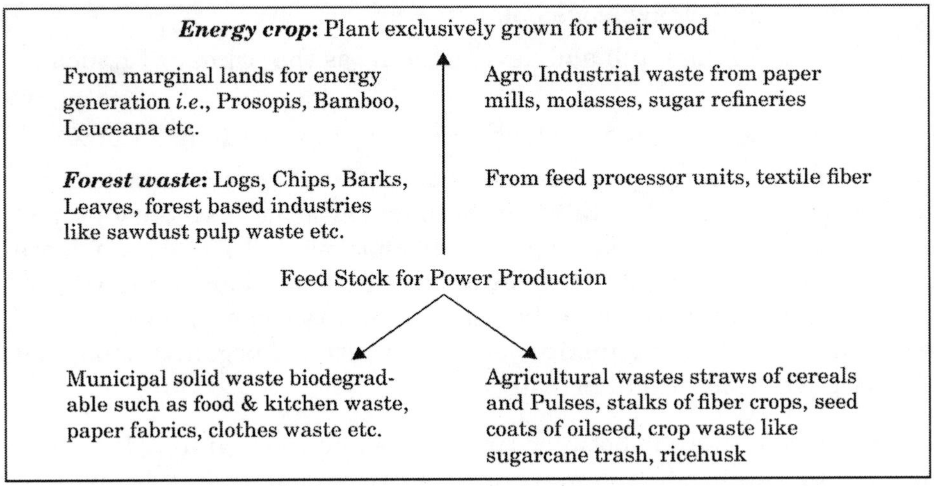

Fig. 2: Biomass energy development in India

CONCLUSIONS

In general, the biomass energy industry produces clean energy from various materials that are currently considered waste, and their utilization produces significant environmental benefits in the form of avoided pollution.

Government and project financing are the two main sources of funding for the proposed plan to develop India's biomass energy. While government investment is especially necessary for projects directly related to farmers, private financing is critical for securing self-sustaining financing cycles and maturing of the biomass industries. The government should still subsidize other types of projects, but they should be largely co-financed by farmers and both domestic and international financing institutions. A joint implementation plan will coordinate the programs and funding levels of the government and each partner in the development of biomass energy in India. A major premise of the plan is that each partner will continue to support those areas it considers a priority based on expertise and funding history.

Government investment should come in the forms of

1. Grant subsidies for constructing household biogas digesters, deploying straw pellet fuel, and constructing an energy crop base;
2. Grant subsidies and interest subsidies for mid-to-large biogas plants, centralized straw gasification

3. Plants, and straw biogas plants;
4. Grants for research and development (as the primary financier); and
5. Co-financing pilot and demonstration projects (at about 50%)

A broad partnership is needed. International finance institutions, such as ADB, can be a catalyst in helping raise the necessary project financing, which will be a significant challenge. To ensure a more independent, sustainable, and competitive field for biomass industries, government should open investments to industry owners, domestic and international financing institutions, international organizations, and foreign governments and private sectors.

Strong coordination between national policy and planning is always needed to establish a comprehensive investment, taxation, and industrial policy framework and to establish goals for the development of rural biomass energy. In the case of the PRC, strong coordination is particularly important between:-

1. The Ministry of Science and Technology and the Ministry of Agriculture to ensure technological advancement of the research plan and the achievement of expected results in pilot testing,
2. Ministry of Agriculture and National Energy Departments to strengthen coordination within the industrial sector. For example, straw to-power projects should consider the situation and development of agriculture, and ensure the construction of the energy crop base is directly integrated with liquid fuel production plants. More emphases to be given on Research and development should be carried out to enhance the potential for these of plants and purposes.

Considering rural biomass energy is one of the key areas to fulfill the gap of energy requirement. Rural Biomass energy has great potential to make a significant impact on two of the India's most pressing development challenges: rural poverty and environmental damage. By harnessing these valuable but wasted resources, biomass energy offers a triple opportunity at advancing:-

a. Rural access to energy.
b. Environmental protection.
c. Rural development.

REFERENCES

Aggarwal, P.K., Singh, A.K., Samra, J.S., Singh, G., Gogoi, A.K., Rao, G.G.S.N. and Ramakrishna, Y.S. (2009b). Introduction. *In*: Global Climate Change and Indian Agriculture, Aggarwal, P.K. (*ed.*), ICAR, New Delhi, pp. 1–5.

Ang, A. (1997). Present Status of Biomass Energy Technologies in Malaysia, Presented in Regional Consultation on Modern Biomass Energy Technologies, Regional Wood Energy Development Programme, FAO, Kuala Lumpur, Malaysia.

Annual Report 2009–2010 (2011). Ministry of Petroleum and Natural Gas (MoPNG), Govt. of India.

B.P. (2015). Statistical Review of World Energy. London: BP.

Balachandra, P. (2011). Modern energy access to all in rural India: An integrated implementation strategy. *Energy Policy*, 39: 7803–14.

Baofen, L. and Xiangjun, Y. (1997). Development and Utilization of Biomass in China, Presented in Regional Consultation on Modern Biomass Energy Technologies, Regional Wood Energy Development Programme, FAO, Kuala Lumpur, Malaysia.

Bawagan, P.V. and Semana, J.A. (1980). Dendrothermal Power Plants: Prospects and Problems, Paper presented at seminar-workshop on Forestry and the Energy Crisis, 26-28 November, Forest Research Institute Central Office, Los Banos, Philippines.

Bioenergy India (2011). A quarter magazine, issue 7, UNDP, India.

CMIE (1996). India's Energy Sector. Centre for Monitoring Indian Economy, Bombay.

Diwedi, G., Sharma, P.P. and Kumar Mukesh (2014). Status and policy of biodiesel development in India. *Journal of Renewable Research*, 4(2).

Durst, P.B. (1986). Dendrothermal dream threatened in the Philippines. *Journal of Forestry*, 84(8).

Dwivedi, G., Sharma, M.P. and Jain, S. (2013). Diesel engine performance and emission analysis using biodiesel from various oil sources –Review. *Journal of Material and Environmental Science*, 434–447.

FAO (1997). Review of Wood Energy Data in RWEDP Member Countries, Field Document No. 47, Bangkok.

Grubler, A. and Nakicenovic, N. (1988). The Dynamic Evolution of Methane Technologies, *In*: Lee, T.H., Linden, H.R., Dryefus, D.A. and Vasko, T. (*eds.*), The Methane Age, Kluwer Academic Publishers, Dordrecht.

Hall, D.O., Rosillo-Calle, F., Williams, R.H. and Woods, J. (1993). Biomass for Energy - Supply Prospects, in Renewable Energy- Sources of Fuels and Electricity - Johansson, T.B., Kelly, H., Reddy, A.K.N. and Williams, R.H. (*eds.*), Island Press, Washington D.C., U.S.A.

Johansson, T.B., Williams, R.H., Ishitani, H. and Edmonds, J.A. (1996). Options for reducing CO_2 emissions from the energy supply sector. *Energy Policy*, 24(10).

Joshi, V., RamanaSinha, C.S., Karuppaswamy, M., Srivastava, K.K. and Singh, P.B. (1992). Rural Energy Data Base, TERI, New Delhi.

Kaushik, D. and Appleby, P. (2015). India's Primary Energy Evolution: Past trends and Future Prospects. India Policy Forum, NCAER, New Delhi. India Policy Forum.

Keyun, D. (1995). Renewable Energy Benefits Rural Women in China. *In*: Energy as an Instrument for Socio-Economic Development. Goldemberg, J. and Johansson, T.B. (*eds.*), UNDP/SEED/BPPS, UNDP, New York.

Kishore, V.V.N., Bhandari, P.M. and Gupta, P. (2004). Biomass energy technologies for rural infrastructure and village power–opportunities and challenges in the context of global climate change concerns. *Energy Policy*, 32: 801–10.

Mekhilef, S., Saidur, R., Safgari, A. and Mustaffa, W.E.S.B. (2011). Biomass energy in Malaysia: Current state and prospects. *J. Elsevier*, 15(7): 3360–3370.

Panyathanya, W. (1997). Modern Industrial Biomass Technology in Thailand, Presented in Regional Consultation on Modern Biomass Energy Technologies, Regional Wood Energy Development Programme, FAO, Kuala Lumpur, Malaysia.

Patel, T. and Raiyani, C.V. (1997). Indoor Air Quality: Problems and Perspectives, in Energy Strategy and Greenhouse Gas Mitigation, Shukla, P.R. (*ed.*), Allied Publishers, New Delhi, India.

Pathak, H., Aggrawal, P.K. and Singh, H.D. (2012). Proceedings of "Climate Change Impact adaptation and Mitigation in agriculture: Methodology for assessment and Application". Indian Agriculture Research Institute, New Delhi.

Pathak, H., Ladha, J.K., Aggarwal, P.K., Peng, S., Das, S., Yadvinder Singh, Bijay-Singh, Kamra, S.K., Mishra, B., Sastri, A.S.R.A.S., Aggarwal, H.P., Das, D.K. and Gupta, R.K. (2003). Climatic potential and on-farm yield trends of rice and wheat in the Indo-gangetic Plains. *Field Crops Res.*, 80(3): 223–234.

Ramana, P.V., Sinha, C.S. and Shukla, P.R. (1997). Environmental Issues in Rural Energy: Policy Responses for the Future, in Energy Strategies and Greenhouse Gas Mitigation, Shukla, P.R. (*ed.*), Allied Publishers, New Delhi, India.

Ravindranath, N.H. and Hall, D.O. (1995). Biomass Energy and Environment- A Developing Country Perspective from India, Oxford University Press, Oxford.

Reddy, A.K.N., Williams, R.H. and Johansson, T.B. (1997), Energy after Rio - Prospects and Challenges, UNDP, New York, USA.

Renjish Kumar, V. K. and Vimala, M. (2016). Energy Consumption in India: Recent Trends. *Asia Pacific Journal of Research*, 1(XXXVI). ISSN: 2320-5504

Shuhua, G., Daxiong, Q. and Xiliang, Z. (1997). Evaluation of the biogas and improved cook stove programme in China, in the Proceedings of the International Conference on Biomass Energy Systems, Ramana, P.V. and Srinivas, S.N. (*eds.*), TERI, New Delhi.

Shukla, P.K. (2007). Biomass energy strategies for aligning development and climate.

Smith, K.R. (1987). Bio-fuels, Air Pollution and Health: A Global Review, New Delhi: Plenum Publishing Corporation {Private}.

Smith, K.R. (1993). Household Fuels and Health, *Urja Bharati*, 3(3).

Sylvain, L., Karthikeyan, N., Erik, D., Ian, M. and Michael, O. (2009). Optimizing biodiesel production in India. *Applied Energy*, 86: 125–131.

Verapong, P. (1997). Status of Biomass Energy and Electricity in Thailand, Presented in Regional Consultation on Modern Biomass Energy Technologies, Regional Wood Energy Development Programme, FAO, Kuala Lumpur, Malaysia.

Woods, J. and Hall, D.O. (1994). Bioenergy for Development - Technical and Environmental Dimensions, FAO Environment and Energy Paper 13, Rome, Italy.

7

Utilization of *Jatropha curcas* L. for Biofuel Production and Sustainable Environmental Development

Ashwani Kumar[1*], Shivani Chaturvedi[2], Ram Chandra Pradhan[3], Lata Nain[4] and Satyawati Sharma[5]

ABSTRACT

Jatropha curcas is commonly known as physic nut and plant of significant potential. Jatropha is cultivated in tropical areas as commercial crop for liquid fuel production and also for many other applications. Its plantation has a potential to provide large-scale rural employment. It is a source of various alkaloids and its cultivation can improve the environment and develop alternative sources of carbon neutral fuels. The non-edible oil produced by Jatropha seeds doesn't compete with commercial oil seed crops, which reduces food versus fuel

[1]Metagenomics and Secretomics Research laboratory, Department of Botany, Dr. Harisingh Gour University (Central University), Sagar-470003, (MP), India.

[2]Department of Chemistry, Indian Institute of Technology Delhi, New Delhi-110016, India.

[3]Food Process Engineering Department, National Institute of Technology, Rourkela, Odisha, India.

[4]Division of Microbiology, ICAR- Indian Agricultural Research Institute, New Delhi, India.

[5]Centre for Rural Development and Technology, Indian Institute of Technology Delhi, New Delhi-110016, India.

**Corresponding author*: E-mail: ashwaniiitd@hotmail.com

crisis. Along with oil, Jatropha produces many useful things. Its seed oil has properties like palm oil and is utilized in rural areas as fuel for lantern generators and cooking medium. By doing rigorous research for its marketing and cost effectiveness we can conclude that for real and potential markets all types of by-products generated from this crop should be considered. From last decades there is a controversial discussion going on for its cultivation, expansion, and various applications. But its proven application as phyto-remediator, sequestration of carbon, soil erosion controller, wasteland reclamation, medicinal and therapeutic uses strongly support Jatropha cultivation in many countries. Although it's essential to explore Jatropha potential risks and benefits before going for mass cultivation's.

Key words: *Jatropha curcas*, Energy, Wasteland reclamation, Biofuel

1. INTRODUCTION

For the development of any civilization energy is utmost important, particularly which is generated from biomass. In the global energy production, biomass to bioenergy conversion plays a significant role; well planned agricultural practices not only improve soil health but also provide good biomass (Vasudevan *et al.*, 2005; Kumar *et al.*, 2005; Kumar *et al.*, 2010a; Kumar *et al.*, 2011; Kumar *et al.*, 2014; Singh *et al.*, 2014; Kumar *et al.*, 2015; Kumar *et al.*, 2016). To cope up with increasing fuel demands, *Jatropha* crop has been planted extensively in tropical and subtropical areas of the world. According to Siang (2009), around 32.72 million hectares land has been involved in *Jatropha* plantation all around the world which produced 160 million tons of seeds. Most of the production of oil, that is about 95% of total production, was from Asia particularly India and China. This shows that the biodiesel industries can generate >80 million tons of seed cake in the coming years. *Jatropha* is a multipurpose plant with many attributes and considerable potential to be commercially exploited (Kumar *et al.*, 2010b; Kumar *et al.*, 2015). It is a drought resistant shrub or tree that grows on semi-cultivated or non-agricultural wastelands in Central and South America, Africa, India and South East Asia (Cano-Asseleih, 1986; Cano-Asseleih *et al.*, 1989). The wood and fruit of *Jatropha* can be used for numerous purposes including fuel. *Jatropha* has high protein content and if its toxicity is removed, it can also be utilized as highly nutritious and cost effective protein supplement in animal feed (Becker and

Makkar, 1998). The plantations are sturdy and can also be applied for soil conservation, land reclamation and also as natural fence (Heller, 1996). Different parts of *Jatropha* can be used in various applications (Fig. 1), including rural areas as well as in the industrial sector. These applications includes, honey production potential, manufacture of soaps and candles and also in cosmetic industry. This plant has been used in recent past for the production of biodiesel as alternative of petroleum to meet the demand for rural energy services and also for reducing GHG emission (Kumar *et al.*, 2008; Kumar *et al.*, 2015). Table 1 is showing potential advantages and disadvantages arise due to long term plantation of *Jatropha*.

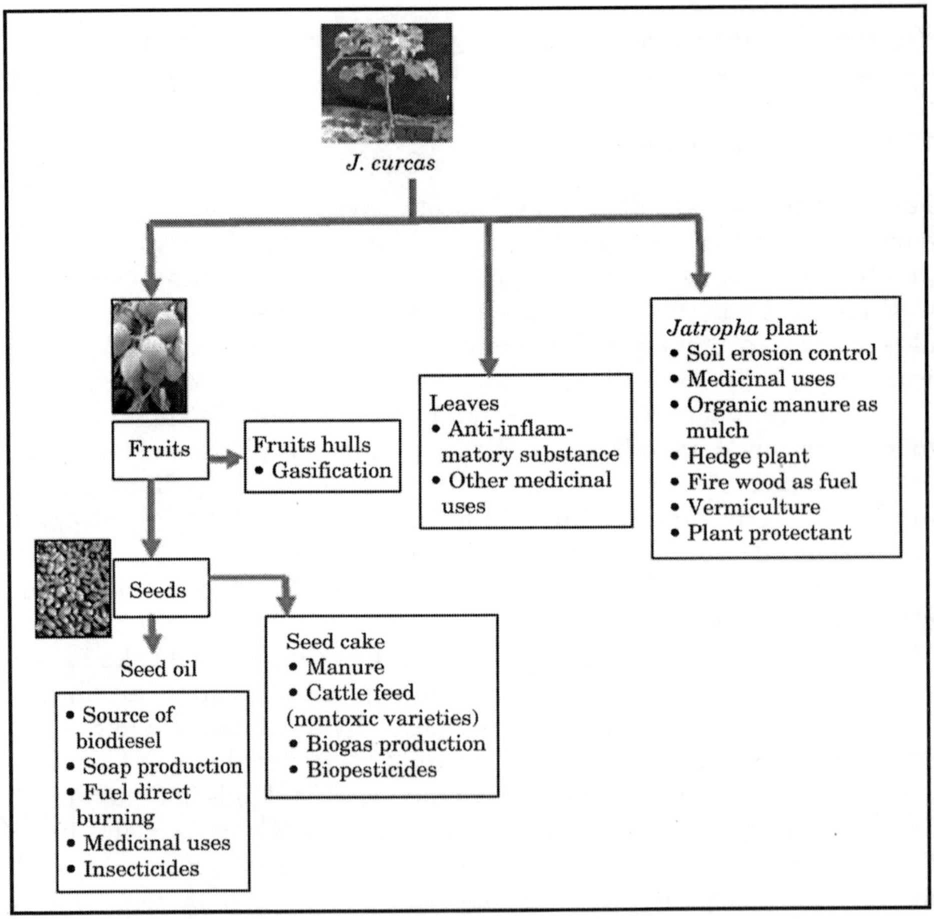

Fig. 1: Potential uses of *J. curcas*

The purpose of this chapter is to provide information about its potential also as a crop particularly, energy production and waste land reclamation. However, few demerits of *Jatropha* limit its utilization. The yield of oil and seeds depends upon the different climatic and soil conditions, so the expected returns in terms of cost is very less. However, its toxicity is a great cause of concern for the public. If we take one kilogram of seed, 240–300 g oil and 700–760 g of seed cake will be generated. This oil of *Jatropha* is non edible and used as biodiesel after trans-esterification and glycerol is the byproduct, which is having high commercial value. After oil extraction the seed cake is left which is rich in nitrogen content and can be used as fertilizer. The cake is having toxicity because of the constituents present in it like phorbol esters (PEs), which is a very active toxic component of *Jatropha*. This compound (PEs) present in 2–3 mg g^{-1} in a defatted kernel meal and 2–8 mg g^{-1} in its oil. High quality genotype can give high yield and low yielding variety may be improved using genetic engineering.

Table 1: Potential advantages and disadvantages of long term cultivation of *Jatropha*

Advantages	*Disadvantages*
Source of renewable energy	Food crisis-increase in price
Environmental conservation	Land use change
Combating global warming and reduction of GHG	Threat to Biodiversity due to invasive species
Employment generation	Chances of soil contamination
Socio-economic and rural development	Production is low
Wasteland reclamation	Nutrient imbalance in soil

2. BOTANICAL ASPECTS

Euphorbiaceae family has genus *Jatropha* which comprises of 170 species. The genus name *Jatropha* derives from the Greek word *jatr´os* (doctor) and *troph´e* (food), it means this plant has medicinal uses. This small tree or shrub is generally 3–5 metres tall but with the support of environmental condition it can attain the height of 8–10 metres. Vegetatively propagated plants do not have tap root system. It has five to seven lobed leaves with hypostomatic condition and stomata are paracytic or rubiaceous type. Deciduous trees shed their leaves in dry season. The regions where general humidity exists throughout the year flowering also occurs whole year. *Jatropha* plant is monoecious with unisexual flowers, which are occasionally hermaphrodite and common pollinator are honey bee (Dehgan and Webster, 1979). The inflorescence changes into a bunch (approximately 10) of ovoid fruits. If enough rain falls the nursery plants can bear fruits in first season itself. While the plants are directly sown in to the ground fruits are produced in the second year. When seed matures their bivalve coccus with dried fleshy exocarp is formed. The maturation of seed takes at least 3–4 months from flowering. The approximate number of seeds per kilogram is 1375; they are black or dark brown in colour. Fig. 2 is showing different parts of *J. curcas*.

Fig. 2: Different parts of *J. curcas*

3. *JATROPHA* PRODUCTION

3.1. Ecological Environment Needed

Portuguese farmers help in distribution of *Jatropha via* the Cape Verde Islands and Guinea Bissau to Asia and Africa (Heller, 1996). *Jatropha* can grow in arid, semiarid, gravelly, sandy, saline soils, stony soil and in crevices of rocks. Water logging hinders its growth and development. In winter season sheded leaves from mulch around it, which enhances organic matter which in turn improves soil health and promotes earthworm activity. *Jatropha* can be found in tropical and subtropical climate. Generally it grows well in heat but withstands in light frost. It grows from 250 to more than 1200 mm per annum of rain fall (Katwal and Soni, 2003). In adverse climate like prolonged drought it sheds its leaves as a counter to reduce transpiration. The requirement of water is very low. It grows well in good aerated soil and adapts in low nutrient marginal soil. Heavy soil may hinder its root growth. The capability to adapt in poor and dry areas shows its strength. The analysis of *Jatropha* canopy is an important structural characteristic for monitoring of crop growth and yield.

3.2. Methods of Propagations

Jatropha can be propagated either from seed or cuttings but plants grown by the seeds show better disease and drought resistance which could be due to true taproot. The plants grown from cuttings have pseudo taproots which may penetrate to only 1/2 to 2/3rd the depth of the soil as compared to true taproot. There are other methods of propagation like direct seedling, precultivation of seedling or nursery raising, transplanting of spontaneous wild plants and direct planting of cuttings. The spacing between the plants (3m × 3m) produces good fruits especially in early years of crop (Heller, 1996). In Cape Verde, Kobilke (1989) had conducted research on different methods of propagation of *Jatropha*. The sowing time, depth of sowing, trial year and cultivation methods are the factors which play an important role. For the sprouting of the vegetative cuttings, the age of the plant from which cuttings are taken and the position within plant are taken into consideration. The ability of root formation may decline with age and distance from stock axis. Cuttings from lower or juvenile regions generally have higher capacity of rooting than upper part of plant (Hartmann and Kester, 1983). Biofertilizers containing beneficial microbes showed growth pro-moting effects on *Jatropha*. The biomass yield was slightly higher with vermi compost than farm yard manure while opposite pattern with respect to stem length was observed (Kumar and Sharma, 2005). As far as yield of seed per plant is concerned it increased with the selection of marker assisted quality planting material (a biotechnological tool).

For the propagation and storage of selected genotypes, tissue cultures have been developed (Engelmann, 1991). The benefits of using tissue culture techniques for propagation are - higher rate of multiplication; minimum risk of infections by microbes, insect and pest, reduction in genetic erosion, or space requirements and reduction in the expense in labor costs. Scientists from India and Nicaragua worked together for genetic improvement of species and their results were very promising (Sujatha and Dingra, 1993; Sujatha and Mukta, 1996; Machado *et al.*, 1997; Sujatha and Prabakaran, 2003; Wei *et al.*, 2004).

Effective regeneration system, amenable to transformation, is a pre requisite for genetic manipulation of *Jatropha* plant. It is done by organogenesis from various explants that includes mature leaves (Sujatha and Mukta, 1996), petiole (Sujatha and Mukta, 1996), hypocotyle (Sujatha and Mukta, 1996), axillary node and by somatic embryogenesis by the mature leaf explants. *Jatropha* crop is of prime importance as a biofuel crop but still there is considerable work to be done on population genetics and plant genomes. The cloning and characterization of gene controlling key steps of FA biosynthesis and lipid breakdown pathways are useful for the understanding of production and accumulation of oil in seeds. However in Central India, Ginwal *et al.* (2005) reported variability in seed source, and from Haryana state 24 accessions were collected on the basis of variation in seed trait and content of oil (Kaushik *et al.*, 2007). Utilization of molecular markers is of high significance in genetic techniques to improve crop quality such as introduction of new strain, marker assisted selection, DNA fingerprinting of germplasm and targeted map based cloning etc. The RFLP markers came as first advancement. It can be used in breading program and also helps in molecular diversity assessment. RTPCR and RACE techniques have been used in developing seeds of *Jatropha* to generate new full length cDNA of stearoyl-acyl carrier protein desaturase and the gene was functionally expressed in *E. coli* (Tong *et al.*, 2006). This gene produce enzyme for fatty acid biosynthesis in higher plants and responsible for deciding the ratio of saturated to unsaturated fatty acids. Ying *et al.* (2007) isolated a new full length cDNA encoding aquaporin (JcPIP2) from seedling of *Jatropha* that is responsible for salt and drought tolerance mechanism. They also found that the abundance of JcPIP2 was induced by heavy drought stress and it played a key role in rapid growth of *Jatropha* under dry conditions. If production of *Jatropha* for oil requirement increases, it will also increase the cake production. Accumulation of cake will be hazardous for the environment because of its toxicity (Becker and Makkar, 1998). With the help of genetic engineering we can not only reduce the toxicity level but also improve the resistance to biotic stresses and also modify the seed oil for better engine efficiency (Hossain and Davies, 2010).

3.3. Biomass Production of *Jatropha*: Agronomic Practices

Significant progress has been made in biomass utilization, yield projection model, selection and diversity assessment but on agronomic properties it is still lacking behind. *Jatropha* has deciduous stem and having drought avoidance strategy with high water use efficiency and characterized by relatively low water footprint (Omobuwajo, 2000; Baryeh, 2002).

3.4. Plantation Practices

Jatropha curcas is easily propagated by generative and by vegetative methods (Heller, 1996; Openshaw, 2000). *Jatropha* shows high survival rate under growth limiting conditions (Kumar *et al.*, 2008). To control soil erosion, soil plantation of this crop through direct cutting is popular (Heller, 1996), except the fact that by this method plants do not develop tap root. This cause uprooting due to the force of wind, also in absence of taproot other roots cannot access extra nutrients present deep in the soil as adventitious roots other than tap root are unable to reach deeper inside (Heller, 1996). Some other researchers recommended that percentage of survival depends on the part of the plant which is cut and used for plantation; whether it is top, middle or base part of the branch. The survival percentages are –42% (top), 72% (middle) and 88% (base) respectively. Length and diameter also affect the survival percentage. It has positive impact on the survival of the cut part. The length and diameter combination of 105 to 2.5, 45 to 3.5 and 45 to 4.5 cm onwards have shown higher than 80% of survival percentage (Kaushik and Kumar, 2006). For their tap root development cuttings can be planted directly in the fields, or nursery beds or polythene bags (Heller, 1996; Kaushik and Kumar, 2006). In the rainy season, 10–20 cm of inter space is preferred in the soil for plantation. One can prepare nursery in polythene bags or tubes or in seed beds. Presoaking in cow dung slurry for 12 hrs shows best result in germination *i.e.*, upto 96% while 72% germination occurs in traditional 24 hrs cold water treatment. However, use of hot water or H_2SO_4 (0.5 M) does not improve germination. Seedlings can be planted at the onset of rains. For seedling plantation, common practices are to keep distance between the seedlings as 2 × 2 m^2 (2500 plants ha^{-1}), 2.5 × 2.5 m^2 (1600 plants ha^{-1}) or 3 × 3 m^2 (1111 plants ha^{-1}) (Heller, 1996). Generally after 90 days of flowering, fruit gets matured (Heller, 1996), its maturity is also displayed by change in colour from green yellowish brown. When seed gets fully matured it yields maximum oil but the fruits do not mature all at the same time, these fruits have to be harvested manually at regular intervals (Heller, 1996). This labor intensive harvesting varies in time and period because of the variation in seasonal conditions. In semi arid zone a period of two month time is very busy for manual harvesting; in persistent humid conditions, weekly harvesting is necessary. Average rainfall and age show positive trends in yield, specially the upper boundaries of the yield. Waste land reclamation and growth of *Jatropha* in marginal soil and low rainfall requirement is its strength (Spaan *et al.*, 2004). Though, its yield depends on rainfall, soil type and soil fertility, genetics, plant age and management like propagation method, spacing, fertilizing, irrigation and pruning etc (Heller, 1996; Openshaw, 2000; Ginwal *et al.*, 2004; Francis *et al.*, 2005). Its production has great variability, which needs careful breeding and systematic selection and development of suitable germplasm. It yields 0.2 kg to more than 2 kg per plant (Francis *et al.*, 2005). If we evaluate ratio of shell and kernel on mass basis it is on an average of 37:63, in which kernel is made up of crude fat and protein is of 30.4 MJ kg^{-1} average calorific value. The shell part is of fiber with 19.4 MJ kg^{-1} of its calorific value. On the basis of these figures the average oil content is 34.4% of dry seed on mass basis. In Egypt an irrigated field's total biomass production is 80t dry m ha^{-1} or 11t dry m ha^{-1} including seeds. They keep spacing of 2.5 × 2.5 m and represent about 5.5t CO_2 ha^{-1} yr^{-1} (Benge, 2006). In Indian wasteland average CO_2 production rate from standing biomass are 72.25t CO_2 ha^{-1} yr^{-1} excluding the 2–2.5t dry seed yields ha^{-1} yr^{-1} (Francis *et al.*, 2005). The density of wood is reported to be from 0.33 to 0.37 cm^3 (Benge, 2006), however, 116 days old young plant has wood density of 0.26 g cm^3 in the base part of young stem.

4. CHEMICAL COMPOSITIONS AND TOXICITY OF PLANT

Previous observations on utilization of *Jatropha* seeds as feed are reported. In Mexico the white winged dove (*Zenaida asiatica*) feed on the seeds of *Jatropha* (Rivera-Lorca and Ku-vera, 1997b). In Mexico it was noticed that pigs and chickens also consumed them. Humans also utilized these *Jatropha* seeds after boiling them in their traditional dishes (Neuwinger, 1994; Rivera-Lorca and Ku-vera, 1997a). In Veracruz, Mexico mice gained weight after they fed on *Jatropha* seeds. It is known that in general *Jatropha* is reported as toxic to animals and humans. In some feeding experiments raw or cooked seeds killed rats within 2–3 days, while raw or cooked defatted seed meal or oil caused death within 6–8 days. In case of seeds which were roasted and cooked, death occurred after 14–16 days (Liberalino *et al.*, 1988). With the constant observation the maximum lethal dose of *Jatropha* for three different species were determined (Felke, 1913; Behme, 1988). During these tests, animals showed abdominal pain, respiratory problems, and diarrhea. They also displayed some histopathological problems like necrosis of liver, heart and kidney, gastrointestinal inflammation and haemorrhages in the liver (Adam, 1974; Adam and Magzoub, 1975; Abdu *et al.*, 1986; Liberalino *et al.*, 1988; Becker, 1995). In contrast to this finding, mice feeding on *Jatropha* oil showed no toxic signs in Mexico. While others reported inflammation of intestine and diarrhea when fed with Indian variety of *Jatropha* oil (Gandhi *et al.*, 1995). In general ingestion of 3–5 seeds can cause gastrointestinal irritation, vomiting, diarrhea, nausea and abdominal pain etc. It can cause dehydrated condition which can be recovered by intravenous fluid replacement (Joubert *et al.*, 1984; Abdu-Agoye, 1986; Mampane *et al.*, 1987). The achiote seeds (*Bixa orellana L.*) are used as antidote for toxicity of *Jatropha curcas* in Mexico (Rivera-Lorca and Ku-Vera, 1997a), whereas extract of *Peltophorum africanum* acts as an antidote in South Africa as traditional healer (Mampane *et al.*, 1987).

The decorticated seed varieties contain 40–60% of oil (Liberalino *et al.*, 1988; Gandhi *et al.*, 1995; Sharma *et al.*, 1997; Wink *et al.*, 1997; Makkar *et al.*, 1997; Openshaw, 2000), this oil is useful in lighting, lubrication, soap making (Rivera-Lorca and Ku-Vera, 1997a) and as a biodiesel. Oil of *Jatropha* is mainly composed of crude protein (24.60%), crude fat (47.25%) and moisture (5.54%) (Akintayo, 2004). Oil from different origins has different fatty acid composition. This oil contains saturated fatty acids mainly palmatic acid (16:0) with 14.1% and stearic acid (18:0) with 6.7%. The unsaturated fatty acids consist of 47% oleic acid (18:1) and 31.6%, linolenic acid (18:2). The composition of *Jatropha* oil has high value of monounsaturated oleic acid and polyunsaturated linolenic acid which shows semi-drying property. It means that it gets partially hardened when the oil is exposed to air. This quality of semi hardening is helpful in substituting the diesel fuel. With the help of growth regulators, plant of *Jatropha* gets induced and the production of hydrocarbons is increased. In this treatment ethephon and morphactin induce the production of hydrocarbon with 5 and 5.4% respectively (Augustus *et al.*, 2002).

5. OIL PRODUCTION AND ITS USES

In production of biodiesel, *Jatropha* oil seed cake is produced as an important by-product and waste water is also produced. Many researchers worked on the procedures of oil extraction. For *Jatropha* mainly two process gained importance mechanical and chemical extraction (Aderibigbe *et al.*, 1997; Forson *et al.*, 2004).

In the first step of extraction, the seed has to be dried, for this, oven may be used at 105°C temperature or sun drying is also popular (3 weeks). Whole seed or kernels or a mixture of both can be fed in mechanical expellers, but commonly whole seeds have been used. For chemical extraction kernels of *Jatropha* seeds are used in expellers. As combustible by-product or feed stock of gasification, shells of *Jatropha* can be used.

5.1. Decortication of *Jatropha* Fruits

Post harvest technology plays an important role in increasing the oil production by reducing the post production losses and increasing the recovery of oil from the seed (Sahay, 1996). The post-harvest processing of *Jatropha* to get raw oil is shown in Fig. 3.

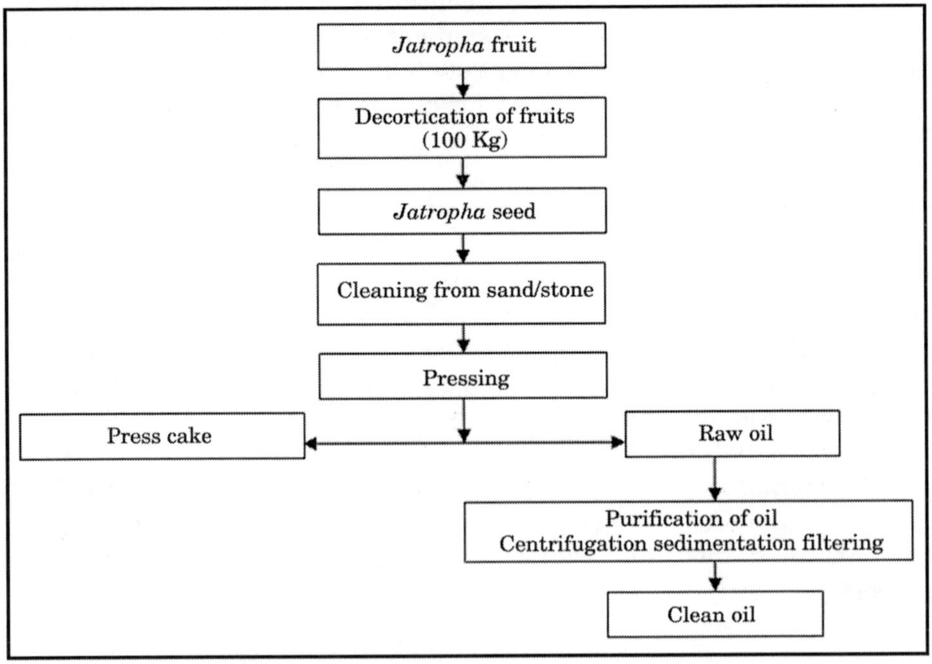

Fig. 3: Post harvest processing of *Jatropha*

Presently, processing is done by seed collector at rural level with stone-aged technology. Decortication by seed collectors is done with either stones or wooden mallets. A table or a concrete floor with small board of wood moving over it are used and dry fruits of *Jatropha are* placed in a thin layer. Due to pressing down, the shell of fruit gets split and a seed comes out. At some places in India, seed decortication is done by a chakki and the crushed kernel is cooked to expel oil. The oil from decorticated seed is highest as compare to under corticated seeds (Sharma and Kumar, 2010). This process of decortication or shelling of fruits is done mechanically by bricks, stone mallets or with the help of fingers. To increase the efficacy of the *Jatropha* system, we have to develop standard method as compared to the traditional methods which is not harmful to human. Reasonably low cost technology is required for the decortication of seeds from the fruits, which is of significance in Indian context and other developing countries (Aviara *et al.*, 2005; Akcali *et al.*, 2006). Pradhan *et al.* (2010) have developed the technologies for rapid and safe decortications of seeds from the fruits. The whole process is finished in two steps, first step involve the compression of fruit in the rotating blades against a stationary concave sieve, there by the pliable endosperm to elastic deformation, while its shell is cracked. Due to the repeated rotational movement, shearing or rubbing action against the stationary concave sieve, the cracked shell is detached from the fruit. Second step, the mixture of seed, shell, unshelled or partially shelled fruits, broken seed and dust, if any, are immediately forcefully passed down through the discharge outlet on to a vibrating separator. This separator separates the shell, seed etc. Small modification in decorticator is always required due to the variable size of the fruits as suggested by Pradhan *et al.* (2010).

5.2. By Chemical Methods

During testing of different ways of chemical extraction methods, the reaction temperature, time consumed, reaction pH and oil yield vary. The results showed highest oil yield in n-hexane method but it also consumed most time. When alkaline protease is used in aqueous enzymatic oil extraction it gave the best result (Winkler *et al.*, 1997; Shah *et al.*, 2005). In this aqueous oil extraction method if material is pretreated with ultra-sonication it improves the process.

The solvent extraction method is economical when the production is large that is more than 50t biodiesel per day. The other important conventional method n-hexane solvent extraction process is not good for the environment because of the fact that it generates waste water, consumes higher specific energy and results in high emission of volatile organic compounds. It also affects the human health since one has to work with inflammable and hazardous chemicals. If the aqueous enzymatic oil extraction method is used it reduces the above problems, as do the use of super critical solvents (mainly super critical CO_2) or bio-renewable solvents as bio-ethanol and isopropyl alcohol. For enhancing the market performance new generation n-hexane extraction units are very efficient and ecofriendly than the older units. With the use of supercritical isopropanol or CO_2 in continuous mechanical extraction system, only 0.3% by weight oil is left in the cake.

5.3. By Direct Use of Oil

In comparison to petroleum fuel, the use of pure *Jatropha* oil showed very good results in low heat rejection diesel (LHR) engine which showed higher brake specific energy consumption (BSEC), lower brake thermal efficiency (BTE), higher exhaust gas temperature (EGT) and lower NOx emissions. Preheating and increasing the injection pressure decreased BSEC, increased BTE, increased EGT and increased NOx emissions only marginally. Kumar *et al.* (2003) have done comparison of *Jatropha* oil with fossil diesel, and observed that the soot (hydrocarbon) emission is higher with *Jatropha* oil as compared to fossil diesel. At maximum output an increase from 100 ppm, for fossil diesel, to 130 ppm, for *Jatropha* oil, was measured and similar trends of CO emissions were observed. Smoke level was higher with *Jatropha curcas* oil (4.4 BSU) in comparison to fossil diesel (3.8 BSU) as well. Furthermore it also been observed that there is an increase in ignition delay and combustion duration with *Jatropha* oil in comparison to fossil diesel (Kumar *et al.*, 2003).

5.4. By Transesterification

Trans-esterification is a very simple and straight forward process but it's always need modification in some parameters to improve the efficiency particularly, in input ratio of reagent (alcohol), reaction catalyst, temperature and time to achieve maximum biodiesel yield. Fig. 4, is showing typical biodiesel production process. For the production of biodiesel from *Jatropha* the optimum parameters are 20% methanol (by mass on oil basis), 1.0% NaOH by mass on oil basis (Kumar *et al.*, 2014). To achieve high ester yield 90 min reaction time at 60°C temp is required (Chitra *et al.*, 2005).

Prasad *et al.* (2000) demonstrated in their study that *Jatropha* methyl ester is having higher brake specific energy consumption than for fossil diesel. Prasad *et al.* (2000) reported that the use of trans-esterified *Jatropha curcas* oil causes less NOx emissions. Kaul *et al.* (2007) reported that *Jatropha curcas* oil after transesterification shows little or no additional engine corrosion as compared to diesel.

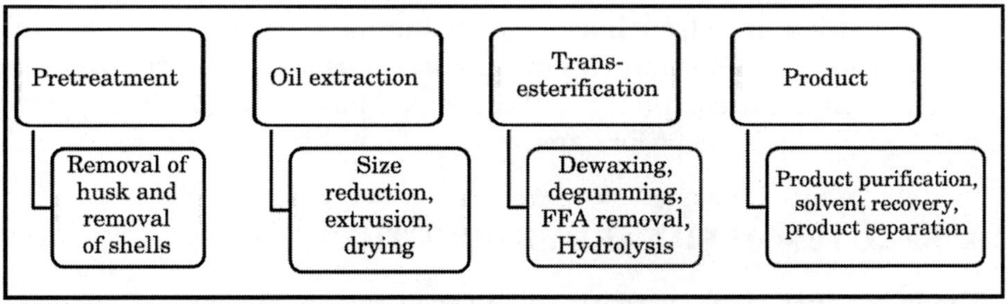

Fig. 4: Steps in biodiesel production from *Jatropha.*

5.5. By Blending

The use of *Jatropha* oil with fossil diesel blends has been tested in a single cylinder, water-cooled open combustion chamber diesel engine (Pramanik, 2003). Brake specific fuel consumption (BSFC) and EGT of the blends were shown to be higher compared to fossil diesel and it also showed increase with increasing proportion of *Jatropha curcas* oil in the blend. The opposite applied for the BTE. However, blends of 30% *Jatropha curcas* oil and 70% fossil diesel by volume and of 40% *Jatropha* oil and 60% fossil diesel by volume displayed brake specific energy consumption and BTE close to the values of fossil diesel.

In the oil extraction procedure of the used *Jatropha* oil, a small portion of water was added which emulsified the *Jatropha curcas* oil. This addition of water resulted in a reduction of EGT and, as such, a reduction in NOx emission with increasing portion of *Jatropha curcas* oil in the blend (Forson *et al.*, 2004) Attention is to be paid to the water content of the used fuel as it causes oxidation inside the injection equipment (FACT, 2006).

5.6. By Alternative Engine Technology

In tropical, areas the use of pure *Jatropha* oil, showed great potential and can be used as straight or blend, (FACT, 2006). The use of pure form of *Jatropha* oil in pre-chamber diesel engines are more suitable than direct injection engines, but simple conversion to a direct injection two-tank system can overcome their problems (FACT, 2006). The diesel engines in old four-wheel-drive vehicles, relying on older technology, are easier to adapt to the characteristics of the used fuel, the temperature of tropical area helps in lowering the viscosity of the oil (FACT, 2006). One can use pure *Jatropha* oil in the stationary diesel engines running at low speed, such as irrigation pumps and electricity generators, without causing any environmental harm. Despite the reported difficulty in the use of pure vegetable oils (Meher *et al.*, 2006; Agarwal, 2007), there is still scope for improvement, and which needs more research.

6. SEED DE-OILED CAKE PRODUCTION

The byproduct of oil extraction is its seed cake which has 58.1% crude protein by weight and 18.2 MJ kg^{-1} average gross energy contents. When extraction of oil is done with the help of expeller it shows that the remaining part of oil left in seed cake is high. The whole seed contains average oil 34.4% on a mass basis and seed cake will contain 9-12% oil by weight. By the oil content the gross energy value of the cake will also change. The cake has various toxins which restrict its utilization as fodder (Francis *et al.*, 2005). The organic nutrient content of *Jatropha* seed cake or its raw kernel is higher than chicken and cattle manure (Francis *et al.*, 2005). The cake possess biodegradable toxins, phorbol ester is the main toxin. This makes the cake as biofertilizer (Chaturvedi *et al.*, 2008, 2009a, b, c), biopesticide, bioinsecticide and bio-molluscicide (Francis *et al.*, 2005; Rug *et al.*, 1997). The phorbol ester decomposes within 6 days (Rug *et al.*, 1997), but it is still advisable to check the phorbol ester content in the produce especially when it is for human consumption. This relation of temperature and moisture to phorbol ester degradation is directly proportionate. If temperature and moisture increase, it enhances the speed of degradation of phorbol ester. If phorbol ester concentration is high it affects snail (*Physa fontinalis*) as mortality rate increases. The cake displays phytotoxicity when applied in high amount (5t ha^{-1}), it causes reduction in germination of tomato. The seed cake can be used as feed stock to biogas plant and slurry of this could be utilized as soil amendment. When pig mannure is used as inoculum to the seed cake, it can produce 0.446 m^3 of biogas, 70% CH_4 per kg of seed cake (Staubmann *et al.*, 1997a,b). With some specific microbial inoculums (Radhakrishna, 2007), 0.5 m^3 biogas $kg^{-1}$of solvent extracted kernel cake and little higher production (0.6 m^3) of biogas kg^{-1} was achieved. Different parts of *Jatropha curcas* can be digested like its fruit husks. Lopez *et al.* (1997) showed that biogas could be produced from husks.

The multi-dimensional utilization of *Jatropha* seed cake makes it an important by-product. Using the cake as a fertilizer can help to reduce inputs needed for both *Jatropha* cultivation and other agricultural crops (Chaturvedi *et al.*, 2010, 2011, 2012, 2013) or it can be used to produce biogas. From environmental perception, digestion of cake and bringing the effluent back to the field is thought to be the best practice at present. The long-term and cumulative impacts of *Jatropha curcas* seed cake on soils have not been addressed yet. Table 2 is showing case studies using *Jatropha curcas* seed cake as fertilizer.

Table 2: Case studies using *Jatropha curcas* seed cake as fertilizer

Country	*Crop*	*Dosage*	*Comments*	*References*
India	JCL	0.75–3t ha^{-1}	Increase of 13–120% yield over zero-input	(Ghosh *et al.*, 2007)
Zimbabwe	Cabbage	2.5–10t ha^{-1}	Increase of 40–113% yield over zero-input	(Ngoma, 2007)
Nepal	Rice	10t ha^{-1}	Increase of 11% yield over zero-input	(Heller, 1996)
Mali	Pearl millet	5t ha^{-1}	Increase of 46% yield over zero-input	(Henning *et al.*, 1995)

Result based on field data showed, one ton seed cake can be produced on 0.27–0.54 ha *Jatropha curcas* plantation. This calculation which is based on an expected seed yield of 2.5–5t dry seed ha^{-1} yr^{-1} with an oil content of 34.4% wt and a mechanical extraction efficiency of 75%.

7. CHEMICAL COMPOSITION AND APPLICATION OF DIFFERENT PARTS OF *JATROPHA*

7.1. Leaves and Twigs

The leaves of *Jatropha* have chemical components which includes, the cyclic triterpenes (stigmasterol, stigmast-5-en-3fl, 7fl-diol, stigmast-5-

en-3fl, %t-diol, cholest-5-en-3fl, 7fl-diol, cholest-5-en-3fl, 7~t-diol, campesterol, fl-sitosterol, and 7-keto-fl-sitosterol). These parts also contain some other compounds such as flavonoids such as apigenin, vitexin and isovitexin (Mitra *et al.*, 1970; Khafagy *et al.,* 1977; Hufford and Oguntimein, 1987; Neuwinger, 1994). Hufford and Oguntimein (1987) have found that an ethanolic extract of the defatted leaves showed activity against P-388 lymphocytic leukaemia. Staubmann *et al.* (1997b) has isolated an anti-inflammatory substance from the leaves. Generally *Jatropha* leaves is not a preferred substrate for silkworm, but some silkworm species (*Samia cynthia ricini*), feed *Jatropha* leaves. Authors conducted some studies on feeding habit of silkworm regarding *Jatropha* and observed that, silkworm showed tendency of feeding of mulberry leaves as compared to *Jatropha*. Fig. 5 is showing various medicinal uses of *Jatropha*.

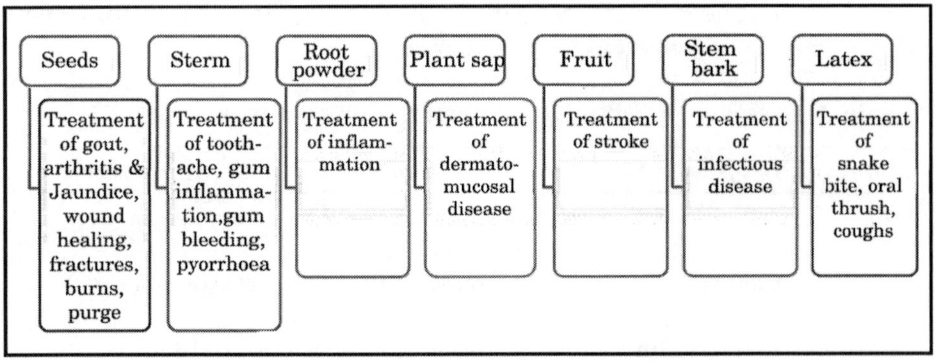

Fig. 5: is showing various medicinal uses of *Jatropha*

7.2. Chemical Composition of Stem and Root

Aerial parts of this plant are rich in various chemicals like o and p-coumaric acid, p-OH-benzoic acid, resorcylic acid, protocatechuic acid, and t-me-cinnamic acid. These parts also contain iridoid, saponin and tannin (Hemalatha and Radhakrishnaiah, 1993). Its stem contains Friedelin, epi-friedelinol, the tetracyclic triterpene ester jatrocurin and scopoletin methyl ester (Talapatra *et al.,* 1993). Underground part like roots contain fl-sitosterol and its fl-D-glucoside, propacin, marmesin, the curculathyranes A and B, the curcusones A-D, jatropholone A and B, diterpenoids jatrophol, the coumarin tomentin, the coumarino-lignan jatrophin and taraxerol (Naengchomnong *et al.,* 1986; Naengchomnong *et al.,* 1994).

7.3. Latex

Latex from *Jatropha* has cyclic octapeptide, which displays proliferation of human t-cells and a moderate inhibition of classical pathway activity of human complement (Vandenberg *et al.,* 1995). *Jatropha* also has proteolytic enzyme (curcain) which have 22 kDa MW and isoelectric point at pH 5.6–6.0 has also been extracted and utilized in various medicinal applications (Nath and Dutta, 1988; Nath and Dutta, 1989; Nath and Dutta, 1991; Dutta, 1997). The curcain is nontoxic on oral administration and its LDs 0 on intraperitoneal administration in mice is 0.96 g kg^{-1} body weight. Another compound, curcacycline B a cyclic nonapeptide that showed to increase the rotamase activity of human cyclophilin B (Auvin *et al.,* 1997). The virus water melon mosaic is inhibited by the latex (Tewari and Shukla, 1982).

7.4. Seeds

In *Jatropha* seed shells, besides the main constituents, a positively inotropic cardioactive principle has been found. The kernel of *Jaropha* possesses dulcitol, sucrose and the fl-D-glucoside of fl-sitosterol (Mitra *et al.,* 1970). Some toxic and anti-nutritional components in the kernel and the press cake include phytates, saponins and a trypsine inhibitor (Aregheore *et al.,* 1997; Makkar and Becker, 1997; Wink *et al.,* 1997). Though, the presence of tannins, amylase inhibitors, glucosinolates and cyanogens were not detected in the seed (Makkar *et al.,* 1997). The chemical composition of the different part of seed is given in Table 3.

Table 3: Chemical composition of kernel, shell and meal

	Kernel	***Shell***	***Meal***
Dry matter (%	94.2–96	89.8–90.4	100
Constituents (% in DM)			
Crude protein	22.2–27	4.3–4.5	56.4–63.8
Lipid	56.8–58	0.5–1.4	1.0–1.5
Ash	3.6–4.3	2.8–6.1	9.6–10.4
Neutral detergent fibre	3.5–3	83.9–89.4	8.1–9.1
Acid detergent fibre	2.4–3	74.6–78	5.7–7.0
Acid detergent lignin	0.0–0.2	45.1–47.5	0.1–0.4
Gross energy (MJ kg^{-1})	30.5–31	19.3–19.5	18.0–18.3

7.5. Other Uses: Utilization of Bio-diesel Fuel and Wasteland Reclamation

Biodiesel is produced from different seed oils from long time and generation of bio-diesel from vegetable oils is well documented, but still there is a need of maximization of the trans-esterification efficiency at minimal cost. An important issue in the biodiesel production is the recovery and the reuse of the catalyst (Francis *et al.*, 2005). There is always a need to develop processing units that has low-cost and robust design for small-scale trans-esterification (Francis *et al.*, 2005). The choice of using methyl esters or the *Jatropha* oil depends on the goal of the use and the available infrastructure. Previous studies on *Jatropha* transesterified oil showed better results than the use of pure oil, in unadjusted diesel engines. When the engine runs on pure oil after changing engine parameters showed considerable improvements in terms of performance and emission. Although, it requires further study.

Several reports have suggested that *Jatropha* has no negative impact on soil and adjoining biodiversity. Ogunwole *et al.* (2007) reported that *Jatropha* have positive impact on soil. Although, land use impact assessment of *Jatropha* has not been executed so far. Several reports proved the role of *Jatropha* in controlling soil erosion and carbon sequestration, and not many studies have been conducted on role of *Jatropha* in nutrient cycling and the impact on soil and biological life. *Jatropha* is being an exotic species and its impact on land use change towards biodiversity is expected to be negative, although this will largely depend on the mix of land use which is replaced by *Jatropha curcas* and on how *Jatropha curcas* is cultivated. Although studies proved the role of *Jatropha* in restoring the local biodiversity when its grown as live fence or in intercropping. However, not many studies have been performed to quantify the allelopathic effects of *Jatropha curcas* on native vegetation.

8. GLOBAL ISSUES

Life cycle assessment showed low GHG emission from *Jatropha* over the production of fossil fuel (Tobin and Fulford, 2005; Prueksakorn and Gheewala, 2006). The largest GHG contributors of the production process are irrigation, fertilizer and transesterification and 70% depending on the applied cultivation intensity.

Prueksakorn and Gheewala (2006) reported that 90% of the total life-cycle GHG emissions are caused by the end-use. According to their calculation, global warming potential of the production and use of *Jatropha curcas* bio-diesel is 23% of the global warming potential of fossil diesel.

The global warming potential of N_2O is 296 times higher than an equal mass of CO_2 (Crutzen *et al.*, 2007). According to the IPCC the N_2O release is equal to about 1% of the nitrogen input from mineral fertilizer or biologically fixed N (IPCC, 2006). The removal of (semi) natural forest for the introduction of *Jatropha curcas*, on the other hand, is expected to have a significant negative effect on the GHG balance of the whole life cycle.

9. CONCLUSIONS

In this article we reviews the cultivation practices of *Jatropha* its botanical and morphological description, ecological role, use of various parts medicines etc, production of biodiesel. This article supports the cultivation of *Jatropha* on marginal land for the production of biodiesel and fertilizers from its cake. Though, policy changes and negative campaigning of *Jatropha*, demoralize the *Jatropha* grower. Several knowledge gaps associated with *Jatropha* need to be bridged before larger scale cultivation can be undertaken. Best practice for *Jatropha* include selection of better germplasm and genetically engineered plant that produce high yield. It is true that the *Jatropha* system cannot meet all expectations at the same time; the plant cannot perform all its functions together at the best level. The impact of *Jatropha* on the soil seems to be positive, depending on used practice and used soil types, but this contribution to soil restoration might find trade-offs in biodiversity loss.

ACKNOWLEDGEMENTS

AK would like to acknowledge CSIR, New Delhi and UGC-BSR Startup grant for providing financial support for the study.

REFERENCES

Abdu-Aguye, I., Sannusi, A., Alafiya-Tayo, R.A. and Bhusnurmath, S.R. (1986). Acute toxicity studies with *Jatropha curcas* L. *Hum. Toxicol.*, 5: 269–274.

Adam, E.I. (1974). Toxic effects *of Jatropha curcas* in mice. *Toxicology,* 2: 67–76.

Adam, E.I. and Magzoub, M. (1975). Toxicity *of Jatropha curcas* for goats. *Toxicology,* 4.

Aderibigbe, A.O., Johnson, C.O.L.E., Makkar, H.P.S., Becker, K. and Foidl, N. (1997). Chemical composition and effect of heat on organic matter-and nitrogen-degradability and some anti nutritional components of *Jatropha* meal. *Animal Feed Science and Technology,* 67: 223–43.

Agarwal, A.K. (2007). Biofuels (alcohols and biodiesel) applications as fuels for internal combustion engines. *Progress in Energy and Combustion Science,* 33: 233–71.

Akcali, I.D., Ince, A. and Guzel, E. (2006). Selected physical properties of peanuts. *International Journal of Food Properties,* 9: 25–37.

Akintayo, E.T. (2004). Characteristics and composition of *Parkia biglobosa* and *Jatropha curcas* oils and cakes. *Bioresour. Technol.,* 92: 307–310.

Aregheore, E.M., Makkar, H.P.S. and Becker, K. (1997). Lectin activity in toxic and non-toxic varieties of *J. curcas* using a latex agglutination test. *In*: Biofuels and Industrial Products from *Jatropha curcas. In*: Giibitz, G.M., Mittelbach, M. and Trabi, M. (*eds.*), DBV Graz. pp. 65–69.

Augustus, G.D.P.S., Jayabalan, M. and Seiler, G.J. (2002). Evaluation and bioinduction of energy components of *Jatropha curcas. Biomass Bioenergy,* 23: 161–164.

Auvin, C., Baraguey, C., Blond, A., Lezenven, F., Pousset, J.L. and Bodo, B. (1997). Curcacycline B, a cyclic non-apeptide from *Jatropha curcas* enhancing rotamase activity of cyclophilin. *Tetrahed. Lett.,* 38: 2845–2848.

Aviara, N.A., Oluwole, F.A. and Haque, M.A. (2005). Effect of moisture content on some physical properties of shea nut. *Int. Agrophysics,* 19: 193–198.

Barnwal, B.K. and Sharma, M.P. (2005). Prospects of biodiesel production from vegetables oils in India. *Renewable and Sustainable Energy Reviews,* 9: 363–78.

Baryeh, E.A. (2002). Physical properties of millet. *J. Food Eng.,* 51: 39–46.

Becker, K. (1995). Einfluss von *Jatropha curcas* aufjunge Karpfen (*Cyprinus carpio* L.). Forschungsbericht, Universitaet Hohenheim, Hohenheim.

Becker, K. and Makkar, H.P.S. (1998). Toxic effects of Phorbol esters in carp (*Cyprinus carpio* L.). *Vet. Human Toxicol.,* 40: 82–86.

Behme, H. (1988). Meglichkeiten der Verwendung von Pressriickst~inden der Purgiernuss in der Tierern~ihrung auf den Kapverden. Bericht, Institut fiir Tierern~ihrung der Bundesanstalt fiir Landwirtschaft, Braunschweig-Ve1kenrode, BRD.

Benge, M. (2006). Assessment of the potential of *Jatropha curcas,* (biodiesel tree) for energy production and other uses in developing countries (*www.echotech.org*).

Cano-Asseleih, L.M. (1986). Chemical investigation of *Jatropha curcas* L. seeds. Ph.D. Thesis. University of London, U.K.

Cano-Asseleih, L.M., Plumbly, R.A. and Hylands, P.J. (1989). Purification and partial characterization of the hemagglutination from seeds of *Jatropha curcas. Jour. Food Biochem.*, 13: 1–20.

Chaturvedi, S. (2011). Influence of composted bio-diesel cake on growth, yield and micro-nutrient composition of tomato. *Communications in Soil Science and Plant Analysis,* 42(21): 2642–2653. DOI: 10.1080/00103624.2011.614039.

Chaturvedi, S., Chel, A., Satya, S. and Kaushik, G. (2008). Viable tailored organic fertilizer alternatives from waste produced by bio-diesel extraction process-an addition in the value chain. ASME 2nd Inter. Conf. on energy sustainability. Florida, USA.

Chaturvedi, S., Kumar, A. and Singh, B. (2009c). Utilizing composted *Jatropha* and Neem cake and Tobacco waste to sustain garlic yields in Indo-gangetic plains. *Journal of Phytology,* 1(6): 353–360.

Chaturvedi, S., Kumar, A., Singh, B., Nain, L., Joshi, M. and Satya, S. (2013). Bioaugmented composting of *Jatropha* de-oiled cake and vegetable waste under aerobic and partial anaerobic conditions. *Journal of Basic Microbiology,* 53(4): 327–335.

Chaturvedi, S., Kumar, V. and Satya, S. (2009a). Composting effects of *Pongamia pinnata* on tomato fertilization. *Archives of Agronomy and Soil Science* 55(5): 535–546.

Chaturvedi, S., Satya, S. and Tiwari, S.K. (2009b). Treatment of biowaste (generated in biodiesel process)- A new strategy for green environment and horticulture crop. *Int. J. of Envi. Sci. and Engineering*, 1(3): 163. (France) *http://www.waset.org/journals/ijese/vl.php.*

Chaturvedi, S., Singh, B., Lata, Khare, S.K., Pandey, A.K. and Satya, S. (2010). Evaluation of hydrolytic enzymes in bioaugmented compost of *Jatropha* cake under aerobic and partial anaerobic conditions. *Annals of Microbiology,* DOI: 10.1007/s13213-010-0113-5.

Chitra, P., Venkatachalam, P. and Sampathrajan, A. (2005). Optimization of experimental conditions for biodiesel production from alkali-catalysed transesterification of *Jatropha curcas* oil. *Energy for Sustainable Development,* 9: 13–8.

Crutzen, P.J., Mosier, A.R., Smith, K.A. and Winiwarter, W. (2007). N_2O release from agro-biofuel production negates global warming reduction by replacing fossil fuels. *Atmospheric Chemistry and Physics Discussions*, 7: 11191–205.

Dehgan, B. and Webster, G.L. (1979). Morphology and intrageneric relationships of the genus *Jatropha* (Euphorbiaceae), vol. 74. University of California Publications in Botany.

Dutta, S.K. (1997). Curcain – A protease isolated from the latex of *Jatropha curcas* L. *In*: Biofuels and industrial products from *Jatropha curcas. In*: Giibitz, G.M., Mittelbach, M. and Trabi, M. (*eds.*), DBV Graz, pp. 75–81.

Engelmann, F. (1991). *In vitro* conservation of tropical plant germplasm. A review. *Euphytica,* 57: 227–243.

FACT Foundation, URL: *http://www.fact-fuels.org/en/FACT Knowledge Centre/ FACT Publications.*

Felke, J. (1913). ,~Ber die giftstoffe der samen von *Jatropha curcas. Landw. Vers. Stat.,* 82: 427–463.

Foidl, N. and Mayorga, L. (2007). Produccion de ester metilico del tempate (EMAT) como sustituto del combustible diesel. *In*: FACT seminar on *Jatropha curcas* L. agronomy and genetics, Wageningen, The Netherlands, March 26–28. Wageningen: FACT Foundation.

Forson, F.K., Oduro, E.K. and Hammond-Donkoh, E. (2004). Performance of *Jatropha* oil blends in a diesel engine. *Renewable Energy,* 29: 1135–45.

Francis, G., Edinger, R. and Becker, K. (2005). A concept for simultaneous wasteland reclamation, fuel production, and socio-economic development in degraded areas in India: Need, potential and perspectives of *Jatropha* plantations. *Natural Resources Forum,* 29: 12–24.

Gandhi, V.M., Cherian, K.M. and Mulky, M.J. (1995). Toxicological studies on Ratan jot oil. *Food Chem. Toxicol.,* 33: 39–42.

Ghosh, A., Patolia, J.S., Chaudhary, D.R., Chikara, J., Rao, S.N., Kumar, D. *et al.* (2007). Response of *Jatropha curcas* under different spacing to *Jatropha* de-oiled cake. *In*: FACT seminar on *Jatropha curcas* L. agronomy and genetics, Wageningen, The Netherlands, March 26–28. Wageningen: FACT Foundation; 2007 Article no. 8.

Ginwal, H.S., Phartyal, S.S., Rawat, P.S. and Srivastava, R.L. (2005). Seed source variation in morphology, germination and seedling growth of *Jatropha curcas* Linn. in Central India. *Silvae Genetica,* 54: 76–80.

Ginwal, H.S., Rawat, P.S. and Srivastava, R.L. (2004). Seed source variation in growth performance and oil yield of *Jatropha curcas* Linn. in Central India. *Silvae Genetica,* 53: 186–92.

Hartmann, H.T. and Kester, D.E. (1983). Plant Propagation. Principles and Practices, fourth ed. Prentice-Hall, Inc., Englewood Cliffs.

Heller, J. (1996). Physic Nut. *Jatropha curcas* L. Promoting the Conservation and use of Underutilized and Neglected Crops. I. Institute of Plant Genetics and Crop Plant Research, Gatersleben/International Plant Genetic Resources Institute, Rome.

Hemalatha, A. and Radhakrishnaiah, M. (1993). Chemosystematics of *Jatropha*. *J. Econ. Taxonom. Bot.,* 17: 75–77.

Henning, R., Samake´, F. and Thie´ro, I. (1995). La valeur fertilisante du tourteau du pourghe‘re. Bamako, Mali: Projet Pourghe‘re DNHE-GTZ.

Hossain, A.K. and Davies, P.A. (2010). Plant oils as fuels for compression ignition engines: A technical review and life-cycle analysis. *Renewable Energy,* 35: 1–13.

Hufford, C.D. and Oguntimein, B.O. (1987). Non-polar constituents of *Jatropha curcas. Lloydia,* 41: 161–165.

Joubert, P.H., Brown, M.M., Hay, I.T. and Sebata, D.B. (1984). Acute poisoning with *Jatropha curcas* (purging nut tree) in children. *S. Afric. Med. J.,* 65: 729–730.

Katwal, R.P.S. and Soni, P.L. (2003). Biofuels: An opportunity for socio-economic development and cleaner environment. *Indian Forester,* 129: 939–949.

Kaul, S., Saxena, R.C., Kumar, A., Negi, M.S., Bhatnagar, A.K., Goyal, H.B. *et al.* (2007). Corrosion behavior of biodiesel from seed oils of Indian origin on diesel engine parts. *Fuel Processing Technology,* 88: 303–7.

Kaushik, N. (2006). Quality planting material and seed standards in *Jatropha curcas*. *In*: Singh, B., Swaminathan, R. and Ponraj, V., (*eds.*), Proceedings of the biodiesel conference toward energy independence–focus of *Jatropha*, Hyderabad, India, 9–10 June. New Delhi: Rashtrapati Bhawan, pp. 179–96.

Kaushik, N. and Kumar, S. (2006). *Jatropha curcas* L. silviculture and uses. 2nd revised ed. Jodhpur: Agrobios.

Kaushik, N., Kumar, K., Kumar, S., Kaushik, N. and Roy, S. (2007). Genetic variability and divergence studies in seed traits and oil content of *Jatropha* (*Jatropha curcas* L.) accessions. *Biomass and Bioenergy,* 31: 497–502.

Khafagy, S.M., Mohamed, Y.A., Abdel, N.A. and Mahmoud, Z.F. (1977). Phytochemical study *of Jatropha curcas. Plant. Med.*, 31: 274–277.

Kobilke, H. (1989). Untersuchungen zur Bestandesbegru¨ndung von Purgiernuß (*Jatropha curcas* L.). Diploma thesis. University Hohenheim, Stuttgart.

Kumar, A. and Sharma, S. (2005). Potential of *Jatropha* and cultural practices to maximize its yield. ICPQR, December, IIT, New Delhi.

Kumar, A. and Sharma, S. (2008). An evaluation of multipurpose oil seed crop for industrial uses (*Jatropha curcas* L.): A review. *Industrial Crops and Products*, 28: 1–10.

Kumar, A. and Sharma, S. (2011). Potential non-edible oil resources as biodiesel feedstock: An Indian perspective. *Renewable and Sustainable Energy Reviews*, 15: 1791–800.

Kumar, A., Kumar, K., Kaushik, N., Sharma, S. and Mishra, S. (2010a). Renewable energy in India: Current status and future potentials. *Renewable and Sustainable Energy Reviews,* 14: 2434–42.

Kumar, A., Sharma, S. and Mishra, S. (2010b). Influence of arbuscular mycorrhizal (AM) fungi and salinity on seedling growth, solute accumulation, and mycorrhizal dependency of *Jatropha curcas* L. *Journal of Plant Growth Regulation*, 29: 297–306.

Kumar, A., Sharma, S. and Mishra, S. (2016). Evaluating effect of arbuscular mycorrhizal fungal consortia and *Azotobacter chroococcum* in improving biomass yield of *Jatropha curcas. Plant Biosystems*, pp. 1–9.

Kumar, A., Sharma, S., Mishra, S. and Dames, J.F. (2015). Arbuscular mycorrhizal inoculation improves growth and antioxidative response of *Jatropha curcas* (L.) under Na_2SO_4 salt stress. *Plant Biosystems*, 149: 260–9.

Kumar, M.S., Ramesh, A. and Nagalingam, B. (2003). An experimental comparison of methods to use methanol and *Jatropha* oil in a compression ignition engine. *Biomass and Bioenergy*, 25: 309–18.

Liberalino, A.A., Bambirra, E.A., Moraes-Santos, T. and Vieira, E.C. (1988). *Jatropha curcas* L. seeds: Chemical analysis and toxicity. *Arq. Biol. Technol.* 31: 539–550.

Lindqvist, Y., Huang, W., Schneider, G. and Shanklin, J. (1996). Crystal structure of delta 9 stearoyl-acyl carrier protein desaturase from castor seed and its relationship to other di-iron proteins. *EMBO J.,* 15: 4081–4092.

Lopez, O., Foidl, G. and Foidl, N. (1997). Production of biogas from *J. curcas* fruit shells. *In*: Gu¨bitz, G.M., Mittelbach, M. and Trabi, M. (*eds.*), biofuels and industrial products from *Jatropha curcas* Proceedings from the symposium "*Jatropha* 97," Managua, Nicaragua, Graz, Austria: Dbv-Verlag, pp. 118–22.

Ma, F. and Hanna, M.A. (1999). Biodiesel production: A review. *Bioresource Technology,* 70: 1–15.

Machado, A.D.C., Frick, N.S., Kremen, R., Katinger, H. and Machado, M.L.D.C. (1997). Biotechnological approaches to the improvement of *J. curcas. In*: Giibitz, G.M., Mittelbach, M. and Trabi, M. (*eds.*), Biofuels and Industrial Products from *Jatropha curcas.* DBV Graz, pp. 22–27.

Makkar, H.P.S. and Becker, K. (1997). Potential of *J. curcas* seed meal as a protein supplement to livestock feed; constraints to its utilization and possible strategies to overcome constraints. *In*: Giibitz, G.M., Mittelbach, M. and Trabi, M. (*eds.*), Biofuels and Industrial Products from *Jatropha curcas.* DBV Graz, pp. 190–205.

Makkar, H.P.S., Becker, K., Sporer, F. and Wink, M. (1997). Studies on nutritive potential and toxic constituents of different provenances of *Jatropha curcas. Journal of Agricultural and Food Chemistry,* 45: 3152–7.

Mampane, K.J., Joubert, P.H. and Hay, I.T. (1987). *Jatropha curcas*: Use as a traditional Tswana medicine and its role as a cause of acute poisoning. *Phytother. Res.,* 1: 50–51.

Meher, L.C., VidyaSagar, D. and Naik, S.N. (2006). Technical aspects of biodiesel production by transesterification–A review. *Renewable and Sustainable Energy Reviews,* 10: 248–68.

Mitra, C.R., Bhatnagar, S.C. and Sinha, M.K. (1970). Chemical examination of *Jatropha curcas. Ind. J. Chem.,* 8: 1047.

Naengchomnong, W., Tarnchompoo, B. and Thebtaranonth, Y. (1994). *Jatropha,* marmesin, propacin and jatrophin from the roots of *Jatropha curcas* (Euphorbiaceae). *J. of the Sci. Soc. of Thail.,* 20: 73–83.

Naengchomnong, W., Thebtaranonth, Y., Wiriyachitra, P., Okamoto, K.T. and Clardy, J. (1986). Isolation and structure determination of four novel diterpenes of *Jatropha curcas. Tetrahed. Lett.,* 27: 2439–2442.

Nath, L.K. and Dutta, S.K. (1988). Extraction and study of certain physico-chemical properties of a new proteolytic enzyme from the latex of *Jatropha curcas* L. *Ind. J. Pharm. Sci.,* 50: 125–127.

Nath, L.K. and Dutta, S.K. (1989). Kinetic studies on curcain, a protease from the latex of *Jatropha curcas* L. *Ind. J. Pharm. Sci.,* 51: 43–47.

Nath, L.K. and Dutta, S.K. (1991). Extraction and purification of curcain, a protease from the latex of *Jatropha curcas* L. *J. Pharm. Pharmacol.,* 43: 111–114.

Nath, L.K. and Dutta, S.K. (1997). Acute toxicity studies and wound healing response of curcain, a proteolytic enzyme extract from the latex of *Jatropha curcas* L. *In*: Biofuels and Industrial Products from *Jatropha curcas.* Gubitz, G.M., Mittelbach, M. and Trabi (*eds.*), DBV Graz. pp. 82–86.

Neuwinger, H.D. (1994). Afrikanische Arzneipflanzen und Jagdgifte. WV GesmH, pp. 450–457.

Ngoma, T. (2007). Fertilizer, manure and *Jatropha* cake application trial, nodes and leaf sections of non-toxic *Jatropha curcas* L. *Plant Growth Regulation,* 47: 83–90.

Ogunwole, J.O., Patolia, J.S., Chaudhary, D.R., Ghosh, A. and Chikara, J. (2007). Improvement of the quality of a degraded entisol with *Jatropha curcas* L. under an Indian semi arid condition. *In*: FACT seminaron *Jatropha curcas* L. agronomy and genetics, Wageningen, The Netherlands, Wageningen: FACT Foundation.

Omobuwajo, T.O., Sanni, L.A. and Olajide, J.O. (2000). Physical properties of Ackee apple (*Blighia sapida*) seeds. *J. Food Eng.,* 45: 43–45.

Openshaw, K. (2000). A review of *Jatropha curcas*: An oil plant of unfulfilled promise. *Biomass and Bioenergy*, 19: 1–15.

Pletsch, M. and Charlwood, B.V. (1997). Accumulation of diterpenoids in cell and root organ cultures of *Jatropha* species. *Pl. Physiol.,* 150: 37–45.

Pradhan, R.C., Naik, S.N., Bhatnagar, N. and Vijay, V.K. (2010). Design, development and testing of hand-operated decorticator for *Jatropha* fruit. *Applied Energy,* 87: 762–768.

Pramanik, K. (2003). Properties and use of *Jatropha curcas* oil and diesel fuel blends in compression ignition engine. *Renewable Energy,* 28: 239–48.

Prasad, C.M.V., Krishna, M.V.S.M., Reddy, C.P. and Mohan, K.R. (2000). Performance evaluation of non-edible vegetable oils as substitute fuels in low heat rejection diesel engines. Proceedings of the Institution of Mechanical Engineers Part D. *Journal of Automobile Engineering,* 214: 181–7.

Prueksakorn, K. and Gheewala, S.H. (2006). Energy and greenhouse gas implications of biodiesel production from *Jatropha curcas* L. *In*: Proceedings of the second joint international conference on "Sustainable energy and environments (SEE2006)," Bangkok, Thailand.

Radhakrishna, P. (2007). Contribution of de-oiled cakes in carbon sequestration and as a source of energy, in Indian agriculture need for a policy initiative. *In*: Proceedings of the fourth international biofuels conference, New Delhi, India, New Delhi: Winrock International India, pp. 65–70.

Reddy, J.N. and Ramesh, A. (2006). Parametric studies for improving the performance of a *Jatropha* oil-fuelled compression ignition engine. *Renewable Energy,* 31: 1994–2016.

Rivera-Lorca, J.A. and Ku-Vera, J.C. (1997a). Chemical composition of three different varieties of *J. curcas* from Mexico. *In*: Biofuels and Industrial Products from *Jatropha curcas*. Gobitz, G.M., Mittelbach, M. and Trabi, M. (*eds.*), DBV Graz. pp. 47–52.

Rivera-Lorca, J.A. and Ku-Vera, J.C. (1997b). Rumen digestion of raw, roasted and boiled seeds of *J. curcas* from Chiapas, Mexico. *In*: Biofuels and Industrial Products from *Jatropha curcas*. Giibitz, G.M., Mittelbach, M. and Trabi, M. (*eds.*), DBV Graz. pp. 167–172.

Rug, M., Sporer, F., Wink. M., Liu, S.Y., Henning, R. and Ruppel, A. (1997). Molluscicidal properties of *J. curcas* against vector snails of the human parasites *Schistosoma mansoni* and *S. japonicum*. *In*: Biofuels and Industrial Products from *Jatropha curcas*. Giibitz, G.M., Mittelbach, M. and Trabi, M. (*eds.*), DBV Graz. pp. 227–232.

Sahay, K.M. and Singh, K.K. (1996). Unit Operation of Agricultural Processing. Vikas Publishing House Pvt. Ltd., New Delhi, India.

Schmook, B. and Serralta-Peraza, L. (1997). *J. curcas*: Distribution and uses in the Yucatan Peninsula of Mexico. *In*: Biofuels and Industrial Products from *Jatropha curcas*. Giibitz, G.M., Mittelbach, M. and Trabi, M. (*eds.*), DBV Graz. pp. 53–57.

Sharma, G.D., Gupta, S.N. and Khabiruddin, M. (1997). Cultivation of *Jatropha curcas* as a future source of hydrocarbon and other industrial products. *In*: Biofuels and Industrial Products from *Jatropha curcas*. Giibitz, G.M., Mittelbach, M. and Trabi, M. (*eds.*), pp. 19–21.

Sharma, S. and Kumar, A. (2010). *Jatropha curcas*: A Source of Energy and Other Applications. Energy Crop. Royal Society of Chemistry, Europe. ISBN 13: 9781849730327; ISBN 10: 1849730326.

Siang, C.C. (2009). *Jatropha curcas* L. development of a new oil crop for biofuel. [Online]. The Institute of Energy Economics, Japan.

Singh, B., Swaminathan, R. and Ponraj, V. (*eds.*) (2006). Proceedings of the biodiesel conference toward energy independence–focus of *Jatropha*, Hyderabad, India, New Delhi: Rashtrapati Bhawan, pp. 223–51.

Singh, N.B., Kumar, A. and Rai, S. (2014). Potential production of bioenergy from biomass in an Indian perspective. *Renewable and Sustainable Energy Reviews*, 39: 65–78.

Spaan, W.P., Bodnar, F., Idoe, O. and De Graaff, J. (2004). Implementation of contour vegetation barriers under farmer conditions in Burkina Faso and Mali. *Quarterly Journal of International Agriculture,* 43: 21–38.

Staubmann, R., Foidl, G., Foidl, N., Gubitz, G.M., Lafferty, R.M., Arbizu, V.M. and Steiner, W. (1997a). Biogas production from *Jatropha curcas* press cake. *Appl. Biochem. Biotech.,* 63: 457–467.

Staubmann, R., Ncube, I., Gubitz, G.M., Steiner, W. and Read, J.S. (1999). Esterase and lipase activity in *Jatropha curcas* L. seeds. *J. Biotechnol.,* 75: 117–126.

Staubmann, R., Schubert-Zsilavecz, M., Hiermann, A. and Kartnig, T. (1997b). The anti-inflammatory effect of *J. curcas* leaves. *In*: Biofuels and Industrial Products from *Jatropha curcas*. Giibitz, G.M., Mittelbach, M. and Trabi, M. (*eds.*), DBV Graz. pp. 60–64.

Stirpe, F., Pession, Brizzi, A., Lorenzoni, E., Strochi, P., Montanaro, L., Sperti, S. (1976). Studies on the proteins from the seeds of *Croton tiglium* and *Jatropha curcas*. *Biochem. J.,* 156: 1–6.

Sujatha, M. and Dingra, M. (1993). Rapid plant regeneration from various explants of *Jatropha integerrima*. *Plant cell, Tissue Organ Cult.,* 35: 293–296.

Sujatha, M. and Mukta, N. (1996). Morphogenesis and plant regeneration from tissue culture of *Jatropha curcas*. *Plant cell, Tissue Organ Cult.,* 44: 135–141.

Sujatha, M. and Prabakaran, A.J. (2003). New ornamental *Jatropha* hybrids through interspecific hybridization. *Genet. Resour. Crop Evol.,* 50: 75–82.

Talapatra, S.K., Mandal, K. and Talapatra, B. (1993). Jatrocurin, a new tetracyclic triterpene from *Jatropha curcas. J. of the Ind. Chem. Soc.,* 70: 543–548.

Tewari, J.P. and Shukla, I.K. (1982). Inhibition of infectivity of two strains of watermelon mosaic virus by latex of some angiosperms. *Geobios*, 9: 124–126.

Tobin, J. and Fulford, D.J. (2005). Life cycle assessment of the production of biodiesel from *Jatropha*. MSc dissertation, The University of Reading.

Tong, L., Peng, S.-M., Deng, W.-Y., Ma, D.-W., Xu, Y., Xiao, M. and Chen, F. (2006). Characterization of a new stearoyl-acyl carrier protein desaturase gene from *Jatropha curcas*. *Biotechnol. Lett.,* 28: 657–662.

Vandenberg, J.J., Horsten, A.J., Kettenesvandenbosch, J.J., Kroes, B.H., Beukelmann, C.J., Leeflang, B.R. and Labadie, R.P. (1995). Curcacycline A – a novel cyclic octapeptide isolated from the latex of *Jatropha curcas* L. *FEBS Lett.,* 358: 215–218.

Vasudevan, P., Sharma, S. and Kumar, A. (2005). Liquid fuel from biomass: An overview. *J. Sci. Ind. Res.*, 64: 822–831.

Wei, Q., Lu, W.D., Liao, Y., Pan, S.L., Xu, Y., Tang, L. and Chen, F. (2004). Plant regeneration from epicotyl explants of *Jatropha curcas. Plant Physiol. Mol. Biol.* 30: 475–478.

Wink, M., Koschmieder, C., Sauerwein, M. and Sporer, F. (1997). Phorbol esters of *J. curcas* – biological activities and potential applications. *In*: Biofuels and Industrial Products from *Jatropha curcas.* Giibitz, G.M., Mittelbach, M. and Trabi, M. (*eds.*), DBV Graz. pp. 160–166.

Winkler, E., Foidl, N., Giibitz, G.M., Staubmann, R. and Steiner, W. (1997). Enzyme supported oil extraction from *Jatropha curcas* seeds. *Appl. Biochem. Biotech.,* 63(5): 449–456.

8

Castor Oil and Alkyl Esters as Lubricity Enhancer

LALIT PRASAD[1,2*], L.M. DAS[1] AND S.N. NAIK[3]

ABSTRACT

Energy is an essential input for economic growth, social development, human welfare and improving the quality of life. Ever since their exploration, the fossil fuels continued as the major conventional energy source. The rising prices of petroleum fuels, depletion of oil reserves and stiff regulations on exhaust emissions are the major driving forces to search for alternative fuels from renewable resources with minimum environmental impact. The emission norms are becoming stringent day by day this has led to the upgradation of the fuel quality to meet these norms. However, this upgradation comes at a cost of reduced lubricity due to progressive reduction in sulfur content. Biodiesel (alkyl esters) of triglycerides derived from vegetable oils has the potential to compensate for this reduction. It has been recognized for many years that vegetable oils can be used as food, fuel and lubricants. In view of the several advantages, vegetable oils have a great potential to replace petroleum-based fuels in the long run and to generate economic benefits. In this

[1]Centre for Energy Studies, Indian Institute of Technology, Delhi, New Delhi, India
[2]School of Basic and Applied Sciences, Galgotias University, Greater Noida, India
[3]Centre for Rural Development and Technology, Indian Institute of Technology, Delhi, New Delhi, India
**Corresponding author*: joshi_lp@yahoo.co.in; lalit.prasad@galgotiasuniversity.edu.in

present study castor (Ricinus communis L.) oil alkyl esters were found to be effective lubricity enhancer for low lubricity diesel fuel apart from an alternative to diesel fuel. The lubricity properties such as wear scar diameter (WSD), friction coefficient (μ), film thickness of castor oil (CO), castor oil methyl ester (COME) and castor oil ethyl ester (COEE) were studied as alone and as a lubricity enhancer for low lubricity diesel fuel (LLDF) by high frequency reciprocating rig (HFRR).

***Key words*:** Transesterification, Low lubricity diesel fuel, Alkyl esters, Lubricity enhancer.

1. INTRODUCTION

Biodiesel has been accepted as a renewable liquid biofuel worldwide on account of its strong lubrication properties it is used as an additive to improve the lubricity of petroleum fuels (Stoldt *et al.*, 1998; Ball *et al.*, 1999). It has been reported that polyol esters of vegetable oils are most suitable lubricants for many drilling conditions (Kania *et al.*, 2015).

It has been recognized for many years that vegetable oils can be used as food, fuel and lubricants (Rudnick, 2006). In view of the several advantages, vegetable oils have a great potential to substitute petroleum-based fuels in the long run. However, the use of vegetable oil in direct injection diesel engines is restricted by some unfavorable physical properties, particularly their viscosity. Transesterification also called alcoholysis is the replacement of alcohol from an ester by another alcohol in a process similar to hydrolysis of vegetable oils this process appears to be more suitable for bring down the viscosity of vegetable oil (Ramadhas *et al.*, 2005; Sarin and Sharma, 2007). The transesterification reaction of triglyceride is presented in Fig. 1.

Efforts are under way in many countries, including India, to search for suitable alternative diesel fuels that are environment friendly. The need to search for these fuels arises mainly from the standpoint of preserving the global environment and the concern about long-term supplies of conventional hydrocarbon-based diesel fuels. Among the different possible sources, diesel fuels derived from triglycerides (vegetable oils/animal fats) seems to be a promising alternative to substitute diesel fuels. Although triglycerides can fuel diesel engines, their high viscosities, low volatilities and poor cold flow properties have led to the investigation of various derivatives (Srivastava and Prasad,

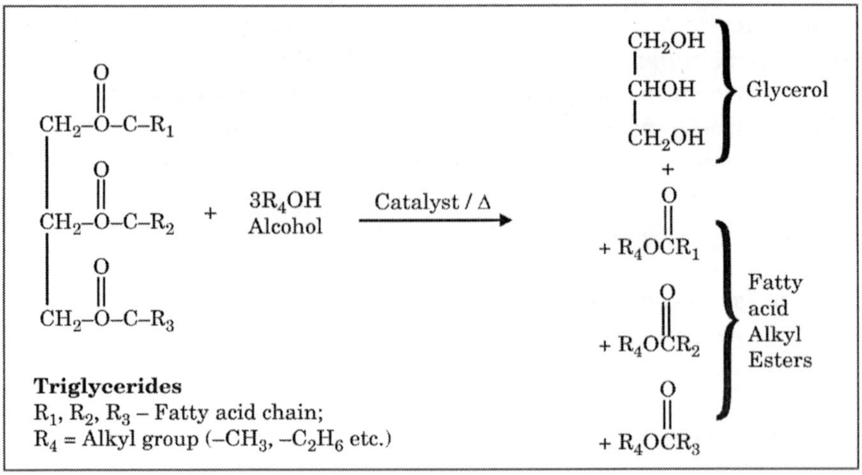

Fig. 1: Transesterification of triglycerides

2000). Fatty acid esters (methyl/ethyl) known as biodiesel, derived from triglycerides by transesterification with methanol have received the most attention (Singh and Singh, 2010). American Society for Testing and Materials (ASTM-D6751) and the European standard (EN 14214) were extensively used test methods for assessing the quality of biodiesel worldwide. The main advantages of using biodiesel are its renewability, lower exhaust gas emissions, its biodegradability and given that the entire organic carbon present is photosynthetic in origin, it does not contribute to a rise in the level of carbon dioxide in the atmosphere (Life cycle analysis) and consequently to the greenhouse effect (Srivastava and Prasad, 2000; Singh and Singh, 2010). Biomass is one of the better sources of energy. Fuels from renewable biomass have the potential to reduce the amount of CO_2, particulate matter, and GHG emissions. This is because the carbon contained in biomass-derived fuel is biogenic and renewable in nature (Kulkarni and Dalai, 2006). Biodiesel has the potential to improve the lubricity of the low/ultra low sulfur diesel fuel as required for the stringent emission norms for diesel fuel. Continuous reduction of sulfur as required for regulations in the country and elsewhere will lead to a decrease in the fuel lubricity (Knothe and Steidley, 2005). However, quality of biodiesel itself needs to meet emerging norms without compromising on performance and emission characteristics. The lubricity issue is significant, because the low sulfur diesel fuel has led to the failure of engine parts such as fuel injectors and pumps, because they are lubricated by the fuel itself.

The use of fats and oils by man dates back to antiquity. Their chemical composition and specific properties have allowed them to find use as foods, fuels and lubricants. Vegetable oils are already in use as lubricant due to their superior lubricity, good anticorrosion, better viscosity-temperature characteristics, and low evaporation loss. Moreover, these oils are readily biodegradable and environmentally safe as these contain readily biodegradable constituents such as fatty acids. There are numerous source of natural oil and fats encompassing vegetable, animal and marine sources. As it is with all matter, their usefulness to man is determined by their chemical nature; and all fats and oils have certain characteristics in common (Aluyor *et al., 2009;* Rudnick, 2006).

Lubricants based on renewable raw materials and their derivatives are drawing increased attention in various applications. The use of such rapidly biodegradable materials is especially favourable in loss-lubrication and hydraulic system. In this respect vegetable based raw materials and their derivatives can play a major role, as vegetable oils have particularly good tribological properties. Due to their specific structure, they combine good boundary friction lubricity and general wear protection with stable viscosity temperature behaviour and very low evaporation. In assertion, they are easily biodegradable. All in all they are excellent raw materials for the formulation of eco-friendly lubricants (Rudnick, 2006; Mobarak *et al.*, 2014).

In the present study, the lubricity study was carried out for pure castor oil and its methyl/ethyl esters as alone and as lubricity improvers for LLDF. The lubricity study was carried out by HFRR.

2. INDIAN SCENARIO: CASTOR (*RICINUS COMMUNIS* L.)

Castor is indigenous to the southeastern Mediterranean Basin, Eastern Africa and India, today it is widespread throughout tropical regions. In areas with a suitable climate, castor establishes itself easily as an apparently "native" plant and can often be found on wasteland. Castor is a non-edible oilseed, belonging to the Euphorbiaceae family. India is largest producer of castor seed and constitutes about 64% of total global production. India is the leader in global castor production and international castor oil trade. India is the world's leading castor seed producer with 62.42% share, followed by China and Brazil with 19.22% and 12.68% share respectively in 2005. The Indian varieties of castor have an oil content of about 40-48%. The genus *Ricinus* is nonspecific, but its evolution under natural selection in a wide range of agro-climatic areas has resulted in numerous wild and semi-wild types with wide genotypic and phenotypic diversity. India is one of the centers of origin. Castor plants either with green, red or purple-coloured stem, green leaves, green or red capsules are common (Anjani, 2005). It is a fast-growing, suckering perennial shrub which can reach the size of a small tree (around 12 metres/39 feet). The glossy leaves are 15–45 centimetres (5.9–18 in) long, long-stalked, alternate and palmate with 5–12 deep lobes with coarsely toothed segments. The wild castor plant common in Indian continent is shown in Fig. 2. In Brazil, castor oil (locally known as mamona oil) is now being used to produce biodiesel. Castor plants are cultivated in almost all of the tropical and subtropical zones. China is the major consumer of castor oil, 31% which is about one third of total production. Other major consumers are Europe, Brazil and US (Ogunniyi, 2006; Report on Castor, 2008; Patel *et al.*, 2016).

Fig. 2: Castor (*Ricinus communis* L.) (a) Tree (b) Tree bearing seeds (Prasad, 2013)

3. ECONOMIC IMPORTANCE AND EMPLOYMENT GENERATION

India, with diverse soil and climate types comprising several agro-ecological regions, provides ample opportunity for growing a variety of crops. Castor plant grows in waste lands without any proper care. Cultivation and processing of castor seed to oil is labour intensive and as such they generate varied employment opportunities for the rural population. Non-edible vegetable seeds have the potential to provide economic benefits at the local level as well as at national level. It minimizes the expenditure of India's foreign-currency reserves for lubricant allowing India to increase its growing foreign currency reserves. Since no food producing farmland is required for producing non-edible feedstocks for the production of alkyl esters, it is considered the most politically and morally acceptable choice in India. Since non edible vegetable plants such as castor are carbon-neutral, large-scale production will improve the country's carbon emissions profile. The processing of seeds to extract oil and production of different alkyl esters and application as a lubricity enhancer will create employment generation in rural areas. Due to the increase in demand of oil/alkyl esters production, huge amount of by-product such as cake and glycerol will be produced during transesterification reaction. Utilization of these by-products into value added products would reduce the cost of alkyl esters/biolubricant production. During oil extraction process about 50% of cake is generated if it is further utilized for value added products that will complement the economics of alkyl ester/biolubricant production. The oil extraction process of castor seeds for producing biodiesel and biolubricant generates castor cake as waste. The cake is rich in protein and is produced at the ratio of 1.2 tons per ton of extracted oil. However, the cake contains anti-nutritional components, ricin and ricinine which are toxic to vertebrates. So the cake cannot be utilized as animal feed. On the other hand the cake contains high concentration of starch, which can be used as potential feedstock for bioethanol production. Thus this process can establish a self-sufficient and integrated biolubricant and bioethanol process and reduce the cost of biolubricant production. Different decentralized units and industries need to be established to produce value added products from the by-products of biolubricant production which generates employment both for rural and urban people.

4. LUBRICITY PROPERTY OF VEGETABLE OIL/ALKYL ESTER

Biodiesel has been accepted as a renewable liquid biofuels worldwide (Stoldt *et al.*, 1998; Ball *et al.*, 1999). Much research has been done to evaluate the impact of biodiesel prepared from different vegetable oils such as rapeseed oil, canola oil, soyabean oil, sunflower oil etc. and different alcohols such as methyl, ethyl, propyl and butyl alcohols, on the lubricity of low-sulfur diesel (Hu *et al.*, 2005).

Drown *et al.* (2001) determined the effects of type of alcohol (methanol and ethanol) and fatty acid profile of vegetable oil on the lubricity properties of biodiesel when used as jet turbine fuel additive. They found that the type of alcohol influenced the lubrication properties of biodiesel, but fatty acid profile did not have any impact. Munson *et al.* (1999) evaluate the lubricity of biodiesels derived from different vegetable oils. They also used methyl, ethyl, butyl and 2-propyl alcohols to make biodiesels and tested their lubricating properties in a base diesel.

5. SIGNIFICANCE OF CASTOR OIL/METHYL/ETHYL ESTER AS LUBRICITY IMPROVERS

Lubricity is the ability of a liquid to provide boundary lubrication to prevent wear between moving parts. The low sulfur level of diesel fuels led to decrease in fuel lubricity, that may lead to wear and damages the diesel engines and primarily fuel injection system. Previous literature states that low blend levels of biodiesel can restore lubricity of low/ ultra sulfur petroleum-derived diesel (petro-diesel) fuels, which has poor lubricity. This feature has been a major technical advantage of biodiesel.

Reduction of 'S' to 0.005% shall drastically reduce lubricating properties of diesel fuel. The reason for the poor lubricity of low sulfur petro-diesel is not only the removal of sulfur containing compounds since in severe hydro-cracking process in refineries also reduced other heteroatom's in the fuel such as oxygen and nitrogen etc. are also reduced (Drown *et al.*, 2001; Schumacher *et al.*, 2005).

Esters of vegetable oils with hydroxyl fatty acids (*i.e.*, castor oil) improve the lubricity at lower levels than the esters of non-hydroxylated vegetable oils since structure of the fatty acid plays important role as lubricity enhancers (Drown *et al.*, 2001; Schumacher *et al.*, 2005; Refaat *et al.*, 2009). Esters of castor and *Lesquerella* oils improve the lubricity of low sulfur petro-diesel at lower blending levels than the esters of non-hydroxylated vegetable oils (Geller and Goodrum, 2004; Goodrum and Geller, 2005). The lubricity behaviour of esters produced from castor oil was found to be most effective even in blending upto 2% in ultra-low sulfur diesel fuels, which have poor lubricity (Moser, 2008).

5.1. Lubricity Study

The lubricity study involving castor oil, methyl ester, ethyl ester and low lubricity test was carried out by using high frequency reciprocating rig (HFRR). Castor oil methyl/ethyl ester were tested as alone and as a fuel additive/lubricity enhancer with low lubricity diesel fuel (LLDF).

The major properties such as wear scar diameter, friction coefficient and film thickness were evaluated for each samples.

There are three ways to evaluate the lubricity of a fuel: a) vehicle test, b) fuel injection test equipment bench test, and c) a laboratory test. The laboratory lubricity test provides accurate evaluation.

The tests available to evaluate lubricity included the:

(i) High Frequency Reciprocating Rig (HFRR)

(ii) Scuffing Load Ball On Cylinder Lubricity Evaluator (SLBOCLE),

(iii) Ball on Three Discs (BOTD) (Schumacher and Howell, 1994) and

(iv) Optimal Reciprocating Rig (SRV)

According to European engine manufacturers, the HFRR gives the best results and therefore this test procedure was adopted as the Coordinating European Council (CEC) standard in 1996. Europe has amended EN 590 to include HFRR test procedure with a maximum wear scar diameter of 460 microns as lubricity standard. Therefore, lubricity analysis done by employing HFRR has become the benchmark for several fuel and lubricant test methods including ASTM D 6079, BS ISO 12156-1, CEC F-06-A-96, EN 590, and JPI-5S-50-98, and IP 450/2000. The HFRR method proved to be sensitive to additives and their structure at the levels below 100 ppm sulfur content (Knothe and Steidley, 2005; Holser and Harry-O'Kuru, 2006).

The lubricity test was carried out individually for low lubricity diesel fuel (LLDF), castor oil, and castor oil methyl and ethyl ester by HFRR. The castor oil/alkyl esters have been used as a lubricity enhancer (additives/blends) for LLDF. The significant change in WSD, friction coefficient and film thickness of LLDF has been tested and correlated. The lubricity test was conducted in triplicate to ensure the repeatability in the data. However the wear scar images taken are close approximate to in between the rest two test results.

5.2. Lubricity Study of LLDF, Oil and Methyl/Ethyl Ester of Castor by HFRR

The average wear scar diameter (WSD), friction coefficient (μ), and film thickness of LLDF, castor oil, COME, and COEE were tested and correlated. From the experimental results (Table 1), it was observed that LLDF showed WSD (654 μm) value above the maximum allowable limit of WSD (450 μm). Fig. 3 shows that the WSD is quite low for oil and methyl/ethyl ester of *Castor* as compared to LLDF. Among CO, COME and COEE, it has been shown that COEE shows minimum WSD value. The comparative fiction coefficient as shows that, oil and methyl/ ethyl ester has minimum friction compare to LLDF. Comparing film percentage indicates high film formation has taken place with oil and methyl/ethyl ester compared to LLDF. The WSD is quite high for LLDF, while for *Castor* oil, methyl and ethyl ester it is quite low. It has been observed that compared to oil, methyl and ethyl esters; ethyl ester possess comparatively better lubricating properties (Prasad *et al.*, 2012). The possible reason for the increase of the good lubrication property of the ethyl ester compared to methyl ester and oil is due to the fact that the electron donating nature of the ethyl ester chain is higher than methyl ester chain; this may lead to higher electron density on the polar oxygen of the ethyl ester chain. However, the lubricity-enhancing effect of double bonds is greater than that of extended chain length, as stated in previous study (Knothe *et al.*, 2005). The results imply that alkyl ester, which is mainly composed of fatty acid esters, is responsible for this lubricity enhancement.

Table 1: Lubricity property of LLDF, Castor oil and methyl/ethyl ester (HFRR test results)

Blend / additive	***Wear scar (μm)***			***Film thickness (%)***	***Friction coefficient***
	X-axis	***Y-axis***	***Average***	***Average***	***Average***
LLDF	635	673	654	2	0.383
Castor oil	163	139	151	99	0.102
COME	138	108	123	78	0.089
COEE	105	87	96	99	0.045

5.3. Lubricity Study of Additive/Blends of Oil and Methyl/Ethyl Ester of *Castor* with LLDF

The effects of oil and methyl/ethyl ester of 2% of castor were tested as lubricity enhancers with LLDF. The effect of addition of castor oil, COME and COEE, with respect to WSD, friction coefficient and film thickness were tested as a lubricity improver and shown in Fig. 4–6. The wear scar generated on the steel ball after the addition of additive dose as tested by HFRR and measured by microscope showed that the WSD of LLDF decreased at the greater extant. The overall effects of additive in

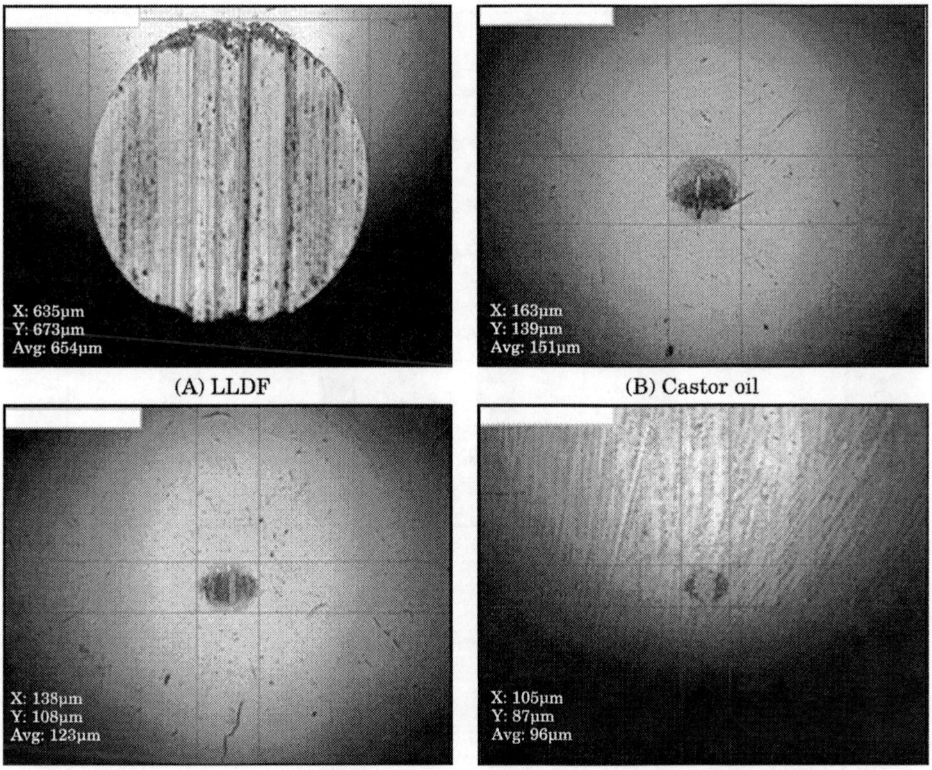

Fig. 3: The Images of wear scar diameter (WSD), of LLDF, Castor oil, COME, and COEE (Prasad, 2013)

LLDF are summarized in Table 2. Among oil, methyl and ethyl esters; ethyl ester was found to be comparatively quite effective additive as a lubricity enhancer for LLDF. Wear scar diameter significantly decreased with addition of a 2% of an additive dose. In the cases of additives, the film averages were observed to be comparatively high for all additives bearing LLDF. The high film average means that there is very little contact between metals when oscillated at a particular frequency; therefore, it leads to minimum wear scar and coefficient of friction (ì). The wear scar diameter reduced with appreciable limit (Table 2) by the addition of oil alkyl esters, which was quite low compared to the maximum allowable limit of ASTM *i.e.*, 450 µm at 60°C for diesel fuel. Adding free fatty acids or other selected oxygenated compounds to low lubricity petro-diesel at additive levels to enhance lubricity has also been reported in the literature (Knothe and Steidley, 2005; Margaroni, 1998).

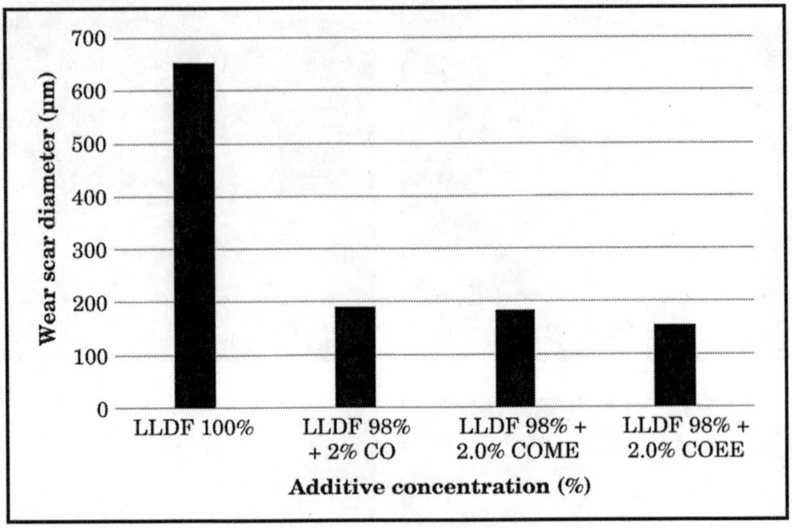

Fig. 4: Effect of additive conc. on wear scar diameter

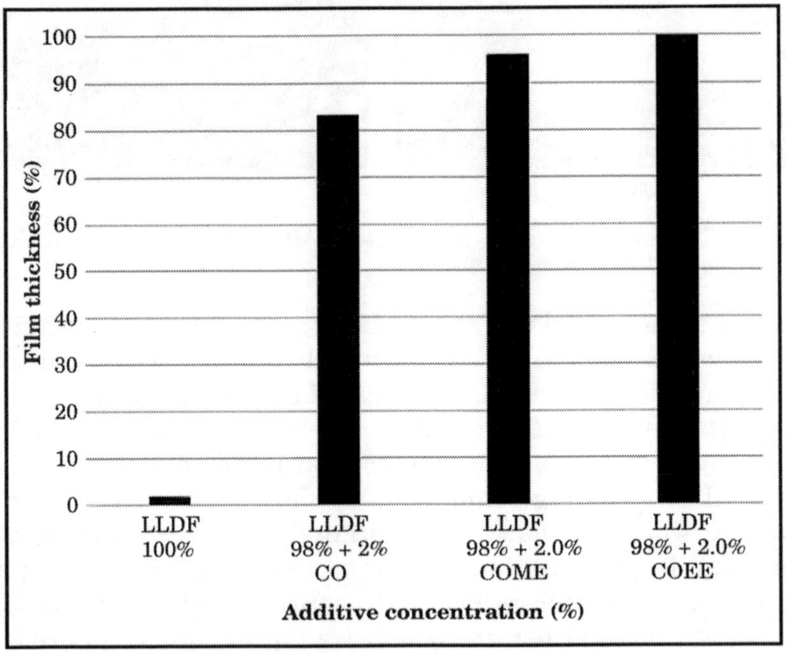

Fig. 5: Effect of additive conc. on film thickness

6. CONCLUSIONS

Biodiesel (fatty acid methyl/ethyl ester) is now accepted worldwide as a

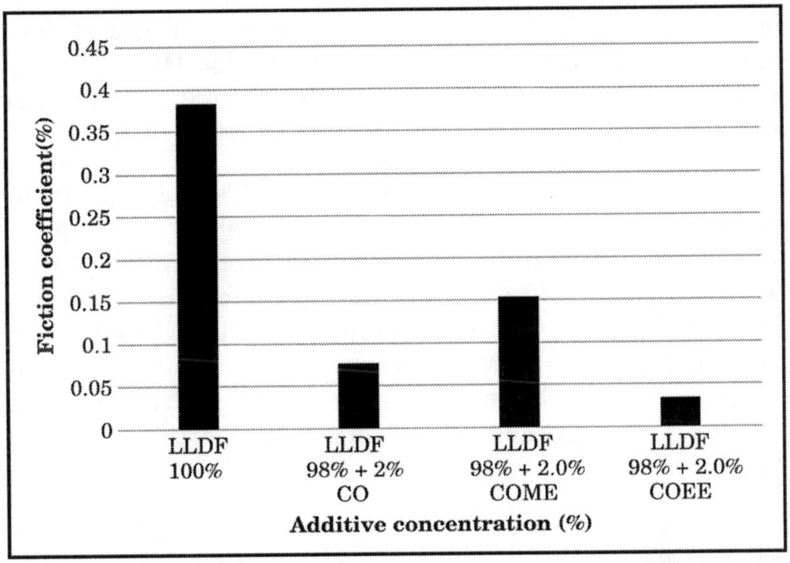

Fig. 6: Effect of additive conc. on friction coefficient

Table 2: Wear scar diameter (WSD), film thickness and friction coefficients of LLDF with addition of Castor oil, COME and COEE (Prasad, 2013)

Additive/Blend with base	*Wear scar (μm)*			*Film thickness (%)*	*Friction coefficient*
LLDF	*X-axis*	*Y-axis*	*Average (WSD)*	*Average*	*Average*
LLDF	635	673	654	2	0.383
2% CO	229	157	193	83	0.077
2.0% COME	222	150	186	96	0.155
2.0% COEE	166	143	155	100	0.034

clean alternative fuel for transport sector. The outcome of the study suggest avenues for establishing feasibility of its application as a lubricity enhancers for low lubricating diesel fuel and a substitute fuel for diesel. Non-edible oil alkyl ester shows excellent tribological properties which is an added advantage since pollution caused by synthetic lubricants is well known. Lubricity study of castor oil and methyl/ethyl ester have been carried out as alone and as an additive/ blends with low lubricity diesel fuel (LLDF) shows tremendous lubricity properties. Fatty ester compounds possess better lubricity than hydrocarbons because of their polarity-imparting O atoms.

The interesting finding of this study is that there is very little difference among oil versus methyl and ethyl ester. However ethyl ester seems to be a comparatively better additive since it effectively reduces the wear scar diameter of LLDF.

Since LLDF causes higher wear/tear (WSD, 654 μm) friction (μ, 0.383) and very low film formation (2%) during its application the percentage reduction in the wear scar diameter (WSD) was observed 70.49%, 71.56% and 76.29% respectively with addition of 2% of castor oil, castor oil methyl ester (COME) and castor oil ethyl ester (COEE) respectively. In the present study it has been observed that indigenous non-edible feedstocks such as castor oil and alky ester (methyl/ethyl) can play significant role by improving the lubricity properties of low lubricity diesel fuel.

REFERENCES

Aluyor, E.O., Obahiagbon, K.O. and Orijesu, M. 2009. Biodegradation of vegetable oils: A review. *Scientific Research and Essay*, 4(6): 543–548.

Anjani, K. 2005. Purple-coloured castor (*Ricinus communis* L.) – A rare multiple resistant morphotype. *Current Science,* 88(2).

ASTM Standard D6079-04 (2005). Standard test method for evaluating lubricity of diesel fuels by the High-Frequency Reciprocating Rig (HFRR).

Ball, K.F., Bostick, J.G. and Brennan, T.J. 1999. Fuel lubricity from blends of a diethanolamine derivative and biodiesel. United States Patent, US5891203.

Drown, D.C., Harper, K. and Frame, E.A. 2001. Screening vegetable oil alcohol esters as fuel lubricity enhancers. *J. Am. Oil Chem. Soc.,* 78(6): 579–84.

Geller, D.P. and Goodrum, J.W. 2004. Effects of specific fatty acid methyl esters on diesel fuel lubricity. *Fuel,* 83: 2351–2356.

Goodrum, J.W. and Geller, D.P. 2005. Influence of fatty acid methyl ester from hydroxylated vegetable oils on diesel fuel lubricity. *Bioresour. Technol.,* 96: 851–855.

Holser, R.A. and Harry-O'Kuru, R. (2006). Transesterified milkweed (*Asclepias*) seed oil a biodiesel fuel. *Fuel,* 85: 2106–2110.

Hu, J., Du, Z., Li, C. and Min, E. (2005). Study on the lubrication properties of biodiesel as fuel lubricity enhancers. *Fuel,* 84: 1601–1606.

Kania, D., Yunus, R., Omar, R., Rashid, S.A. and Jan, B.M. 2015. A review of biolubricants in drilling fluids: Recent research, performance, and applications. *Journal of Petroleum Science and Engineering,* 135: 177–184.

Khatri, K., Sharma, D., Soni, S.L., Kumar, S. and Tanwar, D. 2010. Investigation of optimum fuel injection timing of direct injection C.I. engine operated on preheated karanja-diesel blend. *Jordan Journal of Mechanical and Industrial Engineering,* 4(5): 629–640.

Knothe, G. and Steidley, K.R. 2005. Lubricity of components of biodiesel and petro-diesel. The origin of biodiesel lubricity. *Energy & Fuels,* 19: 1192–1200.

Knothe, G., Gerpen, J.V. and Krahl, J. 2005. The biodiesel handbook, AOCS Press, Champaign, Illinois, p. 302.

Kulkarni, M.G. and Dalai, A.K. 2006. Waste cooking oils-an economical source for biodiesel: A review. *Ind. Eng. Chem. Res.,* 45: 2901–2913.

Margaroni, D.1998. Fuel lubricity. *Ind. Lubr. Tribol.,* 50: 108–118.

Mobarak, H.M., Mohamad Niza, E., Masjuki, H.H., Kalam, M.A., Al Mahmud, K.A.H., Habibullah, M. and Ashraful, A.M. 2014. The prospects of biolubricants as alternatives in automotive applications. *Renewable and Sustainable Energy Reviews,* 33: 34–43.

Moser, B.R., Cermak, S.C. and Isbell, T.A. 2008. Evaluation of castor and *Lesquerella* oil derivatives as additives in biodiesel and ultralow sulfur diesel fuels. *Energy & Fuels,* 22: 1349–1352.

Munson, J.W., Hertz, P.B., Dalai, A.K. and Reaney, M.J. 1999. Lubricity survey of low-level biodiesel fuel additives using the Munson ROCLE bench test. *SAE,* 1999-01-3590.

Ogunniyi, D.S. 2006. Castor oil: A vital industrial raw material. *Bioresource Technology,* 97: 1086–1091.

Patel, V.R.R.R., Dumancas, G.G., Viswanath, L.C.K., Maples, R. and Subong, B.J.J. 2016. Castor oil: Properties, uses, and optimization of processing parameters in commercial production. *Lipid Insights,* 9: 1–12.

Prasad, L. (2013). Studies on formulation, lubricity of alkyl esters and use in a single cylinder C.I. engine, Ph.D. thesis, Indian Institute of Technology Delhi.

Prasad, L., Das, L.M. and Naik, S.N. 2012. *Energy & Fuels,* 26: 5307-5315.

Ramadhas, A.S., Jayaraj, S. and Muraleedharan, C. 2005. Biodiesel production from high FFA rubber seed oil. *Fuel,* 84: 335–340.

Refaat, A.A. 2009. Correlation between the chemical structure of biodiesel and its physical properties. *Int. J. Environ. Sci. Tech.,* 6(4): 677–694.

Rudnick, L.R. 2006. Synthetics, mineral oils and bio-based lubricant chemistry and technology, CRC Press (Taylor & Francis)

Sarin, R., Arora, A.K., Ranjan, R., Gupta, A.A. and Malhota, R.K. 2007. Biodiesel lubricity: Correlation study with residual acidity. *Lubrication Science,* 19: 151–157.

Schumacher, L. 2005. Biodiesel lubricity. Biodiesel Handbook; *In*: Knothe, G., Krahl, J., Van Gerpen, J. (*eds.*), *AOCS Press:* Champaign, IL pp. 137–144.

Singh, S.P. and Singh, D. 2010. Biodiesel production through the use of different sources and characterization of oils and their esters as the substitute of diesel: A review. *Renewable and Sustainable Energy Reviews,* 14: 200–216.

Srivastava, A. and Prasad, R. 2000. Triglycerides-based diesel fuels. *Renewable and Sustainable Energy Reviews,* 4(2): 111–133.

Stoldt, S.H., Dave, H. and Heights, H. 1998. Esters derived from vegetable oils used as additives for fuels. United States Patent, US 5730029.

9

Role of Algae and Their Potential Contribution to Sustainable Energy

A.K. CHATURVEDI[1*], DEEP GUPTA[1] AND SIDDHARTH JAIN[1]

ABSTRACT

The overhanging scarcity of non-renewable energy sources in the near future has led to a higher emphasis on research in the field of renewable resources, particularly algae. Recent research has shown the potential benefits of algal bio-fuels. With sustained algal growth, the algae based bio-fuel can ameliorate the energy deficit that is becoming increasingly alarming with the receding fossil fuel reserves. Extended studies for an improvement of growth in the algae have been conducted concerning variation in multiple parameters governing algal growth. This chapter presents a review of the classification and different aspects of the growth in algae along with their variation and prevailing trends in India. Algae are autotrophic micro-organisms capable of converting CO_2 into carbohydrates and lipids in the presence of light by photosynthesis. Micro algae are cultivated for the treatment of domestic and industrial waste water and also for CO_2 sequestration. Many studies were carried out in the areas of CO_2 sequestration using algae. The present chapter provides review of carbon sequestration using algae as one of the important aspect of algae cultivation. It also discusses the use of algal biodiesel in commercial transportation.

***Key words*:** Algae, Carbon sequestration, Bio-diesel, Bio-mass

[1]College of Engineering Roorkee, Roorkee, India, Pin: 247667

**Corresponding author*: ajay_bengalspr@yahoo.co.in

1. INTRODUCTION

The term “algae” refers to a great diversity of organisms–from microscopic cyanobacteria to giant kelp.

Most algae convert sunlight into energy in a similar manner as plants; however, the genetic diversity of the many different kinds of algae gives researchers an incredible number of unique properties that can be harnessed to develop promising algal biofuel technologies.

The key to assess algae’s potential as a renewable fuel source lies in the high productivity of algal biomass that can be grown in a given area. Some researchers go to the extent to say that the algae could be 10 or even 100 times more productive than the traditional bio-energy feedstock. Achieving the potential for this kind of high productivity in real-world systems is a key challenge to realizing the promise of sustainable and affordable algal bio-fuels. It may be noted that once harvested, algae can readily be processed into the raw material to make fuel/blending agent for fuel for cars, trucks, trains, and planes.

The word ‘algae’ originates from the Latin word for seaweed and is now used for a broad collection of organisms that can be defined both in terms of morphology and general physiology. It is a collective term for all chlorophyll bearing organisms which are thalloid and come under the Sub-Kingdom Thallophyta (not developing embryo). Thallophyta is divided in to two parts: Fungi (non chlorophyll bearing organisms) and Algae.

2. ALGAL HABITATS AND HABIT

Algae form an important part of various ecosystems. These occupy the first trophic level in these ecosystems and have a vast variety of body shapes, bio-chemistries and life cycles. They can, commonly, be found in aquatic conditions both in freshwater and marine environments. They can also be found in damp terrestrial environments and dry environments where they reside in symbiosis with fungus as lichen. Many algae species move themselves through the water column, while others float, attach themselves to objects in water or are terrestrial.

Algae have symbiotic relationship with bacteria in aquatic ecosystem. Algae support aerobic bacterial oxidation of organic matter producing oxygen through photosynthesis, while bacteria releases carbon-dioxide and nutrients in aerobic oxidation used for growth of algal biomass. The study of algae is really a challenging job due to their diverse habitats, some of which are not entirely "fresh". In this connection it is relevant to note that the oceans are clearly saline (35 g salts L^{-1}) and most lakes are relatively dilute (world average <0.1 g L^{-1}; Wetzel, 1983a). Thus there is an enormous variation in the chemical composition of the non marine habitats that algae occupy. The distinction between marine and freshwater habitats is revealed in the variety of algae that occur in these environments. There are no exclusively freshwater divisions of algae, but certain groups exhibit greater abundance and diversity within fresh waters. Conditions in lakes and rivers vary not only in salinity, but also in size, depth, transparency, nutrient conditions, pH, pollution, and many other important factors. Fresh water algae can be further categorised on the basis of their moving and stagnant habitat, namely; Standing (lentic) waters; particularly lakes and wetlands and secondly; Running (lotic) waters; including streams and rivers. The difference between standing and running waters is an important distinction in relation to the algae type which is present there. In this connection it is relevant to note that the *lentic* systems are typically dominated by planktonic algae and *lotic* systems are dominated by benthic (bottom living) organisms. Within freshwater ecosystems, algae occur either as a free-floating (planktonic) or a substrate-associated (largely benthic) organisms. *Planktonic* algae drift freely within the main body of the water (with some species able to regulate their position within the water column), while substrate-associated organisms are either fixed in position (attached) or have limited movement in relation to their substrate. The growth and type of algae depends on the two main factors – the depth of water and the rate of the water flow. Phytoplankton population requires a low rate of flow and adequate light levels for the predominant growth at the surface of lakes and slow moving rivers. *Benthic* algae require adequate light (shallow waters) and can tolerate high rates of water flow and as such can predominate over phytoplankton in fast flowing rivers and streams. Benthic algae also require adequate attachment sites – which include inorganic substrate, submerged water plants and emergent water plants at the edge of the water body. Where the water is clearer, algae can reach up to a depth of 200 m. On the contrary; along the coast, where the water is cloudy, they live on the water surface.

The presence of pigments in the algae influences the distribution: red and brown algae live at greater depth as against green algae because they have pigments capable of absorbing shorter wavelengths. It is important to note that the growth of the planktonic and non planktonic algae is very important with respect to a balanced ecosystem.

2.1. Nourishment

Vast variation in algal population, their shapes and colours can be observed. The factors affecting the type and growth of algae depends upon the presence of some inorganic and organic components like nitrate, sulphite, iron, total hardness and pH of the water. Many algal taxa have particular environmental tolerances or requirements, and are ecologically restricted, but still they are geographically widespread. Limnologists differentiate lakes according to a gradient of productivity (14 C uptake, algal growth) or biomass, from oligo-trophic (poor in nutrients, annual average <50–300 mg C m^{-2} d^{-1}; <0.05–1 µg chlorophyll-*a* L^{-1}) to eutrophic (nutrient-rich >1000 mg C m^{-2} d^{-1}; 15–100 µg chlorophyll-*a* L^{-1})[1] and these levels are influenced, at least in part, by the properties of lake basins (Likens, 1985).

Oligo-trophic lakes are usually deep and steep-sided, and have high transparency, a narrow littoral zone, abundant dissolved oxygen with depth, and larger relative hypolimnion volume. Eutrophic lakes, often shallower with a broad littoral zone, and have depleted summer hypolimnetic oxygen and reduced transparency. Eutrophication of lakes often results when nutrients are added as sewage, detergents, or fertilizers. Growth of planktonic algae, greatly vary among lake basins due to; nutrient availability and bio-geographic regions. Some algae are also able to synthesise oil and protein constituents from the carbohydrates which they manufacture and with the available nitrogen and other nutrients present in the water.

3. CLASSIFICATION OF ALGAE

Every collection of freshwater algae is characterized by a fascinating and confusing diversity of species, many of them are potential competitors for common resources and nutrients. The primary classification of algae is based on certain morphological and physiological features. Chief features are as follows:

---

[1]Algal production is measured as total biomass (the rate of Carbon fixation) it is expressed as Mass of carbon per unit area per year (g C m^{-2} yr^{-1}).

1. Pigment of cell
2. Chemical nature of food stored
3. Relative length, kind, number, point of incorporation of flagella in motile cell
4. Chemical composition of cell wall
5. Presence and absence of definitely organised nucleus in the cell or any other specific cell structure

The current system of classification of algae is based on the following main criteria

1. Types of photosynthetic pigments
2. Type or chemical nature of photosynthetic energy storage products
3. Photosynthetic membranes' (thylakoids) organization and other features of the chloroplasts
4. Cell wall composition and structure
5. The presence or absence of flagella (as well as the number and placement of these flagella, when present)
6. Ultra-structure of the flagella (if any)
7. Scheme and path of nuclear and cell division
8. Sexual cycles
9. Presence of an envelope of endoplasmic reticulum around the chloroplast (see below)
10. Possible connection between the endoplasmic reticulum and the nuclear membrane
11. Recently studies compare the sequence of macromolecules genes and the 5 S, 18 S and 28 S ribosomal RNA sequences
12. Occurrence of any other special features

Major Groups Based on Pigment and Stored Materials are as follows:

Chlorophyta (Green algae)

The Chlorophyta, or green algae, includes a large group of organisms (more than 17,000 identified species) with a great range of somatic differentiation ranging from unicellular flagellates to complex multi cellular thalli. Most species of green algae are aquatic (10% of the algae are of marine origin, whereas 90% are of freshwater origin). Some species inhabit moist terrestrial environments, such as soil and tree trunks. Some green algae live as symbiotic partners with invertebrates, such as corals. Other green algae are in association with fungi as a part of organisms called *lichens*. Pigments are present in definite bodies the chloroplasts, which generally contain chlorophyll a, chlorophyll b, carotenoids and xanthophylls. The reserve food material is in the form of starch. Ex: *Protococcus, Volvox, Spirogyra, Ulothrix* etc.

Chrysophyta (Yellow or Golden green algae)

In this type of algae yellow colour is dominant with chlorophyll. Algae are unicellular, multi-cellular and colonial. Food is stored in the form of oil or insoluble form of carbohydrate Leucosin. The important pigment is phycochrysin.

Cynophyta (Blue Green algae)

No chromatophores, pigments are distributed in cytoplasms. Members appear blue green in colour due to the pigment c, phycocyanin along with phycoerythrin and with chlorophyll a and c and carotenes. Members are characterised by the absence of a definitely organised nucleus. Thus all the cynophyta includes is the; prokaryotic members which are unicellular and filamentous. Reserve food material is in the form of myxo-phycean starch or cyanophycean granules. Ex: *Nostoc, Oscillatoria.*

Phaeophyta (Brown algae)

Contains approximately 1,500 species of multi-cellular algae, few microscopic filamentous forms and many giant forms, mostly seaweed example kelp. Brown algae are mostly marine organisms; they are most common along rocky coasts where ocean water is cool. A few species of brown algae, such as *Sargassum,* can be found for offshore, where they form dense, floating mats. Brown algae contain chlorophylls *a* c_1 and c_2 and fucoxanthin, an accessory pigment that gives the algae their characteristic brown colour. Food stored in the form of Laminarin (β-1,3-linkedglucan), mannitol, glycerol.

Pyrrophyta (Golden Brown algae)

Most important phytoplankton besides diatoms are found in marine waters. Members of this division are typical unicellular flagellates but can also be non-flagellate, ameboid, coccoid, palmelloid, or filamentous. Dino-flagellates possess chlorophylls a, b, c_1, and c_2, fucoxanthin, other carotenoids, and xanthophylls such as peridinin, gyroxanthin diester, dinoxanthin, diadino-xantin, and fucoxanthin. They are readily eaten by large protozoa, rotifer, and planktivorous fishes. Some species produce toxins and red pigments. When the populations of these species explode, they turn the water brownish red, resulting in a phenomenon known as *red tide*.

Rhodophyta (Red algae)

The red algae mostly consist of large, multicellular leaf-like seaweeds but also include the genera of filamentous forms, coccoid unicellular microalgae and encrusting, calcified "coralline" algae (support reefs). Chromatophores characterised by the presence of red pigment 'r' phycobiliproteins (phycoerythrin and phycocyanin), chlorophyll a, chlorophyll d, β-carotenes and xanthophylls. Phycobilins play an important role in absorbing light for photosynthesis. These pigments can absorb the wavelengths of light that penetrate deep into the water. As a result, phycobilins allow red algae to live at depths where algae lacking these pigments cannot survive. Some species of red algae have been found at depths of nearly 270 m.

Freshwater Blue-Green algae

can be divided into four main groups in relation to general morphology, presence/absence of specialized cells and the nature of branching in filamentous forms. These four groups form the basis for current taxonomy of this phylum, as adopted by John *et al.* (2002).

Chroococcales

The simplest of all types; the blue-green algae, occurring essentially as a solitary cell (no filamentous forms), is typically enclosed by a thin layer of mucilage. The cells may remain as a single cell, or be aggregated into plate-like or globular colonies. Typically planktonic with some colonial forms (*e.g.*, species of *Microcystis*) forming massive surface blooms containing individual colonies that are recognizable with the naked eye.

Oscillatoriales

These are relatively simple forms of algae generally referred as Filamentous algae. These occur as planktonic (some bloom forming) or benthic (bottom) population. In some cases they form dense mats on mud or rocky substrata which secondarily detach as meta-phyton into the main body of water.

Nostocales

These are a diverse group of filamentous algae, which are either; planktonic or benthic, with hetero-cysts and akinetes but do not show true branching. Filaments may be un-branched or show false branching. These forms of algae are able to form large colonies by lateral association of filaments (bundles), 3-D tangles or as radiating filaments from the centre of the sphere.

Stigonematales

These filamentous algae, with hetero-cysts and akinetes show true branching. Structurally these are the most complex blue-green algae, with some thalli (*e.g.*, *Fischerella*) differentiated into multi-seriate/uni-seriate basal filaments and uni-seriate erect branches. Largely benthic algae, with genera such as *Stigonema* commonly attached to substrata in standing and flowing waters, gets detached to form planktonic masses.

Climatic conditions

The green algae are one of the most diverse groups of eukaryotes. These shows great variation in morphological forms ranging from flagellated unicells, coccoids, branched or un-branched filaments, to multi-nucleated macrophytes. They are distributed worldwide and can be found everywhere in diverse habitat from Arctic and Antarctic regions to oceans and freshwater lakes as well as in the soil from temperate and arid areas. There are estimated to be at least 600 genera with 10,000 species within the green algae (Norton *et al.,* 1996). The growth and pattern of distribution depends on the climatic conditions like light penetration, temperature, nutrient enrichment, toxic substances, mixing of water, parasites, herbivores and heterotrophic microorganism activities etc. There are certain different types of algae that can grow in unusual habitats. These algae include those growing endophytic in other plants, leaf epiphytes, on mollusca, calcareous rocks, snow algae, thermal algae, epizoic algae and certain others (Smith, 1950). The habitats of algae are mainly classified into three categories *viz.*, aerial habitats, aquatic habitats and unusual habitats. Aerial algae have been defined as algae which obtain their water wholly or in large parts from moisture in the air. Strictly aerial algae are found on the bark and leaves of trees, on wood, stones, and rock and on soil surface. Algal species have a wider distribution among ecosystems and geographical regions. In all ecosystems these remain as a primary producer, and are directly affected by physical and chemical and biological factors. Algal growth is sensitive to some pollutants and they readily accumulate in the presence of these pollutants. The algal metabolism is also sensitive to the variation of environmental and natural disturbances.

Researchers have reported several studies on the phytoplankton distribution with availability of light (Singh and Sharma, 2012), physical, chemical and biological qualities (Munawar, 1974) in freshwater lakes. Today, Indian water bodies are facing terrific ecological stress due to increasing water pollution from rapid industrialization and urbanization. Algal growth chiefly depends on the seasonal variation. For example; summer is the best season for its rapid growth because of long duration of the sunshine period, increased salinity, pH and tropho-tropic activities (Chaturvedi *et al.,* 1999). Climatic conditions not only affect phytoplankton assemblages but also affect watershed features, land use and geochemical features. It may also be noted that the soil or sediments also affect phytoplankton ecology (Giripunje *et al.,* 2013).

During monsoon, wave action of currents and influx of rain water act as limiting factors for the growth of phytoplankton population. The water column of the lake is remarkably stratified to large extent during heavy rainfall and high turbidity, induced due to agricultural and surface runoff and soil erosion. Afterwards, the rate of algal growth steadily increases in post-monsoon period to the late spring. Phytoplankton community progresses in a sequential succession to a climax in a peak sequences with low turbidity and low wind velocity in the lakes (Sugunan, 2000).

4. IMPACT OF CLIMATIC CONDITIONS ON ALGAE

4.1. Biodiversity of Green Algae Based on Climatic Conditions

Not only a climate difference influences the algal communities; but also the geology of the area and the type of soil differences influence the type of algae produced in the area. Further it also needs to be noted that the aquatic ecosystem gets influenced with the regional factors such as chemical weathering and atmospheric precipitation. A number of researchers have published some accounts on phytoplankton ecology in freshwater lakes around India. Phytoplankton distribution is affected by some of the factors, such as; seasonal fluctuations in influent water seasonality; period of sunshine; wind patterns; depth of the lake; temperature; pH; turbidity; dissolved oxygen; nutrient enrichment like dissolved chloride; phosphate; nitrates and organic carbon. This chapter includes a comprehensive review of the interaction between phytoplankton and factors of freshwater lakes of North, South, East and West regions of India (Table 1).

5. ALGAE IN INDIA

5.1. Algae in River Ganges in Uttar Pradesh

Freshwater algae from Uttar Pradesh comes; under the category of Chlorophycean that are chiefly dependent on the physiochemical characteristics of the River Ganga at different time intervals with nutrient availability. Maximum algal population during the year is noticed in early summer, while the lowest is noticed during the month of August. It also needs to be noted that the amount of water discharge velocity and the transparency influence the occurrence and abundance of algal diversity.

Availability or lack of it of the nutrients is the main limiting factor for the algal growth. In freshwater habitats, like River Ganges, phosphorus is usually the principal and an essential nutrient for all forms of the life. Nitrogen can also prompt algal blooms, but it is a more common factor in salty waters. There are two general sources of nutrients in River Ganga; natural nutrients (Rai and Gaur, 2001), and manmade sources like fertilizer, livestock manure, human sewage, industrial run-off etc. In this area particularly 13 species of algae have been found *viz., Chlorella, Cosmarium, Enteromorpha, Eudorina, Hydrodictyon, Pediastrum, Scenedesmus, Pandorina, Spirogyra,*

Table 1: Biodiversity of green algae based on climatic conditions

Lakes	***Dominant groups of phytoplankton***	***Influencing factors***	***References***
Damadama, badk-hal, Peacock lakes	Blue-green algae	Temperature with seasonality	Dakshini and Gupta, 1979
Badua lake	Blue-green algae, diatoms	Seasonal variations	Verma and Munshi, 1983
Dal, waskur lake, anchor lakes	Diatoms, green algae	Hydrographic properties	Sarwar, 1996
Kitham lake	Green algae, diatoms	Seasonal variations	Tiwari and Chauhan, 2006
Ranchi lake	Bluc-green algae, diatoms	Organic matter	Mukherjee *et al.*, 2010
Bhavanisagar lake	Blue-green algae	Seasonal variations	Sreenivasan, 1964; Sreenivasan *et al.*, 1964.
Red hills lake	Blue-green algae	Climatic factors	Ganapati, 1940
Osman sagar, miralam lakes	Blue-green algae	Temperature, seasonality, water chemistry	Mohan, 1987
Sasthamcottah lake	Green algae	Water chemistry	Chaudhary and Pillai, 2009
Kodaikanal lake	Green algae	Temperature, pH, light, Seasonal variation	Singh and Balasingh, 2011
Nongmahir lake	Green algae	Nutrient enrichment	Sugunan and Yadava, 1991
Khecheopalri lake	Green algae	Hydrology and nutrients	Jain *et al.*, 1999
Krishnasager lake	Diatoms	Temperature, transparency, nutrients enrichment	Chattopadhyay and Banerjee, 2007
Santragachi lake	Blue-green algae	Climatic conditions	Ghosh *et al.*, 2012
Ramgarh lake	Diatoms	Seasonality	Jhingran, 1989
Dhom lake	Green algae	Nutrient enrichment	Trivedi, 1993
Masunda lake	Green algae, blue-green algae	Seasonality, nutrient enrichment	Somani *et al.*, 2007
Phutala, ambazari lakes	Blue-green algae	Nutrient enrichment	Maske *et al.*, 2010

Source: Giripunje *et al.,* 2013

Zygnema, Cladophora, Closterium and *Stigeoclonium* respectively. Higher populations of *Spirogyra* and *Cladophora* were recorded during most of the seasons in the River Ganges, but their population were drastically reduced during monsoon season due to the changes in turbidity (suspended soil and sand particles) (Singh and Chaudhary, 2011).

5.2. Algae in Arunachal Pradesh

This region lies in the Indo Burma hotspots, the diversity of algae in its freshwater habitats is still not fully known. According to a study by Das and Adhikari (2012), 86 freshwater algal species were documented from different locations of Arunachal Pradesh. These include eight cyanophyta (cyanoprokaryota/cyanobacteria), 30 chlorophyta, eight euglenophyta, 40 heterokontophyta, 39 bacillariophyceae and one Xanthophyceae). Among all these taxa, 12 species were reported for the first time in India. Some dominant species of algae were *viz.*, cyanophyta (*Aphanocapsa conferta, Merismopedia glauca, Microcystis wesenbergii, Komvophoron crassum, Planktothrix cryptovaginata, Phormidium chlorinum, Phormidium numidicum, Oscillatoria yamadae*), chlorophyta (*Spirogyra acanthophora, Spirogyra parvula, Closterium moniliferum, Staurastrum bicorne, Staurastrum bieneanum* var. *Ellipticum, Staurastrum gracile, Cosmarium awadhense, Cosmarium miscellum, Cosmarium radiosum, Microspora amoena, Oedogonium platygynum* etc.) Euglenophyta (*Trachelomonas armata, F. punctata, Trachelomonas hispida* var. Granulate, *Trachelomonas volvocina* var. *Punctata*) and Heterokontophyta (*Melosira granulate, Melosira granulate* var. *Muzzanensis, Melosira islandica* subspecies, Helvetica, *Cyclotella stelligera, Tabellaria fenestrata, Fragilaria arcus, Fragilaria capucina, Surirella tenera, Gloeobotrys limneticus, Cymbella tumida* etc.) From this account it appears that the biodiversity of these genera are much more than it is expected. It has a great future and real possibilities in the field of fisheries, oil production and as source of single cell protein in Arunachal Pradesh.

5.3. Algae in Jodhpur, Rajasthan

Biodiversity of microalgae and cyano-bacteria from different freshwater bodies of Jodhpur, Rajasthan was studied and compared (Makandar and Bhatnager, 2010) their variations in terms of physicochemical and diversity indices. A total of 84 species were observed in just 25 samples that were collected from seven water bodies. There were 26 green algae belonging to 16 genera, 9 morpho types of diatoms and 48 morpho types of 13 cyano-bacterial genera. It has been found that the generic diversity was high in green algae. The most frequent genera were *Scenedesmus, Oscillatoria, Chroococcus, Gloeocapsa, Phormidium, Chlorococcum, Cosmarium, Chlorella* and *Plectonema*.

5.4. Algae in West Bengal

As far as flora and fauna is concerned, the state of West Bengal is one of the most bio-diversity rich states in India. The lower indo-gangetic basin a fertile hub for various types of flora and fauna. It is situated in the central portion within state which is enriched with numerous water bodies, wetlands, forest and fertile agricultural lands. Climatic condition is one of the most important factors for the growth of algal population and also for other vegetations. There are districts which support typical tropical monsoon and thus have a magnificent reserve of plant resources. Halder and Sinha, (2015) reported four Chlorophycean algal taxa namely *Characium nasutum* Rabenh, *Characium ambiguum* Herm ex Rabenh, *Oedogonium globosum* Nordst and *Cosmarium undulatum* var. Minutum Wittr under the two orders Chlorococcales and Conjugales of Chlorophyceae.

6. CARBON SEQUESTRATION

Sustainability is a key principle in natural resource management and it involves effective utilization of natural resources. Across the world, industrialization and the consequent emergence of economies reliant on fossil fuels have inevitably resulted in the adverse condition of atmospheric warming known as the greenhouse effect. Anthropogenic carbon dioxide (CO_2) has been identified as one of the major causes. Out of several CO_2 sequestration methods, biological methods of sequestration using algae are highly energy efficient due to less energy requirement. Algae are autotrophic micro-organisms capable of converting CO_2 into carbohydrates and lipids in the presence of light by photosynthesis. Micro algae are cultivated for the treatment of domestic and industrial waste water and also for CO_2 sequestration. Many studies were carried out in the field of CO_2 sequestration using algae. From those studies it was concluded that the algae can be effectively used for the treatment of wastewater and CO_2 sequestration. Hence, algal photosynthetic reaction may hold the key to reduce CO_2

emission in a sustainable manner. Algae can be extensively used to capture CO_2 from power plants, steel, cement, oil, automobiles and many other industries and the resulting algal biomass may only be used for bio-fuel production but also for various other industrial products like fertilizers and pharmaceuticals. Besides giving environmental and economic benefit, large scale algae cultivation can create a large number of jobs at different levels in the society.

Pollution has been found to be present widely in the environment. It causes acid rain, eutrophication, haze, bio-magnifications and global climate change etc. The Earth's atmosphere contains a delicate balance of naturally occurring gases that trap some of the sun's heat near the Earth's surface. This "greenhouse effect" keeps the Earth's temperature stable. Unfortunately, evidence is mounting that humans have disturbed this natural balance by producing large amounts of some of these greenhouse gases, including carbon dioxide and methane. As a result, the Earth's atmosphere appears to be trapping more of the sun's heat, causing the Earth's average temperature to rise; a phenomenon known as global warming. Global climatic change is considered to be a highly sensitive issue and global warming could have significant impact on human health, agriculture, water resources, forests, wildlife, and coastal areas (Wetzel, 1983; Likens, 1985). Therefore, the use of fossil fuel is now widely accepted as unsustainable due to depleting resources and the accumulation of greenhouse gases in the environment. Due to industrial revolution, the developed countries have emitted most of the anthropogenic greenhouse gases into the atmosphere. On the other hand the developing countries are the most vulnerable to climate change impact because they have fewer resources to adapt socially, technologically and financially (Norton *et al.,* 1996).

6.1. Greenhouse Gases and their Contributions

The effect of greenhouse gas not only depends on the quantity but also on their chemical nature. For example, the influence of methane and CFC are very high though they are present in lower quantities. Though water vapour dominates the greenhouse effect among the contributing gases, CO_2 plays a major role as it influences the increase of water vapour concentrations. The concentration of nitrous oxide began to rise at the beginning of the industrial revolution. Greenhouse gases such as CFC's are mostly synthesized by the use of refrigerants, aerosol propellants and cleaning solvents.

Table 2: Greenhouse gases and their contribution (Smith, 1950)

Sl. no.	*Greenhouse gases*	*Greenhouse effect (in %)*	*Natural (in %)*	*Man-made (in %)*
1	Water vapour	95.000	94.999	0.001
2	Carbon dioxide	3.618	3.502	0.117
3	Methane	0.360	0.294	0.066
4	Nitrous oxide	0.959	0.903	0.047
5	Misc. gases	0.072	0.025	0.047
6	Total	100	99.72	0.28

Table 2 shows the various greenhouse gases and their contribution. From the above table, it is observed that, among all the greenhouse gases, contribution of carbon dioxide to the environment is considered to be the maximum (Smith, 1950).

6.2. CO_2 Sequestration Using Microalgae

Algae are recognized as one of the oldest life-forms (Giripunje *et al.*, 2013). They are primitive plants which lack roots, stems and leaves, have no sterile covering of cells around the reproductive cells and have chlorophyll a as their primary photosynthetic pigment (Suguman and Yadav, 1991). Depending on the species, algae can range from 1 micron (μm) to a few hundred microns in size. As a point of reference, 1 micron (μm)=1/1000 of a millimetre (mm). Microalgae are one of the most important groups of organisms on the planet. It is estimated that microalgae produce approximately half of the atmospheric oxygen on earth, while consuming vast amounts of the greenhouse gas carbon dioxide. Microalgae have a rapid growth potential and many species have oil content in the range of 20–50% dry weight of biomass (Dakshini and Gupta, 1979; Verma and Munshi, 1983; Sarwar, 1996). With respect to air quality maintenance and improvement, microalgae biomass production can effect bio-fixation of waste CO_2 (One kg of dry algal biomass utilize about 1.83 kg of CO_2) (Dakshini and Gupta, 1979). Nutrients for microalgae cultivation can be obtained from wastewater as a growth medium, there by treating the organic effluent from the agro-food industry (Tiwari and Chauhan, 2006). Algal cultivation does not require herbicides or pesticides application (Mukharjee *et al.*, 2010). They can also be used to produce valuable co-products such as proteins and also biomass residue after oil extraction which may be used as fermented to produce ethanol or methane (Sreenivasan *et al.*, 1964) or fertilizer (Tiwari and Chauhan, 2006). The potential application of microalgae is biofuel production, CO_2 fixation, bio hydrogen production and bio-treatment of wastewater.

6.3. Microalgae Tolerance of CO_2

Direct utilization of power plant flue gas has been considered for CO_2 sequestration systems (Maske *et al.,* 2010). The advantage of direct utilization of flue is the reduction of the cost of separating CO_2 gas. Since power plant flue gas contains higher concentration of CO_2, identifying high CO_2 tolerant species is important. Although CO_2 concentrations vary depending on the flue gas source, 15–20% v/v is typically assumed. Several species have been tested under CO_2 concentrations of over 15%. For example, *Chlorococcum littorale* could grow under 60% CO_2 using the stepwise adaptation technique (Singh and Chaudhary, 2011). The following Table (Table 3) shows the CO_2 tolerance of various species.

Microalgae can assimilate CO_2 within various ranges of concentration from ambient (0.04%) to 100% v/v CO_2 by selecting adequate species.

From the above table it is observed that at a temperature of 60ºC, 100% CO_2 tolerance is observed for *Cyanidium caldarium*. From the results certain thermophilic micro-organisms are known to survive till temperatures of 60ºC. It is suggested that the temperature dependent solubility of CO_2 gives an advantage to thermophilic algae to tolerate a higher concentration of CO_2 (Dixit *et al.,* 1992).

7. PRODUCTION OF ALGAE

The commercial production of algal feedstock gives an opportunity to utilize wasteland and different water resources that are currently not suitable for any other commercial use. Algae can be cultivated in areas that are far remove from farm and forestlands, thereby minimizing the damages caused to the food chain, and obviating the food-versus-fuel dilemma.

Algae can be grown in the state of suspension or can be attached to a solid surface. Each system has its own advantages and disadvantages. Currently, there are four algae cultivation technologies in use for commercial algae production and proposed for algal bio-fuel production.

Table 3: CO_2 tolerance of various species (Das and Akhikary, 2012)

Sl. no.	*Species*	*Known maximum CO_2 uptake concentration*	*References*
1	*Cyanidium caldarium*	100	Das and Akhikary, 2012
2	*Scenedesmus* sp.	80	Makandar and Bhatnager, 2010
3	*Chlorococcum littorale*	60	Singh and Chaudhary, 2011
4	*Synechococcus elongatus*	60	Halder and Sinha, 2015
5	*Euglena gracilis*	45	Kolkwitz and Marsson, 1908
6	*Chlorella* sp.	40	Makandar and Bhatnager, 2010
7	*Eudorina* spp.	20	Makandar and Bhatnager, 2010
8	*Dunaliella tertiolecta*	15	Patrick, 1949
9	*Nannochloris* sp.	15	Dixit *et al.*, 1992

These are extensive or open ponds, intensive or raceway ponds, closed photo bio-reactors in many designs and closed fermented systems. The suspension-based open ponds and enclosed photo bio-reactors are commonly used for algal bio-fuel production. In general, an open pond is simply a series of raceways outside, while a photo bio-reactor is indoor sophisticated reactor design and can work as a greenhouse in outdoor environment.

7.1. Open Pond System

Open ponds are the traditional and simplest systems for mass cultivation of algae. This system is generally a shallow pond about one foot deep in which algae are cultured under natural or environmental conditions. The pond is designed in a raceway configuration, in which a paddle wheel provides circulation and mixing of the algal cells and nutrients (Fig. 1). The raceways (Fig. 2) are made up of poured concrete, or they can simply be dug into the earth. It is very essential to line them with plastic to prevent the soaking up of the liquid in the ground. This system has a continuous movement of all the nutrients, water and algae. It generally follows the recirculation method that can help flourish growth of algae. Baffles in the channel guide the flow around bends in order to minimize space. The system is often operated in a continuous mode.

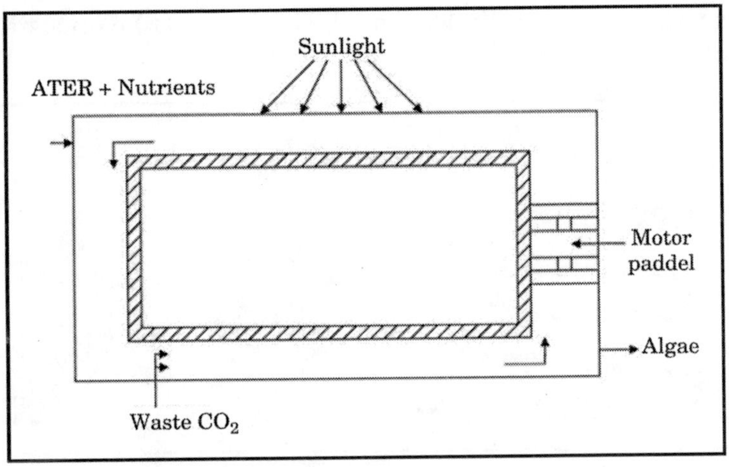

Fig. 1: Open pond system (Yue, 2013)

Fig. 2: Raceway system (*Internet source: http://www2.hci.edu.sg/y11hci01 49/website/algae-open-pond.jpg*)

7.2. Closed Loop Photo Bio-reactor (PBR)

The closed loop photo bio-reactors (Fig. 3) are commonly used for culturing algae for; mainly; the bio-fuel production. The closed vessel may consist of the steel, plastics, or fibres and their size may vary according to the production capacity of the plant. Tubular algae bio-reactors may consist of a vertical or horizontal arrangement of tubes vessels with suspended form of algae in fluid and it should be circulated throughout the system. The introduction of CO_2 gas either at the beginning or at the end of the system is commonly done to speed up algae growth and generally leaves a lack of gas at the opposite end of the circulatory system.

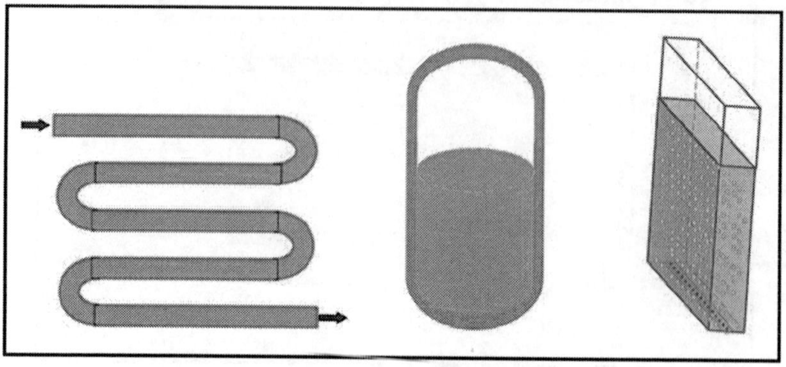

Fig. 3: Closed bio-reactor (Chanakya *et al.,* 2012)

7.3 Hybrid Cultivation of Algae

Hybrid cultivation system utilizes the advantages of both the systems as closed photo-bio-reactor and open pond system. In hybrid system, phosphorous and nitrogen in the form of phosphoric acid, ammonium nitrate and urea are added on regular interval to accelerate the growth of micro algae, this is not the case with other cultivation method (Ryan, 2009). Table 4 presents a short comparison of open pond systems and closed photo bio-reactors. Each system has some benefits and drawbacks with respect to optimal growth conditions.

8. FUTURE PROSPECTS OF ALGAE AS SOURCE OF ENERGY

Algae can utilize solar energy more efficiently than higher plants due to their unicellular unique structure and minimal competition between plant functions. Algae have short life cycle with fast growth, therefore, these are better for biomass production.

Algae simply utilize full sunlight, CO_2 and some nutrients to grow in control and favourable environment, algae have a capability to multiply several times in a day. Algal growth is natural occurrence in all types of water bodies. This unique property makes algae one of the most valuable sustainable and renewable fuel resources all over the world.

Bio-fuels have become more important in recent years because of their environmental benefits over conventional fossil fuels. The use of

Table 4: Comparison of open and closed algae growth systems (Otto Pulz, 2001)

Parameter	***Open pond***	***Closed photobioreactor***
Construction	Simple	More complicated - varies by design
Cost	Cheaper to construct,	Operate more expensive construction, operation
Typical growth rates (g/m^2-day)	Low: 10–25	Variable: 1–500
Water losses	High	Low
Typical biomass concentration	Low: 0.1–0.2 g/L	High: 2–8 g/L
Temperature control	Difficult	Easily controlled
Species control	Difficult	Simple
Contamination	High risk	Low risk
Light utilization	Poor	Very high
CO_2 losses to atmosphere	High	Almost none
Area requirements	Large	Small
Depth/diameter of water	0.3 m	0.1 m
Surface: volume ratio (m^2/m^3)	~6	60–400

bio-diesel causes reduction in carbon and sulphur content in atmosphere which cut down their amount by 30-20% and it also maintains the CO_2 in the atmosphere (Huang *et al.,* 2010).

Algae have some benefits over the conventional fuels. These are as follows:

There is no need for fresh water for the cultivation of algae. It can be grown using either waste water or marine water or even industrial effluents as well. It can be grown on non-arable land and have very less requirement of land for cultivation as compared to the other traditional oil seeds. Algae are capable of synthesizing more oil content per acre than the terrestrial plants which are currently used for the fabrication of bio-fuels. It gets matured within 12-20 hours. Algae biodiesel is non-toxic, highly bio-degradable and easy to get process. Algal bio-diesel can be utilized in any type of engine. There is no requirement of either a new design or a modification of the existing diesel engine design.

In every environment algal growth takes CO_2 as means of nourishment. This implies that the algae farms could be utilized as a carbon absorber and helps in pollution mitigation. There is another benefit of its growth and that is that it utilizes CO_2 that has been created by the algae bio-fuel combustion.

In the production of energy from algal biomass, two basic approaches are employed depending on the particular organism and the hydrocarbons that are produced as a by-product. The first is simply the biological conversion of nutrients into lipids or hydrocarbons. It depends on the species and type of environment in which it grows. The second procedure entails the thermo-chemical liquefaction of algal biomass into lipid or hydrocarbons. Lipids and hydrocarbons can normally be found throughout the algal biomass. They can occur as membrane components, storage products and metabolites. Numerous algal strains have been shown in the laboratory to produce more than 50 percent of their biomass as lipid with much of this as tri acetyl glycerides (TAGs), also called triglycerides, the anticipated starting material for biodiesel fuels. Most of the observations of high lipid content come from algal cultures grown under nutrient (especially nitrogen, phosphorous, or silicon) limitation (Darzins *et al.*, 2010). Microalgae grow quickly and contain high oil content compared with terrestrial crops, which take a season to grow and only contain a maximum of about 5 percent dry weight of oil (Chisti, 2007). During the peak growth phase, some algae can double every three and half hours. Oil content of algae is usually between 20 percent and 50 percent (dry weight, Table 5), while some strains can reach as high as 80 percent.

Table 5: Oil content of various algal species

Sl. no.	*Micro algae*	*Oil content (% dry weight)*
1	*Botryococcus braunii*	25–75
2	*Chlorella* sp.	28–32
3	*Crypthecodinium cohnii*	20
4	*Cylindrotheca* sp.	16–37
5	*Nitzschia* sp.	45–47
6	*Phaeodactylum tricornutum*	20–30
7	*Schizochytrium* sp.	50–77
8	*Tetraselmis suecia*	15–23

Source: National Biodiesel Board, Corn Refiners Association, year?

9. BIODIESEL FROM ALGAE

9.1. Biodiesel Production Methods

***Blending*:** Vegetable oil can be directly mixed with diesel fuel and may be used for running an engine. The blending of SVO with diesel fuel was experimented successfully. It has been proved that the use of 100% SVO was also possible with some minor modifications in the fuel system

(Jain and Sharma, 2010). However, high fuel content caused major problems due to fuel viscosity of pure vegetable oils causes problems in the engins. Micro-emulsification, pyrolysis and trans-esterification are the methods used to solve the problems encountered due to high fuel viscosity.

***Micro-emulsification*:** To solve the problem of high viscosity of SVO, micro-emulsions with solvents such as methanol, ethanol and butanol has been used. A micro emulsion is defined as the colloidal equilibrium dispersion of optically isotropic fluid. These can improve spray characteristics by explosive vaporization of the low boiling constituents in the micelles. All micro-emulsions with butanol, hexanol and octanol can meet the maximum viscosity limitation for diesel engines (Jain and Sharma, 2010).

***Cracking*:** Cracking is the process of conversion of one substance into another by means of heat or with the aid of catalyst. It involves heating in the absence of air or oxygen and cleavage of chemical bonds to yield small molecules. The pyrolyzed material can be vegetable oils, animal fats, natural fatty acids and methyl esters of fatty acids. The pyrolysis of fats has been investigated for more than 100 years, especially in those areas of the world that lack deposits of petroleum. Since World War I, many investigators have studied the pyrolysis of vegetable oil to obtain products suitable for engine fuel application. Tung oil was saponified with lime and then thermally cracked to yield crude oil, which was refined to produce diesel fuel and small amounts of gasoline and kerosene (Jain and Sharma, 2010).

***Transesterification*:** Transesterification is the most common method of converting oil into biodiesel that can be used directly or as blend with diesel in diesel engine. It is also called alcoholysis; which is displacement of alcohol from an ester by another alcohol in a process similar to hydrolysis, except that in this process, an alcohol is used instead of water. This process has been widely used to reduce the viscosity of triglycerides (Jain and Sharma, 2010). A transesterification reaction is represented in Fig. 4. A catalyst is usually used to speed up the reaction that may be basic, acid or enzymatic in nature. The entire transesterification can be represented by three steps. Transesterification is a method of transforming of an ester into another when a vegetable oil is reacted with methanol in the presence of a catalyst to give methyl ester also biodiesel and amount of glycerine.

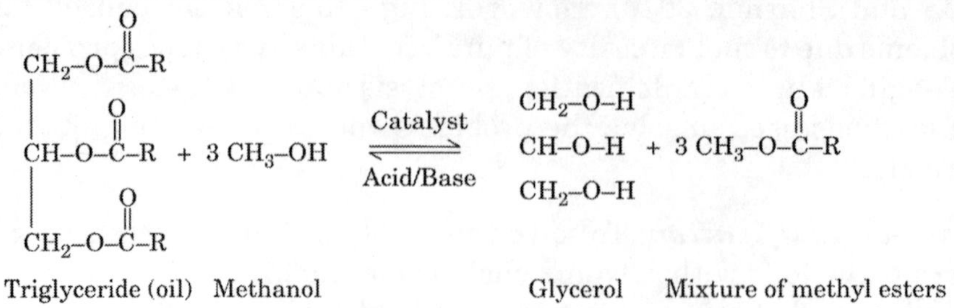

Fig. 4: Trans esterification reaction

Acid catalysed trans-esterification: When the amount of full fatty acid (FFA) content in the oil is more than 1%, acid is used to catalyze the trans esterification reaction. This is known as acid catalysed esterification reaction, used to reduce FFA contents and increase the yield of alkyl esters.

Base catalysed trans-esterification: When base is used to catalyze the Trans esterification reaction, it is known as base catalyst trans esterification reaction. It is used when the amount of FFA contents are less than 1%. In this step, conversion of tri-glycerides into alkyl ester and glycerol occur. The lighter layer (biodiesel) is separated from heavier glycerol phase by separation. The lighter biodiesel is removed, washed with water, dried over anhydrous sodium sulphate and can be used directly as engine fuel in diesel engines.

9.2. Factors Affecting the Trans-esterification Reactions

The most important variables that influence the Trans esterification reaction are:

- Reaction temperature
- Catalyst
- Purity of reactants
- Ratio of alcohol to oil
- Mixing intensity
- Catalyst percentage

9.3. Standards for Comparing Bio-diesel Quality

The countries like Germany, Italy, France, USA, etc., have developed their own biodiesel standards. An Indian standard under the name IS: 15607 is also being developed. Usually ASTM standards are followed all over the world. All above stated standards are reported in Table 6, which shows the limiting values of viscosity, flash point and acid value in ASTM standard slightly higher than BIS standards. The limiting value of carbon residue is almost same for all the above standards. Among the general parameters for biodiesel, viscosity controls the characteristics of the injection from the diesel engine. The viscosity of fatty acid methyl ester can reach very high levels and hence it is important to control it within the acceptable level to avoid negative impacts on fuel injector system performance. Therefore, the viscosity specifications proposed are nearly same as that of the diesel fuel.

9.4. Advantages of Biodiesel

Following are the advantages of using biodiesel as substitute of diesel:

- Biodiesel is a green liquid fuel free from environmental problems as it emits lower emissions compared to diesel.
- Biodiesel may not require engine modification up to B20. However, higher blends may need some minor modifications.

Table 6: Bio-diesel properties as a fuel in different standards (Jain and Sharma, 2010)

Sl. no.	*Parameter*	*Austria ON*	*India BIS*	*France general official*	*Germany DIN*	*Italy UNI*	*USA ASTM*
1	Density at 15°C (g/cm^3)	0.85–0.89	0.87–0.89	0.87–0.89	0.875–0.89	0.86–0.90	0.88
2	Viscosity at 40°C (mm^2/s)	3.5–5	1.9–6	3.5–5	3.5–5	3.5–5	1.96
3	Flash point (°C)	100	130	100	110	100	130
4	Pour point (°C)	–	–	10	–	1–5	15–18
5	Cetane number	≥49	≥40	≥49	≥49	–	≥47
6	Neutralization number (mg KOH/g)	≤0.8	≤0.5	≤0.5	≤0.5	≤0.5	≤0.8
7	Carbon residue (%)	≤0.05	≤0.05	–	≤0.05	–	≤0.05

- Biodiesel is cheaper than diesel and can be a "on farm fuel"; where the farmer can grow the seed oil crops, produce biodiesel and can use it in the field itself.
- Biodiesel can make the vehicle perform better as it has a cetane number of over 100 which is a measure of the quality of the fuel's ignition.

- Owing to the clarity and the purity of biodiesel it can be used without adding additional lubricants unlike diesel.
- Biodiesel reduces the environmental effects of a waste product. As biodiesel is made out of waste products itself, it does not contribute to nature's garbage at all. Biodiesel can be made out of used cooking oils and lards. So instead of throwing these substances away, the ability to turn them into biodiesel becomes more than welcome.
- Biodiesel is energy efficient. If the production of biodiesel is compared with the production of regular type, producing the latter consumes more energy. Biodiesel does not need to be drilled, transported, or refined like petroleum diesel. Producing biodiesel is easier and is less time consuming.
- Biodiesel is produced locally. A locally produced fuel will be more cost efficient. There is no need to pay tariffs or similar taxes to the countries from which oil and petroleum diesel is sourced. Every country has the ability to produce bio-diesel.

The advantages of microalgae in energy production can be found in the following points

Microalgae have chlorophyll and other photosynthetic organs, which can perform photosynthesis. So microalgae use the sunlight, H_2O which is contained in the microalgae cell and CO_2 from the air to convert organic compounds which can be converted into biofuel.

The reproduction of microalgae is generally split type breeding, the cell cycle is relatively short, as such it is easy to carry out large scale cultivation.

Microalgae can be cultivated within sea water, alkaline water and even waste water, so it is a significant method to produce bio-energy in the fresh water shortage areas and barren land areas.

10. CASE STUDY OF AN ALGAE USED BY COLLEGE OF ENGINEERING ROORKEE

Significant research work is being carried out in the field of renewable energy resources such as in Solar Energy, Geothermal Energy, Biomass Energy, and Wind Energy as well as in alternative fuels such as CNG, LPG and Bio-diesel. Algae is a relatively new area which needs to be explored because it is found in abundance in environment and therefore algal bio-diesel is likely to be considered as a good replacement to diesel. Moreover, bio-diesel does not require any separate storage infrastructure and it is less volatile than petroleum diesel. In addition to this, Oil drilling is a complicated process, while extraction of oil from algae and later the formulation of Bio-diesel by trans-esterification is a simpler process. Thus, algal bio-diesel is considered as a cheaper and much cleaner source of energy.

Keeping in mind the increased rate of consumption of fossil fuels and shift of the world towards renewable energy, an initiative was taken by the Research and Development Cell of the College of Engineering Roorkee (COER). It produced biodiesel by making a reaction between algae oil and methanol in the presence of potassium hydroxide (catalyst). The final products obtained were bio-diesel and glycerol. Glycerol was the by product, which was removed with the help of separating funnel. Bio-diesel consisting of long chains of fatty acid and esters was produced by trans-esterification reaction of algae oil with short chain alcohols further blends of B10 and B20 biodiesel were made and several experiments were carried out on an IC engine in the laboratories of College of Engineering Roorkee and Indian Institute of Technology (IIT), Roorkee. Thereafter, the B10 blend has been tested successfully in the college bus which goes from college to Haridwar (Please specify the distance covered by bus and its frequency once/twice).

Bio-diesel has similar properties as diesel but unlike diesel it is not a fossil fuel. It can operate effectively on a diesel engine and is considered as a Clean Fuel due to its complete combustion property. Bio-diesel can be considered as a renewable energy resource as its sources of extraction are plant and algae-based. Thus, as an alternative to diesel, bio-diesel has great scope in the automobile as well as industrial sector.

Algae are the fastest growing and most abundant species found in all types of water. Because of their low cost and availability in abundance in nature, algae are mostly preferred for the production of bio-diesel. When Algae is chemically reacted with methanol in the presence of catalyst, a fatty acid alkyl ester or we can say biodiesel is produced. A study has been opted to optimize the biodiesel production from algae oil using low temperature trans esterification process. Three process variables *viz.*, catalyst (Al_2O_3) concentration (0–2% w/w), reaction time (60–180 min) and methanol/oil ratio (v/v) (20–60%) for the Trans esterification process were optimized to get maximum yield using Boxe Behnken Design with the help of Design Expert 9.0.6.2 software.

A biodiesel yield of 89.98% was achieved with methanol/oil volumetric ratio (3:5) using Al_2O_3 as catalyst (1% w/w) in 160 min time at 50°C temperature. It was observed that catalyst concentration, reaction time and methanol/oil volumetric ratio had a significant effect on blend yield. It is found out that this model can further be tested at industrial level for use in the industry to improve the efficiency of biodiesel production from algae oil thereby, saving time and cost of the process in optimizing the process parameters.

11. BENEFITS OF MICRO-ALGAL BIOFUELS

Microalgae are a diverse group of single-celled organisms that have the potential to offer a variety of solutions for our liquid transportation fuel requirements through a number of avenues. Algal species grow in a wide range of aquatic environments, from fresh water through saturated saline. Algae efficiently use CO_2, and are responsible for more than 40% of the global carbon fixation, with the majority of this productivity coming from marine microalgae. Algae can produce biomass very rapidly, with some species doubling in as few as 6 hours, and many exhibiting two doublings per day. All algae have the capacity to produce energy-rich oils, and a number of micro algal species have been found to naturally accumulate high oil levels in total dry biomass. For example, some *Botryococcus* spp. have been identified that have up to 50% of their dry mass stored as long-chain hydrocarbons (Halder and Sinha, 2015). With potentially millions of species, algal diversity gives researchers various options for identifying production strains and also provides sources for genetic information that can be used to improve these production strains. The micro algal species being investigated as potential biofuel crops originate from groups whose ancestral relationships are significantly broader than the most diverse land plants, providing a wealth of genetic diversity. The groups most often considered when discussing microalgae are diatoms, green Algae, golden brown, prymnesiophytes, eustigmatophytes and cyanobacteria members from all of these groups have been examined as potential fuel production strains. However, it should be noted that cyanobacteria are not algae but a class of photosynthetic bacteria (Makandar and Bhatnager, 2010).

12. CONCLUSIONS

Microalgae have additional advantages over terrestrial plants. Since they are single-celled organisms that duplicate by division, high-throughput technologies can be used to rapidly evolve strains. This can reduce processes that take years in crop plants, down to a few months in algae. Algae have a reduced impact on the environment compared with terrestrial sources of bio-mass used for bio-fuels. They can be grown on land that would not be used for traditional agriculture, and are very efficient in removing nutrients from water. Thus, not only would production of algae bio-fuels minimize land use compared with bio-fuels produced from terrestrial plants but, in the process of culturing these microalgae, waste streams can be remediated. Potential waste streams include municipal wastewater to remove nitrates and phosphates before discharge, and flue gas of coal or other combustible-based power plants to capture sulfates and CO_2. Algae production strains also have the potential to be bioengineered, allowing improvement of specific traits and production of valuable co-products, which may allow algal bio-fuels to compete economically with petroleum. These characteristics make algae a platform with a high potential to produce cost-competitive bio-fuels.

Bio-prospecting is of importance to identify algal species that have desired traits (*e.g.*, high lipid content, growth rates, growth densities and/or the presence of valuable co-products), while growing on low-cost media. Despite the potential of this strategy, the most likely scenario is that the bio-prospecting will not identify species that are cost competitive with petroleum and subsequent genetic engineering and breeding will be required to bring these strains to economic viability. The range of potential for engineering algae is just beginning to be realized, from improving lipid biogenesis and improving crop protection, to producing valuable enzyme or protein co-products. No sustainable technology is without its challenges but blind promotion of those technologies without honest consideration of the long-term implications may lead to the acceptance of strategies whose long-term consequences outweigh their short-term benefits. We have presented the most important current and upcoming challenges of algae bio fuels but, as with any new industry, the more we learn the more we realize that challenges exist that we had not foreseen. Even given these uncertainties, it is believed that fuel production from algae can be cost competitive and widely scalable and deployable in the next 7–10 years, but only if we continue to expand our understanding of these amazing organisms as we expand our ability to engineer them for the specific task of developing a new energy industry.

REFERENCES

Balakumar, K., Sharpudin, J., Mubarak, B.H. and Mohammed, F.H. 2014. Comparative study of carbon sequestration by algae in domestic sewage and industrial effluent. *International Journal of Emerging Technology and Advanced Engineering, Certified Journal* 4(3): 807–809. (ISSN 2250-2459, ISO 9001:2008).

Benemann, J.R. 1993. Utilization of carbon dioxide from fossil fuel burning power plants with biological system. *Energy Conversion and Management,* 34(9/11): 999–1004.

Berndes, G., Hoogwijk, M. and Broek, R.V.D. 2003. The contribution of biomass in the future global energy supply: A review of 17 studies. *Biomass Bioenergy,* 25: 1–28.

Borowitzka, M. 1992. Algal biotechnology products and processes matching science and economics. *Journal of Applied Phycology,* 4(3): 267–79.

Borowitzka, M. 1997. Microalgae for aquaculture: Opportunities and constraints. *Journal of Applied Phycology,* 9(5): 393–401.

Borowitzka, M.A. 1999. Commercial production of microalgae: Ponds, tanks, tubes and fermenters. *Journal of Biotechnology,* 70(1–3): 313–21.

Boussiba, S., Sandbank, E., Shelef, E., Cohen, Z., Vonshak, A., Ben-Amotz, A., Arad, S. and Richmond, A. 1988. Outdoor cultivation of the marine microalga Isochrysisgalbana in open reactors. *Aquaculture,* 72: 247–253.

Cantrell, K.B., Ducey, T., RoK, S. and Hunt, P.G. 2008. Livestock waste-to-bioenergy generation opportunities. *Bioresource Technology,* 99 (17): 7941–7953.

Catie Ryan –Cultivating Clean Energy the Promise of Algae Biofuels, A report to Natural Resources Defense Council, USA, October.

Chanakya, H.N., Mahapatra, D.M., Sarada, R., Chauhan, V.S. and Abitha, R. 2012. Sustainability of large-scale algal bio-fuel production in India. *Journal of the Indian Institute of Science,* 92:1. *journal.library.iisc.ernet.in*

Chattopadhyay, C. and Banerjee, T.C. 2007. Temporal changes in environmental characteristics and diversity of net phytoplankton in a freshwater lake. *Turk. J. Bot.,* 31: 287–296.

Chaudhary, R. and Pillai, R.S. 2009. Algal biodiversity and related physico-chemical parameters in Sasthamcotta lake, Kerala (India). *Journal of Environmental Research and Development,* 3:3.

Chaumont, D. 1993. Biotechnology of algal biomass production: A review of systems for outdoor mass culture. *Journal of Applied Phycology,* 5: 593–604.

Chisti, Y. 2007. Biodiesel from microalgae. *Biotechnology Advances,* 25(3): 294–306.

Dakshini, K.M. and Gupta, S.K. 1979. Preliminary observations of limnological study of three freshwater lakes around Delhi state, India. *Proc. Indian. Nat. Sci. Acad. B.,* 45: 564–570.

Darzins, A.l., Pienkos, P. and Edye, L. 2010. Current Status and Potential for Algal Biofuels Production. A Report to IEA Bioenegy Task 39. *www.Task39.org*

Das, S.K. and Akhikary, S.P. 2012. Diversity of freshwater algae in Arunachal Pradesh and their distribution in different altitudes. *J. Indian bot. Soc.*, 91(1–3): 160–182.

Dixit, S.S., Smol, J.P., Kingston, J.C. and Charles, D.F. 1992. Diatoms: Powerful indicators of environmental change. *Environmental Science and Technology,* 26: 23–33.

Falkowski, P.G. and Raven, J.A. 1997. Aquatic photosynthesis. London: Blackwater Science. p. 375.

Ganapati, S.V. 1940. Geographical aspects of Red Hill Lake. *J. Madras Geogr. Assoc,* 25: 152–254.

Ghosh, S., Barinova, S. and Keshri J.P. 2012. Diversity and seasonal variation of phytoplankton community in the Santragachi Lake, West Bengal, India. *Science Connect,* 3.

Giripunje, M.D., Fulke, A.B., Khairnari, K., Meshram, P.U. and Paunikari, W.N. 2013. A review of phytoplankton ecology in fresh water lakes in India. *Lakes. Reservoirs and Ponds,* 7(2): 127–141.

Grossman, R.V. 2005. Paths toward Algal Genomics. *Plant Physiology,* 137: 410–427.

Guan Hua, H., Feng, C. and Dong, W. 2010. Biodiesel production by micro algal. *Journal homepage: www.elsevier.com / locate / apenergy*, Contents lists available at Science Direct, *Applied Energy,* 87: 38–46.

Halder, N. and Sinha, S.N. 2015. New report of the occurrence of four chlorophycean Algal Species from West Bengal, India. *J. Algal Biomass Utln.*, 6(3): 20–23.

Hanagata, N., Takeuchi, T., Fukuju, Y., Barnes, D.J. and Karube, I. 1997. CO_2 fixation by *Chlorella* sp. KR-1 and its Cultural Characteristics, 68(3): 269–273.

Hirano, A., Ueda, R., Hirayama, S. and Ogushi, Y. 1997. CO_2 fixation and ethanol production with microalgal photosynthesis and intracellular anaerobic fermentation. *Energy,* 22(2–3): 137–42.

Hoffert, M.I., Caldeira, K., Jain, A.K., Haites, E.F., Harveyk, L.D.D., Potter, S. D., Schlesinger, M.E., Schneider, S.H., WattsI, R.G., Wigley, T.M.L. and Wuebble, D.J. 1998. Energy implications of future stabilization of atmospheric CO_2 content. *Nature,* 395: 881.

Internet source. *http: / / www2.hci.edu.sg / y11hci0149 / website / algae-open-pond.jpg.*

Jain, S. and Sharma, M.P. 2010. Prospects of biodiesel in India: A Review. *Renewable and Sustainable Energy Review,* 14 (2): 763–771.

Jain, S., Jain, A., Rai, S.C., Pal, J. and Sharma, E. 1999. Hydrology and nutrient dynamics of a sacred lake in Sikkim Himalaya. *Hydrobiologia,* 416: 13–22.

Jhingran, A.G. 1989. Limnology and production biology of two man-made lakes in Rajasthan (India) with management strategies for their fish yield optimisation. Final Report, IDA Fisheries Management in Rajasthan, Central Inland Capture Fisheries Research Institute, Barrackpore, India. pp. 1-63.

Jorquera, O., Kiperstok, A., Sales, E.A., Embiruçu, M. and Ghirardi, M.L. 2010. Comparative energy life-cycle analyses of microalgal biomass production in open ponds and photo-bioreactors. *Bioresource Technology,* 101: 1406–1413.

Kodama *et al.* 1993. Cloning and Characterization of High-CO_2-Specific cDNAs from a Marine microalga, *Chlorococcum littorale*, and Effect of CO_2 Concentration and Iron Deficiency on the Gene Expression 39(2): 131–138.

Kolkwitz, R. and Marsson, M. 1908. Oekologie der pflanzlichen Saprobien. *Berichte der Deutschen Botanischen Gesellschaft,* 26: 505–519.

Lee, R.E. 1980. *Phycology. New York: Cambridge University Press.*

Likens, G.E. 1985. An ecosystem approach to aquatic ecology, *Mirror Lake and its environment,* Springer-Verlag, New York.

Makandar, M.B. and Bhatnager, A. 2010. Biodiversity of microalgae and cyanobacteria from freshwater bodies of Jodhpur, Rajasthan (India). *J. Algal Biomass Utln.,* 1(3): 54–69.

Maske, S., Sangolkar, L.N. and Chakrabarti, T. 2010. Temporal variation in density and diversity of cyanobacteria and cyanotoxins in lakes at Nagpur (Maharashtra State), India. *Environmental Monitoring and Assessment,* 169: 299–308.

Matsumoto, H., Shioji, N., Hamasaki, A., Ikuta, Y., Fukuda, Y., Sato, M., Endo, N., Tsukamoto, T. 1995. Carbon dioxide fixation by microalgae photosynthesis using actual flue gas discharged from a boiler. *Applied Biochemictory and Biotechnology,* 51–52: 681–692.

Metting, F.B. 1996. Biodiversity and application of microalgae. *Journal of Industrial Microbiology,* 17(5–6): 477–89.

Milledge, J.J. 2010. Commercial application of microalgae other than as biofuels a brief review. *Reviews in Environmental Science and Biotechnology.* 1–11.

Miura, M., Zhu, H., Rotello, R., Hartwieg, E.A. and Yuan, J. 1993. Induction of apoptosis in fibroblasts by IL-1β-converting enzyme, a mammalian homolog of the *C. elegans* cell death gene ced-3. *Cell,* 75(4): 653–660.

Miyairi, S. 1995. CO_2 assimilation in a thermophilic cyanobacterium. *Energy Conver. Mgmt.,* 36: 763–766.

Mohan, K.S. 1987. Bacillariophyceae of two tropical Indian lake of Hyderabad. *Bot. Bull. Academia Sinica,* 28: 13–24.

Mukherjee, B., Nivedita, M. and Mukherjee, D. 2010. Plankton diversity and dynamics in a polluted eutrophic lake, Ranchi. *Journal of Environmental Biology,* 31: 827–839.

Nagase *et al.* 1998. Regulation of phytochelatin synthesis by zinc and cadmium in marine green alga. *Dunaliella tertiolecta,* 62(3): 453–459.

Nakano *et al.* 1996. Regulation of Myosin Phosphatase by Rho and Rho - Associated Kinase (Rho - Kinase), 273(5272): 245–248.

Norton, T.A., Melkonian, M. and Andersen, R.A. 1996. Algal biodiversity, *Phycologia,* 35: 308–326.

Otto Pulz, K.S. 2001. Photo-bioreactors: Production systems for phototrophic microorganisms. *Appl. Microbiol. Biotechnol.,* 57: 287–293.

Patrick, R. 1949. A proposed biological measure of stream conditions, based on a survey of the Conestoga Basin, Lancaster County, Pennsylvania. *Proceedings of the Academy of Natural Sciences of Philadelphia,* 101: 277–341.

Ritschard, R.L. (1992). Marine algae as a CO_2 Sink. *Water, Air and Soil pollution,* 64: 289–303.

Rodolfi, L., Zittelli, G.C., Bassi, N., Padovani, G., Biondi, N., Bonini, G. *et al.* 2008. Microalgae for oil: Strain selection, induction of lipid synthesis and outdoor mass cultivation in a low-cost photo-bioreactor. *Biotechnology and Bioengineering,* 102(1): 100–12.

Sahoo, D., Elangbam, G. and Devi, S.S. 2012. Using algae for carbon dioxide capture and bio-fuel production to combat climate change. *Phykos,* 42(1): 32–38.

Sarwar, S.G. 1996. Epiphytic algal flora attached to *Myriophyllum spicatum* L. in three Kashmir lakes, India. *Environment and Biodiversity: In the Context of South Asia,* 226-231.

Seckbach *et al.* 1971. Growth pattern and isotope fractionation of *Cyanidium caldarium* and hot spring algal mats. 12(3): 161–169.

Singh, A.P. and Chaudhary, B.R. 2011. Phenological diversity of chlorophycean algae from river Ganges at Varanasi, Uttar Pradesh. *J. Algal Biomass Utln.,* 2(1): 21–29.

Singh, R.P. and Balasingh, G.S. 2011. Limnological Studies of Kodaikanal Lake (Dindugal District), in Special Reference to Phytoplankton Diversity. *Indian Journal of Fundamental and Applied Life Sciences,* 1: 112–118.

Smith, G.M. 1950 The fresh water algae of United States- McGraw-Hill Book Company, INC, New York, London.

Somani, V., Gholba, M. and Pejaver, M. 2007. Study of phytoplankton population in lake Masunda, Thane, employing multivariate analysis, *Eco. Env. & Cons.,* 13: 847–848.

Sreenivasan, A., Soundar, R. and Antony, F. 1964. Limnological studies of tropical impoundments II Hydrobiological features and plankton of Bhavanisagar reservoir for 1961–62. *Proc. Indian Acad. Sci.,* 59: 51–71.

Sugunan, V.V. and Yadava, Y.S. 1991. Feasibility studies for fisheries development of Nongmahir reservoir, Central Inland Capture Fisheries Research Institute, Barrackpore. p. 30.

Tiwari, A. and Chauhan, S.V. 2006. Seasonal phytoplanktonic diversity of Kitham Lake, Agra. *Journal of Environmental Biology,* 27: 35–38.

Tredici, M.R. and Materassi, R. 1992. From open ponds to vertical alveolar panels: The Italian experience in the development of reactors for the mass cultivation of photoautotrophic microorganisms. *J. Appl. Phycol.,* 4: 221–231.

Trivedi, R.K. 1993. Water quality of Dhom reservoir Maharashtra, India. Ecology and pollution of Indian lakes and reservoirs, Ashish Publishing House, New Delhi.

Ugwu, C.U., Aoyagi, H. and Uchiyama, H. 2008. Photo-bioreactors for mass cultivation of algae, *Bioresource Technology,* 99: 4021–4028.

UNFCCC 2007. Climate change: Impacts, vulnerability and adaptation in developing countries. Climate change Secretariate, United Nations framework Convention on Climate change.

Verma, P.K. and Munshi, J.D. 1983. Limnology of Badua reservoir of Bhagalpur, Bihar. *Proc. Indian Nat. Sci. Acad. B.,* 49: 598–609.

Wang Yue 2013. Microalgae as the Third Generation Biofuel: Production, Usage, Challenges and Prospects Master thesis in Sustainable Development at Uppsala University, No. 166 31 p. 30. ECTS/hp *https://www.diva-portal.org/smash/get/diva2:665095/FULLTEXT01.pdf.*

Wetzel, R.G. 1983. Limnology, 2nd ed. Saunders, Philadelphia.

Yoshihara, K., Nagase, H., Eguchi, K., Hirata, K. and Miyamoto, K. 1996. Biological Elimination of nitric oxide and carbon dioxide from flue gas by marine microalgae NOE - 113 cultivated in long tubular photo-bioreactor. *Journal of Fermentation and Bioengineering,* 4: 351–354.

10

Algae in a Vat may Power the Future: A Review on Algal Biodiesel Production

FARID AHMAD[1], ANIKA FERDOUS[1] AND MOHIDUS SAMAD KHAN[1*]

ABSTRACT

With the growing energy demand, gradual depletion of fossil fuel, price fluctuation and environmental issues related to conventional fossil fuel, there is a pressing need for alternative fuel sources that are renewable, environment friendly and economically viable. Biofuel derived from edible and non-edible sources is considered to be a promising alternative to conventional fossil fuel. Biofuels are renewable, carbon neutral and cause lower environmental damages than fossil fuel. However, edible biofuel sources such as: Soybean, Corn, Jatropha, or sunflower have raised concerns due to their demand as food crops and edible oils. Non-edible sources like microalgae have received considerable interest since these do not compete with food supply. In recent years, countries like Australia, USA, India and China are investing in extensive research in algal biodiesel production. Microalgae can be cultured in wastelands and water bodies, and the extracted lipid content can be turned into biodiesel which makes algal biodiesel a lucrative energy source for the countries having minimum energy resources. This article provides a technical overview of algae cultivation, harvesting and algal biodiesel production procedures.

[1]Department of Chemical Engineering, Bangladesh University of Engineering and Technology (BUET). Dhaka 1000, Bangladesh.

**Corresponding author*: mohid@che.buet.ac.bd

Key words: Energy demand, Biofuel, Carbon neutral, Renewable, Microalgae, Biodiesel, Lipid, Algae cultivation.

1. INTRODUCTION

Globally, fossil fuels are considered as the primary fuel source. The industrially developed countries are running their economy mainly based on the supply of fossil fuels.

However, the reserves of fossil fuels are depleting, and considering the ever increasing demand it is estimated that by 2050 to 2075, the crude oil reserve is going to be exhausted (*World Fossil Fuel Reserves and Projected Depletion*, The Colorado River Commission of Nevada, 2002). Moreover, combustion of fossil fuel causes irreversible damage to the atmosphere. The greenhouse gases (CO_2, SO_x, NO_x) caused by fossil fuel consumption play a vital role in global warming. Most of the CO_2 emissions are caused by developed countries like USA, China and EU countries which contribute to global warming. In contrast, developing countries like Bangladesh, Maldives and Sri Lanka are at greater risk from the impact of this phenomenon, albeit, for example Bangladesh causes less than 0.1 tons of Carbon emission/Person compared to almost 6 tons of Carbon emission/Person of USA (Kais *et al*., 2011). Considering the ever-growing demand of fossil fuel, energy security and the environmental vulnerability associated with the use of fossil fuel, governments, scientists and researchers all over the world are looking for sustainable alternative fuel sources (Saifullah *et al*., 2014). There are several alternative energy sources, such as solar energy, wind energy, and biofuel production from biomass, which are gaining attention because of their ability to provide clean energy. However, the ecological imbalance caused by wind mills and dams, the economic factors, reservation and longevity associated with solar energy and possible threat to food security associated with biofuel produced from food crops limit the potentiality of these alternative sources (Hossain *et al*., 2008; Hasan *et al*., 2013). Microalgae can be a potential ingredient of future biofuel production due to their higher photosynthetic efficiency, higher biomass production, and faster growth compared to energy crops (Ginzburg, 1993). The prospect of producing biodiesel from microalgae is high since it yields 30 times more energy per acre than land crops such as soybeans. Lipid extracted from algal biomass can be turned into biodegradable and carbon neutral biodiesel (Meher *et al*., 2006; Hossain *et al*., 2008). Biodiesel qualifies as a suitable alternative to fossil fuels due to its qualities of being bio-degradable, non-toxic and renewable; it also does not contribute any net carbon dioxide or sulfur to the atmosphere and emits less gaseous pollutants than normal diesel (Antolin *et al.*, 2002).

Different research groups throughout the world have been working to develop technically sound and economically feasible procedures to produce biodiesel from algae. There are few critical factors of algal biodiesel research: a) algae strain with high lipid content b) high yield culture technique c) lipid extraction technique d) processing of high quality biodiesel from lipid and e) overall production cost. This article presents an overview of biodiesel production from microalgae and discusses different technical issues related to algae culture, algae harvesting, post-harvesting, lipid extraction and transesterification process. The lipid contents of different microalgae strain and the key factors to be considered in cultivation design for algal growth are also discussed here.

2. ALGAE

Algae comprises of one of the most robust organisms on earth, these are large group of primitive, mostly aquatic, photosynthetic chlorophyll-bearing plants, lacking specialized tissues and organs namely roots, stems, leaves, flowers, etc (Qudus and Zohra, 2015). Algae can be broadly categorized into two groups: microalgae and macro algae. Microalgae are very small, their sizes can range from a few micrometres to a few hundreds of micrometres while macroalgae can be up to six order of magnitude larger than microalgae (Sayed, 2012). Algae are usually found in damp places or water bodies. Like plants, most algae require primarily three components to grow: sunlight, carbon-dioxide, and water (Sayed, 2012). Algae contain proteins, carbohydrates, fats (lipids) and nucleic acids. Though the percentages vary with the type of algae, there are algae types that contain as much as 70% lipids (oil). It is this fatty acid/lipid (oil) that can be extracted and converted into biodiesel (Becker, 1994; Mehlitz, 2009).

Microalgae have many different species with widely varying compositions. These live as single cells or colonies without any specialization. The small size of microalgae makes subsequent harvesting more complicated (Bhatnagar *et al.*, 2011). Biologists have categorized microalgae according to their pigmentation, life cycle and basic cellular structure; the important categories are: diatoms (Bacillariophyceae), green algae (Chlorophyceae), blue-green algae (Cyanophyceae), and golden algae (Chrysophyceae) (NREL, 1998). In recent years, different microalgae strains have been receiving high attention because of their less complex structure, fast growth rate, and high lipid content.

Several microalgal strains have been reported to have the ability to contain high amount of lipids. Allard and Templier (2000) extracted lipids from a variety of marine and freshwater algae and reported that the lipid contain varied from 1-26% (Allard and Templier, 2000). Miao and Wu reported heterotrophic growth of *Chlorella protothecoides* is capable of as high as 55% of lipid growth, which can be later converted to biodiesel (Miao, 2004). In spite of low growth rate, *Botryococcus braunii* has received vast attention because of its high lipid content (up to 64%) and high hydrocarbon production ability (Metzger, 1995; Sheehan *et al.*, 1998). Table 1 enlisted different algal strains with their lipid contents (Gouveia and Oliveira, 2009).

Table 1: Lipid content of some microalgae (% dry matter) (Gouveia and Oliveira, 2009)

Species	*Lipid (%dry matter)*
Scenedesmus obliquus	11–22/35–55
Scenedesmus dimorphus	6–7/16–40
Chlorella vulgaris	14–40
Chlorella emersonii	63
Chlorella protothecoides	23/55
Chlorella sorokiana 22	22
Chlorella minutissima	57
Dunaliella bioculata	8
Dunaliella salina	14–20
Neochloris oleoabundans	35–65
Spirulina maxima	4–9

3. BIODIESEL PRODUCTION FROM MICROALGAE

3.1. Biodiesel

Technically, biodiesel is an alternative transportation fuel based on monoalkyl ester building blocks of long-chain fatty acids, which are produced from common vegetable sources like soybean, sunflower, canola, and palm (Aikins *et al.*, 2010; Oncel, 2013). There are several advantages of using Biodiesel (Sharp, 1996; Ma and Hanna, 1999; Huang *et al.*, 2010). It is a renewable resource of energy that could be sustainably supplied. It does not cause any net increase in carbon dioxide release and also has very low sulfur content. By using biodiesel as energy source, the release of oxides of sulfur and carbon monoxide would be cut down by 30% and 10%, respectively. It does not contain any aromatic compound and other chemical substances which are harmful to the environment. Recent studies have showed that use of biodiesel can decrease 90% of air toxicity compared to common diesel source. Biodiesel is better than conventional diesel fuel in terms of flash point and biodegradability. Table 2 presents the fuel properties of microalgal biodiesel and conventional diesel.

Table 2: Comparative assessment of microalgae biodiesel and diesel (Kais *et al.*, 2011).

Properties	***Microalgae biodiesel***	***Diesel***
Density	0.864 kg/L	0.838 kg/L
Viscosity	5×10^{-4} Pa s at 40ºC	1.9×10^{-4} Pa s at 40ºC
Flash point	75ºC	75ºC
Heating value	0.35 MJ kg^{-1}	0.5 MJ kg^{-1}

3.2. Microalgae as a Potential Alternate Source of Fuel

Microalgae are photosynthetic microorganisms that use solar energy in combination with water and CO_2 to create biomass. Depending on the choice of chemical or biological treatment of microalgae, there can be various end products such as-biodiesel, bioethanol, biohydrogen and biomethane.

Use of microalgal lipids has been of interest since Harder and von Witsch proposed the idea of mass growth of diatoms to meet the urgent supply of fat in World War II (Liang *et al.*, 2009). Microalgae can form diverse kinds of cellular lipids including neutral lipids, polar lipids, wax esters, sterols and hydrocarbons. As compared to lipids extracted from vegetables, these lipids from microalgae have various fatty acid compositions with higher levels of unsaturation. The size of the fatty acid chains and their level of unsaturation play an important role in the quality of biodiesel (Hu, 2008; Oncel, 2013). Under optimum growth conditions, algae can synthesize fatty acids principally for esterification into glycerol-based membrane lipids, which constituent about 5-20% of their dry cell weight (DCW). However, under unfavorable and stressful conditions, many algae alter their lipid synthesis pathways towards the formation and accumulation of neutral lipids (20-50% DCW), mainly in the form of triacylglycerol (TAG) (Hu, 2008).

There are several advantages of using microalgae as a biodiesel source over other conventional sources (Chisti, 2007; Hu *et al.*, 2008; Aikins *et al.*, 2010):

Algae has high growth rate (*e.g.*, 1-3 doublings per day) and its oil content (per dry weight) can be as high as 80%. As a result the yield of oil extracted from microalgae can be up to 80–90%. Algal biofuel can be a potential candidate for sustainable alternative fuel. Algal culture does not require high land area and does not compete for land with crops. For example: 1 to 3% of the U.S. cropping land will be sufficient for producing algal biomass that can satisfy 50% of U.S. transport fuel demand. Being able to utilize nutrients such as nitrogen and phosphorus from a variety of wastewater sources, algae provide an additional benefit of bio-remediation of wastewater. It can sequester carbon dioxide from flue gases emitted from fossil fuel-fired power plants and thus contribute in reducing the amount of a major greenhouse gases in the atmosphere. Algae can thrive in saline/brackish water. Value-added by-products as biopolymers, proteins, animal feed, and fertilizers can be produced from algae.

In addition, microalgae can convert 3–6% of solar energy to biomass (Huntley, 2007). Microalgae grow very quickly compared to terrestrial crops; some microalgae can double in just 3.5 hours (Chisti, 2007). Compared to terrestrial crops, which take a season to grow and contain a maximum of about 5% dry weight of oil, microalgae grow quickly and contain high amount of lipid (Wen and Johnosn, 2009).

Fig. 1 represents oil yield from various oilseeds and microalgae; the yield is shown in the Gallons per acre per year.

3.3. Manufacturing of Biodiesel from Microalgae

To make the process of producing biodiesel from microalgae economically sustainable it is very important to select the right kind of microalgae strain. Ideally, algal strains having the criteria such as: ability to capture large quantities of carbon dioxide, high lipid content/capability to produce high levels of algal oil, resistance to contamination and adaptability to thermal conditions, local water conditions and CO_2 source

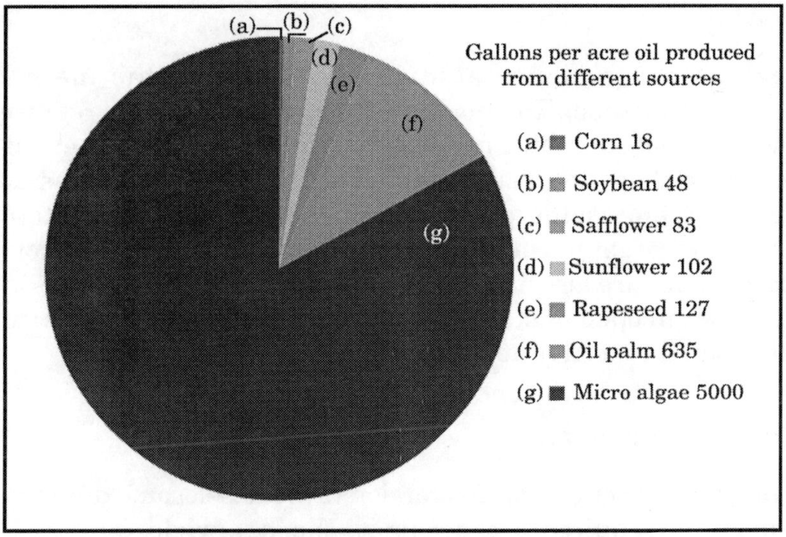

Fig. 1: Comparison of average oil yields (Gallons per acre) from microalgae and oilseeds (Chisti, 2008)
Reference: Biodiesel from Microalgae Beats Bioethanol. Tends in Biotechnology, 26(3): 126–131.

are considered as the best algae strains (Ghosh *et al.*, 2014). Following factors should be considered for selecting microalgae strains to produce biodiesel (Mata *et al.*, 2010):

- ***Growing rate*:** Normally measured by total amount of biomass accumulated per unit time and unit volume
- ***Lipid content*:** Total lipid along with the distribution of fatty acids and triglycerides, a factor that may be relevant in biodiesel production.
- Resistance to environmental condition changes, particularly temperature, nutrients input, light, competition from other microalgae strain, and/or bacteria.
- Nutrient availability, in particular CO_2 sources.
- Ease of biomass separation and processing.
- Possibility of obtaining other valuable chemicals.

3.3.1. *Steps involved in biodiesel production from algae*

Following steps are involved in microalgal biodiesel production (Blinova *et al.*, 2015)

Cultivation

Microalgae can be cultivated in photoautotrophic methods or heterotrophic methods. In photoautotrophic method algae require light to grow and create new biomass, whereas, in heterotrophic method algae are grown without light and are fed with carbon source as sugars. However, heterotrophic process is not as efficient as phototrophic process, since organic carbon sources required for the heterotrophic microorganisms are also derived by photosynthesis (Chisti, 2007). Algae can be grown in open or closed ponds. Photobioreactors (PBRs) are one the most popular cultivation techniques.

Harvesting/de-watering

Prior to lipid extraction and conversion to biodiesel, some preprocessing steps might be required to concentrate the algal yield.

Extraction

Three major products can be extracted from algal biomass: lipids (including triglycerides and fatty acids), proteins and carbohydrates. While lipids and carbohydrates are fuel precursors (*e.g.*, gasoline, biodiesel and jet fuel), proteins can be used as byproducts (*e.g.*, animal/fish feed).

Conversion

Conversion techniques include chemical, biochemical, thermochemical or a combination of these approaches. The end product varies depending on the application of conversion technique(s) (Fishman, 2010). Transesterification can be used to produce biodiesel from lipid. Transesterification is the reaction between triacylglycerol and an acyl acceptor, which can be carboxylic acid (acidolysis), alcohols (alcoholysis) or another ester (interesterification) (Medina *et al.*, 2009). Fig. 2 shows different steps and factors involved in microalgal biodiesel production. The steps are briefly discussed in the following sections.

4. ALGAE CULTURE

Algae can be cultivated in either open ponds or photo-bioreactors (PBRs). Generally, an open pond is a series of outdoor 'raceways'. Open ponds

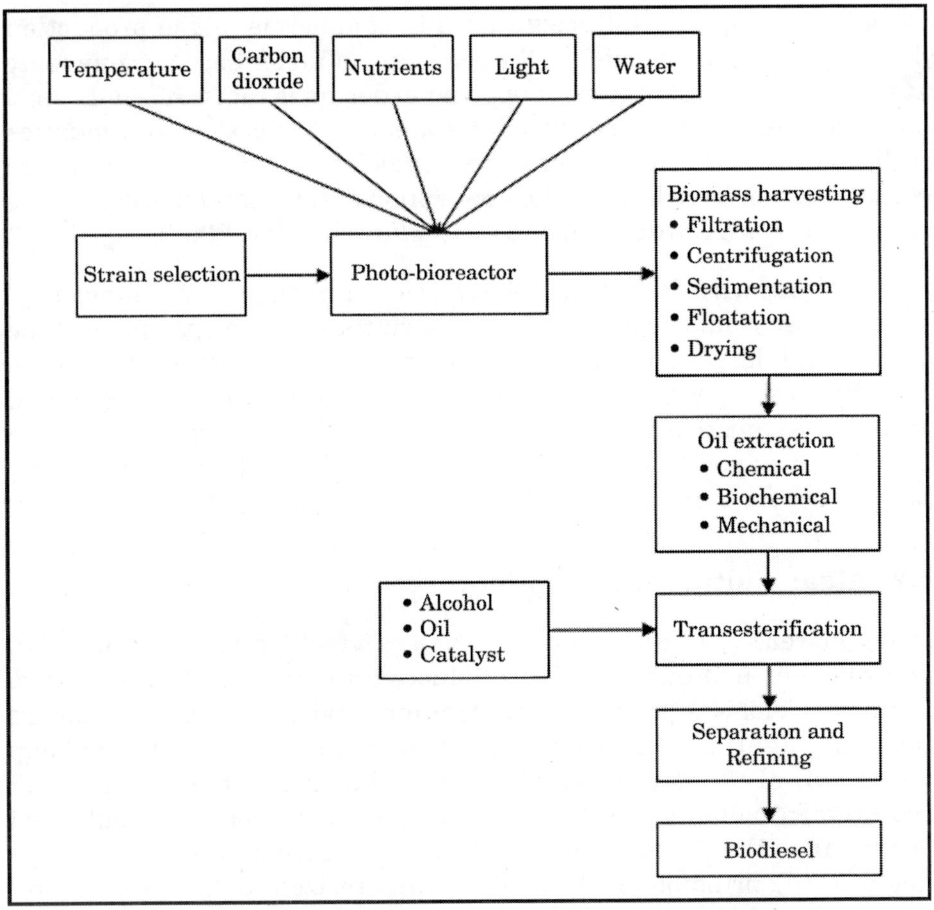

Fig. 2: Microlagal biodiesel production (Oncel, 2013)

are mostly used for cultivation of algae strains of high lipid content. PBRs are closed, controlled systems that provide an ideal environment for high algae productivity (Briggs, 2004). Appropriate technique is selected to maximize algal growth for production of biofuel and/or other products.

4.1. Algae Culture in Open Ponds

Open ponds are the most widespread technique for commercially growing algae commercially. There are three commonly used open pond systems, which are: unmixed, circular and raceways (Sukenik *et al.*, 1991; Lundquist *et al.*, 2010). Unmixed ponds are not controllable, unlikely to be able to efficiently supply CO_2 and are of low productivity.

Circular ponds, the first design used in commercial algae production do not scale above 1000 m^2, as the center pivot mixer becomes unwieldy at this size. The best option for open pond design is raceway pond. Open ponds are easy to construct and operate, and the cost is minimum as well; however, the nature of open ponds and their susceptibility to contamination have limited the application of this method to only three taxa: *Spirulina*, *Dunaliella*, and *Chlorella* (Huntley, 2007).

An alternative to open ponds are closed ponds where the control over the environment is much better than that for the open ponds. The idea behind the closed pond is to cover a pond or pool with a greenhouse. Usually closed ponds are used in *Spirulina* cultivation. These closed systems are constructed using Plexiglas. Closed Pond systems cost more than the open ponds, however, these cost considerably less than photo-bioreactors for similar areas of operation (Parmara, 2011).

4.2. Algae Culture in Photo-bioreactors

Photo-bioreactor (PBR) is a closed system which provides a controlled environment and enables high productivity of algae. PBR is used to promote biological growth by controlling environmental parameters including light, pH, carbon dioxide supply, gas supply rate, makeup water supply, culture density and mixing regime. All growth requirements of algae are introduced into the system and controlled according to the requirements (Hasan *et al.*, 2013). Based on the shape and working principle, PBRs can be characterized as: flat-plate photo-bioreactor, tubular photo-bioreactor, vertical column photo-bioreactor and internally illuminated photo-bioreactor (Ren, 2014). Table 3 presents characteristics of flat plate, tubular and vertical column photo-bioreactors for culturing microalgae:

4.3. Algae Cultivation in Hybrid System

Recent interest in microalgae as a feedstock for biofuel has intrigued researchers to improve the existing cultural techniques and to develop new designs. Researchers have proposed improved designs for the classic tubular reactor and flat panel, and also brought forward new concepts as floating PBR and hybrid systems (Tredici *et al.*, 2010).

Hybrid systems integrate the advantages of both the PBR and open ponds. This can improve the performance of the algae cultivation and

Table 3: Characteristics for culture system of microalgae (Ugwu *et al.*, 2008; Chen *et al.*, 2009; Brennan and Owende, 2010)

Culture systems	*Advantages*	*Limitations*
Flat plate photo-bioreactors	• Large illumination area • Suitable for outdoor culture • Good light path • Good biomass productivities • Relatively cheap • Easy to sterilize and clean up • Low oxygen build-up	• Scale-up requires many compartments and support materials. • Difficult temperature control • Small degree of hydrodynamic stress • Some degree of wall growth
Tubular photo-bioreactors	• Large illumination surface area • Suitable for outdoor cultures • Fairly good biomass productivities • Relatively cheap	• Gradients of pH • Dissolved O_2 • Dissolved O_2 and CO_2 along the tubes • Fouling • Requires large land surface • Some degrees of wall growth.
Column photo-bioreactor	• Compact • High mass transfer • Low energy consumption • Good mixing with low shear stress • Easy to sterilize • Reduced photo-inhibition and photo-oxidation	• Small illumination area • Expensive compared to open ponds • Shear stress • Sophisticated construction

increase yield. During exponential growth, microalgae are cultivated in the photo-bioreactor. In this technique contamination and growth parameters, such as: mixing, pH, and CO_2 diffusion can be easily maintained, hence, higher biomass concentration could be achieved (Yongmanitchai, 1991).

Using an inoculum large enough for the pond reduces the risk of contamination since it allows the strain to establish in the pond before the unwanted species. Cleaning the ponds in between also reduces the risk of contamination (Chenk, 2008; Demirbas, 2010).

A comparison of factors affecting cultivation between open ponds, photo-bioreactors and hybrid system is shown in Table 4 (Nerala *et al.*, 2016).

5. FACTORS TO BE CONSIDERED IN CULTIVATION DESIGN FOR ALGAL GROWTH

Photosynthetic growth of algae requires light, carbon dioxide, sugar and inorganic salts. Growth medium provides the micro-and macronutrients that ensure the unhindered algal growth (Blair *et al.*, 2013). Different physical and chemical factors that must be assured to get satisfactory algal yield are briefly described below:

Table 4: Comparison between open pond, photo-bioreactors and hybrid system (Nerala *et al.*, 2016)

Factors	***Open ponds***	***Photo-bioreactors (PBRs)***	***Hybrid system***
Space required	High	Moderate	High
Contamination risk	High	Low	Low
Setup cost	Low	High	Moderate
Maintenance	Easy	Difficult	Moderate
Maintaining continuous exponential phase	Difficult	Difficult	Easy
Mixing	Very poor	Uniform	Moderate
Operation regime	Batch or semi-continuous	Same	Continuous
Light utilization efficiency	Poor	High	High
Temperature control	Difficult	More uniform temperature	Moderate
Biomass quality	Variable	Reproducible	Reproducible
Evaporation loss of growth medium	High	Low	Moderate
CO_2 spurging efficiency	Low	High	Moderate
O_2 inhibition	Low	High	High during growth Low during stress

5.1. Light

Light helps algae to convert carbon dioxide into carbohydrate *via*, photosynthesis. Microalgae need a light/dark regime for productive photosynthesis. Microalgae need light for a photochemical phase where adenosine triphosphate (ATP) and nicotinamide adenine (NADPH) are produced. The dark phase is necessary for biochemical phases to synthesize essential molecules for growth (Cheirsilp and Torpee, 2012). Research has demonstrated that, both the growth rate and cell density of microalgae increase proportionally with light period duration (Lam and Lee, 2012). Further research on this topic has revealed that a 12 h: 12 h (light: dark) photoperiod resulted in a higher productivity and maximum cell density compared to a 14 h: 10 h (light: dark) photoperiod (Jacob-Lopes *et al.*, 2009).

The intensity of light depends on the depth and density of the algal culture. At higher depth and cell concentrations, the light intensity should be increased high enough so that in can penetrates through the algae culture. Along with natural light, fluorescent tubes can be used as a light source. Generally microalgae use light of wavelengths from 400 to 700 nm for photosynthesis (Blinova *et al.*, 2015).

5.2. Mixing

Mixing is necessary to prevent sedimentation of the algae, to ensure that all cells of the culture are equally exposed to light and nutrients, to avoid thermal stratification (*e.g.*, in outdoor cultures), and to improve gas exchange between the culture medium and the air (Blinova *et al.*, 2015). The rate at which atmospheric carbon dioxide can diffuse into an algal culture will definitely affect the growth rate. Scientists have demonstrated that combination of blue light and carbon dioxide induced higher rate of liquid accumulation in algal strains (Carvalho *et al.*, 2014).

Depending on the scale of the culture system, mixing is achieved by stirring daily by hand (test tubes, Erlenmeyer's), aerating (bags, tanks), or using paddle wheels and jet pumps (ponds). However, it should be noted that not all algal species can tolerate vigorous mixing (Mohsenpour and Willoughby, 2016).

5.3. Nutrient Medium

Media composition has a robust effect on both the growth rate and final concentration of microalgae. Two major nutrients are nitrogen and phosphorus, which play a significant role in controlling growth rates and lipid production. Other essential nutrients are carbon, hydrogen, oxygen, sulfur, calcium, magnesium, sodium, potassium and chlorine. Nutrients needed in trace quantities include iron, boron, manganese, copper, molybdenum, vanadium, cobalt, nickel, silicon, and selenium (Munir *et al.*, 2015). Nutrient requirement must be maintained in an optimal range to ensure maximum growth. However, scarce nutrient can reduce the growth rate and too much of it can be toxic and inhibiting

(Darzins *et al.*, 2010). Table 5 presents the composition of several nutrient media widely used for algae culture.

Table 5: Composition of different nutrients used in algae culture

Name of the medium	*Name of compounds used*
Bold's Basal (BB) (Nicholas, 1973)	$NaNO_3$, $MgSO_4.7H_2O$, NaCl, K_2HPO_4, KH_2PO_4, $CaCl_2.2H_2O$, $ZnSO_4.7H_2O$, $MnCl_2.4H_2O$, MoO_3, $CuSO_4.5H_2O$, $Co(NO_3)_2.6H_2O$, H_3BO_3, EDTA, KOH, $FeSO_4.7H_2O$, H_2SO_4
MLA (Bolch, 1996)	$MgSO_4.7H_2O$, $NaNO_3$, K_2HPO_4, H_3BO_3, H_2SeO_3, Biotin, Vitamin B_{-12}, Thiamin-HCl, Na_2EDTA, $FeCl_3$, $NaHCO_3$, $MnCl_2.4H_2O$, $CuSO_4.5H_2O$, $ZnSO_4.7H_2O$, $CoCl_2.6H_2O$, $Na_2MoO_4.2H_2O$, $CaCl_2.2H_2O$
Blue Green (BG) (Crofcheck, 2012)	$NaNO_3$, $K_2HPO_4..3H_2O$, $MgSO_4.7H_2O$, $CaCl_2.2H_2O$,Citric acid, Ferric ammonium citrate, EDTA, Na_2CO_3, H_3BO_3, $MnCl_2.4H_2O$, $ZnSO_4.7H_2O$, $Na_2MoO_4.2H_2O$, $CuSO_4.5H_2O$, $Co(NO_3)_2.6H_2O$

5.4. Temperature

Temperature influences the rate of chemical reactions related to algal growth and metabolism. Like all other living organisms, algae also have an optimum temperature range for growth. As the temperature deviates from the optimum range, growth is inhibited, then halted, and then cell death occurs (Darzins *et al.*, 2010). Temperature strongly influences the cellular chemical composition specifically lipids and proteins, the uptake of nutrients, carbon dioxide fixation and the growth rate for every species of algae (Evans, 1884). It was demonstrated that, below 4°C the photosynthesis is completely inhibited, photosynthesis is substantially inhibited between 4° and 11°C and above 11°C the rate of photosynthesis is high (Takemura *et al.*, 1985).

5.5. pH

Optimum pH influences the carbon availability, metabolic functions and biochemical composition of algal cells (Richmond, 2000). At $pH>9.0$ some calcium salts may form precipitation (Becker, 1994). Since algae uptakes carbon primarily in inorganic form, pH of the medium can increase significantly. At higher pH, carbon for algae is available in carbonate forms, which limits the availability of carbon in the form of CO_2, and by turn suppresses the algae growth rate (Azov, 1982). Optimum pH may vary for algae strains. Researches demonstrated that

Chlorella vulgaris can tolerate pH as low as 3.0 and as high as 11.0 (Mayo, 1997; Yeh *et al.*, 2010).

6. HARVESTING

Algal harvesting means recovery of biomass from the culture medium, which may contribute to 20-30% of the total biomass production cost (Grima *et al.*, 2003). In order to remove large quantities of water (about 99% by weight of the algae-water mixture is water), and process large algal biomass volumes, a suitable harvesting technique must be selected. It may involve one or more steps. There are several physical, chemical or biological ways to achieve the desired separation. No single technique has proven to be the universal solution to harvesting algae. The preference of technique also depends on how the algae will be treated after harvesting, *i.e.*, oil extraction for biodiesel production or anaerobic digestion. Different downstream production processes have different demands regarding the water content (Ho *et al.*, 2011).

Most common harvesting methods include:

- Sedimentation,
- Centrifugation,
- Filtration and ultra-filtration, and
- Flocculation or a combination of flocculation-floatation.

6.1. Sedimentation

Gravity sedimentation is the most common technique used in wastewater. However, the small size of the microalgae and their low specific gravity result a settling rate which is too low for routine algae harvesting thus, severely affect the overall cost of the procedure (Grima *et al.*, 2003).

6.2. Centrifugation

Centrifugation is suitable for algae used for high value products. While this method is effective in removing the majority of cells from suspension, a high energy input is required to run the centrifuge separation of algal cells from large amounts of culture water (Richmond, 2000) hence, this technique is costly because of high energy input.

6.3. Filtration and Membrane Filtration

Filter presses operating under pressure or vacuum can be used to recover large quantities of biomass. However, for some applications filtration can be relatively slow which may be unsatisfactory. Pressure or vacuum filtration, along with the use of filter aids such as diatomaceous earth or cellulose are suitable for the recovery of large algal cells (>70 µm) (Brennan and Owende, 2010; Mata *et al.*, 2010). Membrane filtration units can also be used to remove algae cells from the culture water without adding any additional materials to the harvested biomass. Membrane microfiltration and ultra-filtration are other possible alternatives for recovering algal biomass, which are more suitable for fragile cells and small-scale production processes. But membrane replacement and pumping make these processes comparatively expensive (Pittman *et al.*, 2011).

6.4. Flocculation

Flocculation is an effective step in the bulk harvesting process and is used to aggregate algal cells, and thus easing subsequent centrifugation, sedimentation or filtration steps (Haifeng, 2008). Alum and ferric chloride are chemical flocculants used to harvest algae. Another flocculating agent is chitosan. Interrupting the carbon dioxide supply can cause algae to flocculate on its own, which is called 'auto flocculation' (Travieso *et al.*, 1992).

6.5. Immobilization

Immobilization of microalgae prior to and during cultivation is an alternative method of harvesting microalgae. It is conducted to retrieve the microalgae easily at the end of the growth period. Artificial attachment or polymer encapsulation of microalgae such as in alginate, are the most frequently used immobilization approaches (Olaizola, 2003). Alginate-immobilized microalgae have been shown to be as effective as free-swimming microalgae in accumulating nitrogen and phosphorous from the nutrient medium (de-Bashan and Bashan, 2010).

7. PROCESSING OF ALGAL BIOMASS AND LIPID EXTRACTION

Algal oil can be extracted using different methods. Dry algae cell retains its oil content, which can be pressed out with an oil press. Algal oil can be extracted using solvents such as benzene, ether or hexane. This method can be combined with cold press. The process together will be able to derive 95% of the total oil present in algae. Enzymatic extraction uses enzymes to degrade the cell walls, which makes fractionation of the oil much easier (Amin, 2009; Morales *et al.*, 1985).

Prior to biodiesel production, algal biomass needs to be processed. There are two types of methods for processing algal biomass (Muhit *et al.*, 2014)

a) ***Wet extraction process*:** Focuses on disrupting the algae cells in solution;
b) ***Dry extraction process*:** Removes the algae from aqueous medium and then disrupt the cell

7.1. Pre-treatment of Microalgal Biomass

Pre-treatment of algal biomass is usually conducted prior to lipid extraction and anaerobic digestion steps. This step reduces the cost and energy requirement of lipid extraction. When solvent-based extraction techniques are applied to wet algae paste after harvesting, the cells tend to remain intact due to their cell envelope (cell wall and membranes), which prevents them from coming into direct contact with the solvent. The algal cell envelope is a thick and rigid layer consisting of complex carbohydrates and glycoproteins with high mechanical strength and chemical resistance (Bai, 2015; Kim *et al.*, 2011). For the efficiency of the lipid extraction process, complete disintegration of cell structure is important (Roesijadi *et al.*, 2010). Successful cell disintegration demands small amount of water in algal biomass (Halim *et al.*, 2012b). Lipid yields increase with the degree of cell disruption, since lipids are discharged from the cellular structures outside cell. Mechanical, chemical and enzymatic methods are widely used for cell disintegration in microorganism biomass (Piasecka *et al.*, 2014).

Mechanical pre-treatment methods generally break the cell wall directly through physical forces, which have been proved to be relatively efficient for cell disruption. This method can be used to algal biomass regardless of its species, which is the key advantage of using this method. However, mechanical method is energy intensive. Microwave, sonication, bead beating, high pressure thermo-hydrolysis and electroporation are some of the mechanical methods (Kim *et al.*, 2011).

In chemical methods, treatments with acids, alkali, surfactants, or heated chemicals can degrade the chemical linkages on the rigid cell envelope (Miranda *et al.*, 2012; Sarada *et al.*, 2006). Compared to mechanical methods, chemical methods are less energy intensive, however, the efficiency of the process can be relatively low or strain dependent (Bai, 2015).

Biological methods refer to the methods that degrade cell envelope using enzymes or other biological reagents (Young *et al.*, 2011; Zheng *et al.*, 2011). While mechanical methods destroy almost every particle existing in the solution, enzymatic methods offer mild reaction conditions and high selectivity. The critical downfall of the biological methods is high cost of the enzymes (Bai, 2015).

7.1.1. *Wet process*

Freezing

The expansion of the water inside the cell as well as the water around the cell will cause the cell walls to rupture from the inside out or be disrupted by the compressive forces (Huang *et al.*, 2010).

Homogenization

To obtain a uniform media, the particle size in a mixture should be reduced. This reducing process is termed as homogenization. This process involves expelling mixture at high pressure through small valves. Homogenizers having nano-tolerance are a potentially viable source of extracting oil on a lab scale but they are unlikely to be appropriate for industrial scale (Greenwell *et al.*, 2010; Huang *et al.*, 2010).

Sonification

The thick cell walls of microalgae block the release of intralipids present inside. Sonication process could be useful to extract intralipides. In the sonication process cell wall collapses when cavitation bubbles create a pressure as well as shock and finally contained oil is released (Greenwell *et al.*, 2010). Ultrasonic waves can be generated by using ultrasonic reactor.

7.1.2. *Dry process*

Dehydration of biomass is a common technique, which also increases its shelf-life. Several methods are commonly used to dry microalgae such as spray drying, drum-drying, freeze-drying and sun-drying (*Handbook of Microalgal Culture: Biotechnology and Applied Phycology*, 2003). Due to high water content of algal biomass, sun-drying is not a very effective method; on the other hand, spray-drying may not be economically viable for low value products as biofuel.

7.2. Lipid Extraction Methods

After drying, metabolite such as lipid is extracted by disrupting algal cell (Mata *et al.*, 2010). Lipid extraction techniques are categorized as mechanical and chemical methods (Halim *et al.*, 2012a). Different lipid extraction techniques are shown in the following flowchart (Fig. 3) (Mubarak, 2015):

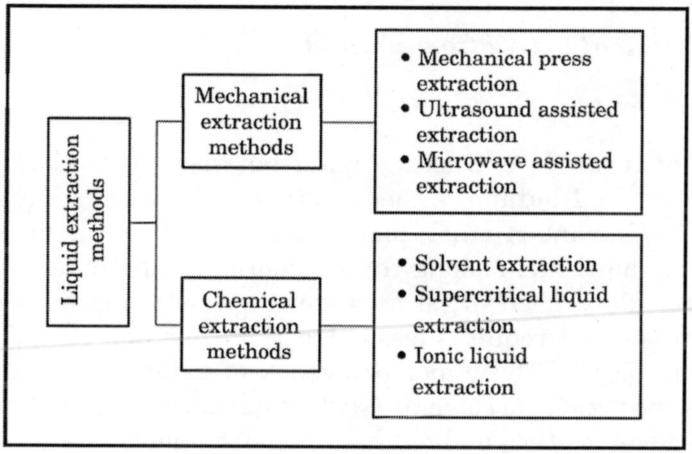

Fig. 3: Flowchart of different lipid extraction methods from microalgae (Mubarak *et al.*, 2015)

7.2.1. *Mechanical extraction method*

Mechanical presses extraction

A mechanical press is used to break the cell walls and compress out the oil from the dry biomass and can extract about 75% of oil. But this method is quite slow and requires a large quantity of biomass (*Handbook of Microalgal Culture: Biotechnology and Applied Phycology*, 2003).

Ultrasound assisted extraction

The intense sonication of liquid generates high frequency (20 KHz) sound waves that propagate into the liquid media and results in alternate high and low pressure cycles. This cycle causes cavitation and the shearing forces break the cell structure during cavitation. This results in improved mass transfer supporting the extraction of lipids (Lee *et al.*, 1998).

Microwave assisted extraction

The microwave (0.3 to 300 GHz) assisted heating uses a non-contact heat source, which can penetrate the biomaterials, interact with polar molecules like water in the biomass, and heat the whole sample uniformly. The higher oil yield with superior quality and reduced extraction time are the main advantages of microwave-assisted extraction (MAE) (Iqbal and Theegala, 2013).

7.2.2. *Chemical extraction method*

Solvent extraction

The solvents used for extracting lipid from microalgae biomass are n-hexane, ethanol, 1-butanol, dimethyl ether and mixtures of chloroform/ methanol, n-hexane/ethanol, methylene chloride/methanol, acetone/ dichloro-methane etc. Among these, chloroform/methanol (1/2 v/v) is the most widely used organic solvent system employed for lipid extraction, since it requires less extraction time and offers increased yield (Neto, 2013). The major drawback of using solvent extraction method is the toxicity of solvents (such as hexane and chloroform) which can have adverse effect on health and environment.

Supercritical fluid extraction

The effectiveness of supercritical fluids as solvent is a function of density. Therefore, the solvent power can be varied by changing extraction pressure and temperature, which makes supercritical fluids suitable as extraction solvents. Moreover, these fluids are capable of producing solvent free crude lipids. Supercritical CO_2 is commonly used for extraction purpose (Santana *et al.*, 2012).

Ionic liquid extraction

Ionic liquids are suitable for the extraction of lipids from algae due to their non-volatility, thermal stability and synthetic flexibility.

8. TRANSESTERIFICATION

The lipid extraction method is followed by fractionation step for purification. In this step the undesired polar lipids and non-acyl-glycerol neutral lipids are removed from the crude lipids or total lipids. For fractionation, different methods such as liquid chromatography, acid precipitation and urea crystallization are used (Medina *et al.*, 1998).

The purified product is then processed into biodiesel by transesterfication method. During this method, the fatty-acids containing lipid fractions are reacted with alcohol and converted to fatty acid alkyl ester. Methanol is widely used in trans-esterfication process due to its low cost and physical advantages. Ethanol, propanol, butanol and amyl alcohol are other options. An acid (such as H_2SO_4) or an alkali (such as NaOH) can be used as a catalyst in this reaction. Since studies have shown that alkali-catalyzed trans-esterification is about 4000 times faster than acid-based one, the former is most frequently used commercially (Fakuda *et al.*, 2001). Fig. 4 shows the general form of trans-esterificaton reaction.

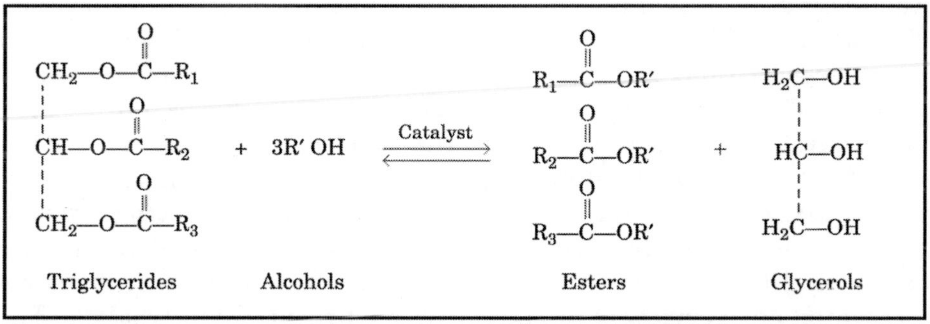

Fig. 4: General form of transesterfication reaction to produce algal biodiesel (Pokoo-Aikins *et al.*, 2010)

Transesterification is a multiple step reaction, including three reversible steps in series, where the triglycerides are first converted to diglycerides, then diglycerides are converted to monoglycerides, and lastly monoglycerides are converted to esters (biodiesel) and glycerol (by-product). Although the alcohol: oil theoretical molar ratio is 3:1, a ratio of 6:1 is generally used to complete the reaction accurately (Mata *et al.*, 2010).

Once trans-esterification is complete, the reaction mixture contains, biodiesel, glycerol, reformed alkali catalyst, excess methanol and remaining un-transesterified lipids. Then the reaction mixture undergoes post-transesterification purification process. The reaction mixture is left to settle down under gravity to induce bio-phasic partitioning. After bio-phasic partitioning biodiesel/un-transesterified lipids are accumulated at the top phase, while glycerol accumulates at the bottom phase. Once the top phase is decanted off, it is washed repeatedly to remove any trace of alkali catalyst and excess methanol (Halim *et al.*, 2012a).

9. CONCLUSIONS

In recent decades the rising demand of global energy is heavily dependent on fossil fuels. However, it is predicted that the supply of fossil fuels will come to an end by 21st century considering a 5% flat increase in demand. In addition, excessive use of fossil fuel is causing increase of greenhouse gas CO_2 in the atmosphere, which is considered responsible for temperature increase, sea level hike, deviation of biodiversity and ecological imbalance. Therefore, many nations are conducting extensive research program to explore future generation fuel which will be greener, cleaner and renewable. The investment in algae research has been increasing significantly in the recent years, since it offers biodiesel as an alternative energy source, requires small footprint, helps reducing CO_2 from the atmosphere and can be used in wastewater treatment. However, the critical challenge is to keep the production cost of algal biodiesel compatible to that of petro diesel. To make it economically competitive, different stages of producing microalgal biodiesel need to be optimized substantially. This article presented an overview of production of biodiesel from microalgae. The basics and advancements of algae culture, harvesting, lipid extraction and processing were discussed here. Different technical and limiting factors related to different stages of algae production were also discussed. In the upcoming years, through extensive research and development, it is expected that cost effective methods and technologies to produce algal biodiesel will be developed.

LITERATURE CITED

Aikins, G.P., Nadim, A., El-Halwagi, M.M. and Mahalec, V. (2010). Design and analysis of biodiesel production from algae grown through carbon sequestration. *Clean Technology and Environment Policy,* 12: 15.

Allard, B. and Templier, J. (2000). Comparison of neutral lipid profile of various trilaminar outer cell wall (TLS)-containing microalgae with emphasis on algaenan occurrence. *Phytochemistry,* 54(4): 11.

Amin, S. (2009). Review on biofuel oil and gas production processes from microalgae. *Energy Conversion and Management,* 50: 1834–1840.

Antolin, G., Tinaut, F.V., Briceno, Y., Castano, V., Perez, C. and Ramirez, A.I.. (2002). Optimisation of biodiesel production by sunflower oil transesterification. *Bioresource Technology,* 83: 111.

Azov, Y. (1982). Effect of pH on inorganic carbon uptake in algal cultures. *Applied and Environmental Microbiology,* 43: 6.

Bai, X. (2015). *Enhancing algal biomass and biofuels recovery from open culture systems.* (Doctor of Philosophy), The University of Queensland, Australia.

Becker, E. (1994). *Microalgae: biotechnology and microbiology* (J. Baddiley Ed. Vol. 10). Cambridge, New York: Cambridge University Press.

Bhatnagar, S.K., Saxena, A. and Kraan, S. (2011). Algae-Based Biofuels: A review of challenges and opportunities for developing countries. *Algae Biofuel,* pp. 1–40.

Blair, M.F., Kokabian, B. and Gude, V.G. (2013). Light and growth medium effect on *Chlorella vulgaris* biomass production. *Journal of Environmental Chemical Engineering,* 10.

Blinova, L., Bartosova, A. and Gerulova, K. (2015). Cultivation of microalgae (*Chlorella vulgaris*) for biodiesel production. *Research Papers, Faculty of Materials Science and Technology in Trnava Slovak University of Technology in Bratislava,* 23(36): 10.

Bolch, C.J.S. and B.S.I. (1996). Isolation and purification of Australian isolates of the toxic cyanobacterium *Microcystis aeruginosa* Kütz. *Journal of Applied Phycology,* 8: 8.

Brennan, L. and Owende, P. (2010). Biofuels from microalgae – A review of technologies for production, processing, and extractions of biofuels and co-products. *Renewable and Sustainable Energy Reviews,* 14: 557–577.

Briggs, M. (2004). Widescale Biodiesel Production from Algae: University of New Hampshire Biodiesel Group.

Carvalho, A.P., Silva, S., Baptista, J,M. and MAlcata, F.X. (2014). Light requirements in microalgal photobioreactors: An overview of biophotonic aspects. *Applied Microbial Biotechnology,* 89: 13.

Cheirsilp, B. and Torpee, S. (2012). Enhanced growth and lipid production of microalgae under mixotrophic culture condition: Effect of light intensity, glucose concentration and fed-batch cultivation. *Bioresource Technology.*

Chen, P., Min, M., Chen, Y., Wang, L., Li, Y., Chen, Q., Wang, C., Wan, Y., Wang, X., Cheng, Y., Deng, S., Hennessy, K., Lin, X., Liu, Y., Wang, Y., Martinez, B.

and Ruan, R. (2009). Review of the biological and engineering aspects of algae to fuels approach. *International Journal of Agricultural and Biological Engineering, 2*(4): 1–30.

Chenk, P.M. *et al.* (2008). Second generation biofuels: High-efficiency microalgae for biodiesel production. *Bio Energy Research,* 1(1): 20–43.

Chisti, Y. (2007). Biodiesel from microalgae. *Biotechnology Advance,* 25: 12.

Crofcheck, C., Shea, E.X., Montross, M., Crocker, M. and Andrews, R. (2012). Influence of media composition on the growth rate of *Chlorella vulgaris* and *Scenedesmus* acutus utilized for CO_2 mitigation. *Journal of Biochemical Technology,* 4(2): 5.

Darzins, A., Pienkos, P. and Edye, L. (2010). Current Status and Potential for Algal Biofuels Production.

De-Bashan, L.E. and Bashan, Y. (2010). Immobilized microalgae for removing pollutants: Review of practical aspects. *Bioresource Technology,* 101(6): 1611–1627.

Demirbas, A. (2010). "Use of algae as biofuel sources". *Energy Conversion and Management,* 51(22): 2238–2749.

Evans, C.T. and C.R. (1884). Effect of nitrogen source on lipid accumulation in oleaginous yeasts. *Journal of General Microbiology,* 130.

Fakuda, H., Kondo, A. and Noda, H. (2001). Biodiesel fuel prodcution by transesterification of oils. *Journal of Bioscience and Bioengineering,* 92: 11.

Fishman, D. *et al.* (2010). National Algal Biofuels Technology Roadmap. U.S. Department of Energy, Office of Energy Efficiency and Renewable Energy, Biomass Program. In U.S. DOE (Ed.).

Ghosh, R., Bhattacharjee, S. and Mitra, A. (2014). Production of Lipids in Photobioreactors Using Microalgae. *International Journal of Scientific & Engineering Research,* 5(8): 16.

Ginzburg, B.Z. (1993). Liquid fuel (iil) from halophilic algae: A renewable source of non- polluting energy. *Renewable Energy,* 3.

Gouveia, L. and Oliveira, A.C. (2009). Microalgae as a raw material for biofuels production. *Journal of Industrial & Microbial Biotechnology,* 36(2): 5.

Greenwell, H.C., Laurens, L.M. L., Shields, R.J., Lovitt, R.W. and Flynn, K.J. (2010). Placing microalgae on the biofuels priority list: A review of the technological challenges. *Journal of the Royal Society Interface,* 7: 703–726.

Grima, M.E., Belarbi, E.H., Fernandez, F.G.A., Medina, A.R. and Chisti, Y. (2003). Recovery of microalgal biomass and metabolites: Process options and economics. *Biotechnology Advance,* 20(7-8): 491–515.

Haifeng, L. *et al.* (2008). Effect of sodium phosphate buffer on horseradish peroxidase thermal stability. *Journal of Thermal Analysis and Calorimetry,* 93(2): 5.

Halim, R., Danquah, M.K. and Webly, P.A. (2012a). Extraction of oil from microalgae for biodiesel production: A review. *Biotechnology Advances,* 30: 21.

Halim, R., Harun, R., Danquah, M.K. and Webley, P.A. (2012b). Microalgal cell disruption for biofuel development. *Applied Energy,* 91: 5.

Handbook of Microalgal Culture: Biotechnology and Applied Phycology. (2003). Wiley-Blackwell.

Hasan, M.F., Halim, S.Z., Shahariar, C.M. and Quddus, N. (2013). Mitigation of Agricultural Energy Requirement. *In*: Bangladesh Using Microalgae Production. *Journal of Mechanical Engineering and Sciences.*

Ho, S.H., Chen, C.Y., Lee, D.J. and Chang, J.S. (2011). Perspectives on microalgal CO_2-emission mitigation systems – A review. *Biotechnology Advance,* 29(2): 189–198.

Hossain, A.B.M.S., Sallen, A., Byee, A.N., Chowdhury, P. and Naqiuddin, M. (2008). Biodiesel fuel production from algae as renewable energy. *American Journal of Biochemistry and Biotechnology*, 4(3): 4.

Hu, Q., Sommerfeld, M., Jarvis, E., Ghirardi, M., Posewitz, M., Seibert, M. and Drazins, A. (2008). Microalgal triacylglycerols as feedstocks for biofuel production: Perspectives and advances. *The Plant Journal,* 54: 18.

Huang, G., Chen, F., Wei, D., Zhang, X. and Chen, G. (2010). Biodiesel production by microalgal biotechnology. *Appied Energy,* 87: 38–46.

Huntley, M.E.R. and D.G. (2007). CO_2 Mitigation and Renewable Oil from Photosynthetic Microbes: A New Appraisal. *Mitigation and Adaptation Strategies for Global Change,* 12.

Iqbal, J. and Theegala, C. (2013). Microwave assisted lipid extraction from microalgae using biodiesel as co-solvent. *Algal Research, 2*(1): 34–42.

Jacob-Lopes, E., Scoparo, C.H.G., Lacerda, L.M.C.F. and Franco, T.T. (2009). Effect of light cycles (night/day) on CO_2 fixation and biomass production by microalgae in photobioreactors. *Chemical Engineering and Processing: Process Intensification,* 48(1): 4.

Kais, M.I., Chowdhury, F.I. and Shahriar, K.F. (2011). Biodiesel from Microalgae as a solution of third world energy crisis. *Bioenergy Technology*, 8.

Kim, D.-G., La, H.-J., Ahn, C.-Y., Park, Y.-H. and Oh, H.-M. (2011). Harvest of *Scenedesmus* sp. with bioflocculant and reuse of culture medium for subsequent high-density cultures. *Bioenergy Technology,* 102(3): 5.

Lam, M.K. and Lee, K.T. (2012). Potential of using organic fertilizer to cultivate *Chlorella vulgaris* for biodiesel production. *Applied Energy,* 94: 5.

Lee, J.S., Yoon, B.D. and Oh, H.-M. (1998). Rapid method for the determination of lipid from the green alga *Botryococcus braunii. Biotechnology Techniques,* 12(7): 553–556.

Liang, Y., Sarkany, N. and Cui, Y. (2009). Biomass and lipid productivities of *Chlorella vulgaris* under autotrophic, heterotrophic and mixotrophic growth conditions. *Biotechnology Letters,* 31: 6.

Lundquist, T.J., Woertz, I.C., Quinn, I.W.T. and Benemann, J.R. (2010). A Realistic Technology and Engineering Assessment of Algae Biofuel Production.

Ma, F. and Hanna, M.A. (1999). Bidiesel production: A review. *Bioresource Technology,* 70: 15.

Mata, T.M., Martins, A. and Caetano, N.S. (2010). Microalgae for biodiesel production and other applications: A review. *Renewable and Sustainable Energy Reviews,* 14: 217–232.

Mayo, A.W. (1997). Effects of temperature and pH on the kinetic growth of unialga *Chlorella vulgaris* cultures containing bacteria. *Water Environment Research,* 69: 8.

Medina, A.R., Grima, E.M., Gimenez, A.G. and Ibanez, M.J. (1998). Downstream processing of algal polyunsaturated fatty acids. *Biotechnology Advance,* 16(3): 63.

Medina, A.R., Moreno, P.A.G., Cerdán, L.E. and Grima, E.M. (2009). Biocatalysis: Towards ever greener biodiesel production. *Biotechnology Advances,* 27: 10.

Meher, L.C., Vidyasagar, D. and Naik, S.N. (2006). Technical aspects of biodiesel production by transesterification- A review. *Renewable and Sustainable Energy Reviews,* 10: 20.

Mehlitz, T.H. (2009). *Temperature Influence and Heat Management Requirements of Microalgae Cultivation in Photobioreactors.* (Master of Science in Agriculture), California Polytechnic State University.

Metzger, P. and Y.P. (1995). N-alkenylpyrogallol dimethyl ethers, aliphatic diol monoesters and some minor ether lipids from *Botryococcus braunii*: A Race. *Phytochemistry,* 40(2).

Miao, X., Wu, Q. and Yang, C. (2004). Fast pyrolysis of microalgae to produce renewable fuels. *Journal of Analytical and Applied Pyrolysis,* 71.

Miranda, J.R., Passarinho, P.C. and Gouveia, L. (2012). Pre-treatment optimization of *Scenedesmus* obliquus microalga for bioethanol production. *Bioresource Technology,* 104(0): 6.

Mohsenpour, S.F. and Willoughby, N. (2016). Effect of CO_2 aeration on cultivation of microalgae in luminescent photobioreactors. *Biomass and Bioenergy,* 85: 9.

Morales, J., Noue, J. and Picard, G. (1985). Harvesting marine microalgae species by chitosan flocculation. *Aquacultural Engineering,* 4(4): 257–270.

Mubarak, M., Shajia, A. and Suchitra, T.V. (2015). A review on the extraction of lipid from microalgae for biodiesel production. *Algal Research,* 7: 117–123.

Muhit, I.B., Baidya, D. and Nurangir, N. (2014). Prospect of algal biodiesel production in Bangladesh: Overview from developed countries. *IOSR Journal of Mechanical and Civil Engineering (IOSR-JMCE),* 11(1): 49–54.

Munir, N., Imtiaz, A., Sharif, N. and Naz, S. (2015). Optimization of growth conditions of different algal strains and determination of their lipid contents. *The Journal of Animal & Plant Sciences,* 25(2): 8.

Nerala, R.R., Garg, S., Sharma, K.K., Thomas-Hall, S.R., Deme, M., Li. Y. and Schenk, P.M. (2016). Comparison of microalgae cultivation in photobioreactor, open raceway pond and a two-stage hybrid system. *Frontiers in Eenrgy Research,* 4.

Neto, A.M.P. *et al.* (2013). Improvement in microalgae lipid extraction using a sonication-assisted method. *Renewable Energy,* 55: 525–531.

Nicholas, H.W. (1973). Growth media – freshwater. *In*: Stein, J. (*ed.*), *Handbook of Phycological Methods, Culture Methods and Growth Measurements,* Cambridge University Press. pp. 7–24

NREL (1998). A Look Back at the U.S. Department of Energy's Aquatic Species Program: Biodiesel from Algae: National Renewable Energy Laboratory, U.S. Department of Energy's, Office of Fuels Development. .

Olaizola, M. (2003). Commercial development of microalgal biotechnology: From the test tube to the market place. *Biomolecular Engineering,* 20(4–6): 459–466.

Oncel, S.S. (2013). Microalgae for a macroenergy world. *Renewable and Sustainable Energy Reviews,* 26: 23.

Parmara, A., Singha, N.K., Pandeyb, A., Gnansounouc, E. and Madamwara, D. (2011). Review Cyanobacteria and microalgae: A positive prospect for biofuels. *Bioresource Technology,* 102(22): 9.

Piasecka, A., Krzemińska, I. and Tys, J. (2014). Physical methods of microalgal biomass pretreatment. *International Agrophysics,* 28: 7.

Pittman, J.K., Dean, A.P. and Osudeko, O. (2011). The potential of sustainable algal biofuel production using wastewater resources. *Bioresource Technology,* 102: 17–25.

Pokoo-Aikins, G., Nadim, A., El-Halwagi, M.M. and Mahalec, V. (2010). Design and analysis of biodiesel production from algae grown through carbon sequestration. *Clean Technology and Environment Policy,* 12: 15.

Qudus, N.A. and Zohra, H.S. (2015). A Techno-economic Assessment of Algae Biofuel Production in Bangladesh.

Halim, R., Danquah., M.K. and Webley, P.A. (2012). Extraction of oil from microalgae for biodiesel production: A review. *Biotechnology Advance,* 30(3): 709–732.

Ren, T. (2014). *Primary Factors Affecting Growth of Microalgae Optimal Light Exposure Duration and Frequency.* (Master of Science), Iowa State University USA.

Richmond, A. (2000). Microalgal biotechnology at the turn of the millenium: A personal view. *Journal of Applied Phycology,* 12: 10.

Roesijadi, G., Jones, S.B., Snowden-Swan, L.J. and Zhu, Y. (2010). Macroalgae as a Biomass Feedstock: A Preliminary Analysis.Macroalgae as a Biomass Feedstock: A Preliminary Analysis. *Pacific Northwest National Laboratory*: Washington.

Saifullah, A.Z.A., Karim Md, A. and Yazid, A.A. (2014). Microalgae: An alternative source of renewable energy. *American Journal of Engineering Research (AJER),* 3(3): 9.

Santana, A., Jesus, S., Larrayoz, M.A. and Filho, R.M. (2012). Supercritical carbon dioxide extraction of algal lipids for the biodiesel production. *Procedia Engineering,* 42: 1755–1761.

Sarada, R., Vidhyavathi, R., Usha, D. and Ravishankar, G.A. (2006). An Efficient Method for Extraction of Astaxanthin from Green Alga *Phaeodactylum tricornutum. Journal of Agricultural and Food Chemistry*, 54(20), 3.

Sayed, M.A. (2012). Algal biofuel: A promising next-generation potential. *The Daily Star.*

Sharp, C.A. (1996). Emissions and lubricity evaluation of rapeseed derived biodiesel fuels [R]. Final Report for Montana Department of Environmental Quality. *Southwest Research Institute.*

Sheehan, J., Dunahay, T., Benemann, J. and Roessler, P. (1998). A Look Back at the U.S. Department of Energy's Aquatic Species Program-Biodiesel from Algae: U.S. Department of Energy's Office of Fuels Development.

Sukenik, A.L., Levy, R.S., Levy, Y., Falkowski, P.G. and Dubinsky, Z. (1991). Optimizing algal biomass production in an outdoor pond: A simulation model. *Journal of Applied Phycology,* 3: 10.

Takemura, N., Iwkume, T. and Rusuno, M. (1985). Photosynthesis and primary production of *Microcystis aeruginosa* Kütz. in Lake Kasumigaura. *Journal of Plankton Research,* 7(2): 9.

Travieso, L., Benitez, F. and Dupeiron, R. (1992). Sewage treatment using immobilized microalgae. *Bioresource Technology,* 40: 183–187.

Tredici, M.R., Chini Zittelli, G. and Rodolfi, L. (2010). *Photobioreactors. In: Encyclopedia of Industrial Biotechnology*: *Bioprocess, Bioseparation and Cell Technology. In*: Flickinger, M.C. (*ed.*), Vol. 6. New York: John Wiley and Sons.

Ugwu, C.U., Aoyag, H. and Uchiyama, H. (2008). Photobioreactors for mass cultivation of algae. *Bioresource Technology,* 99(10): 4021–4028.

Wen, Z. and Johnson, M.B. (2009). Microalgae as a feedstock for biofuel production. *Virginia Cooperative Extension.*

World Fossil Fuel Reserves and Projected Depletion (2002): The Colorado River Comission Nevada.

Yeh, K., Chang, J. and Chen, W. (2010). Effect of light supply and carbon source on cell growth and cellular composition of a newly isolated microalga *Chlorella vulgaris* ESP-31. *Engineering in Life Sciences,* 10: 7.

Yongmanitchai, W. and O.P.W. (1991). Growth of and omega-3 fatty acid production by *Phaeodactylum tricornutum* under different culture conditions. *Applied and Environmental Microbiology,* 57(2): 18.

Young, G., Nippen, F., Titterbrandt, S. and Cooney, M.J. (2011). Direct transesterification of biomass using an ionic liquid co-solvent system. *Biofuels,* 2(3): 5.

Zheng, H., Yin, J., Gao, Z., Huang, H., Ji, X. and Dou, C. (2011). Disruption of *Chlorella vulgaris* cells for the release of biodiesel-producing lipids: A comparison of grinding, ultrasonication, bead milling, enzymatic lysis, and microwaves. *Applied Biochemistry and Biotechnology,* 164(7): 9.

11

Algal Biofuel and Its Potential to Reduce Greenhouse Gas Emissions: A Case Study

K. HUNDT[1], BALE V. REDDY[1*] AND A. ALAM[1]

ABSTRACT

There is a growing interest in algae and production of biofuels from algae. The chapter presents an overview of algae, its benefits over other biofuel sources, the technology and developments in producing algal biofuel. The research advances and the status of algae biofuel systems are also discussed. A case study is undertaken to observe the potential of algal biodiesel, produced using power plant exhaust, and consequently reduce greenhouse gas emissions. The case study results indicate that using 60% of all coal and gas power plants would unable production of significant amount of biofuel, which can be used for reducing greenhouse gas emissions. Research investigations are needed to improve the efficiency of algal fuel production technology, so as to lower the cost of algal biodiesel and thereby make it commercially competitive with petrol and diesel.

Key words: Algae, Greenhouse gas emissions, Biofuels, Petrodiesel, Transportation, Extraction

[1]Department of Automotive, Mechanical and Manufacturing Engineering, Faculty of Engineering and Applied Science, University of Ontario Institute of Technology, Oshawa L1H 7K4, Canada

**Corresponding author*: Bale.Reddy@uoit.ca

1. INTRODUCTION

Fossil fuels still contribute a significant portion of energy supply for electricity generation, transportation sector, industrial applications and in residential and commercial space heating. There is growing concern on greenhouse gas emissions from fossil fuels use for power generation, industrial applications and transportation sector, residential and in other uses. Tables 1 and 2 shows the emissions from coal, natural gas and oil (IEA, 2017; EIA, 2017). For power generation greater attention has to be paid to develop renewable energy systems and enhance their share in total power generation. Currently for transportation, oil is the major fuel source. There are research investigations and efforts from many companies to develop fuel cells for transportation applications.

Interest in hybrid vehicles is also growing in recent years. Compressed Natural Gas (CNG) is also gaining popularity for transportation in some countries due to its clean burning nature. There is also growing approach to use those fuels that do not contribute to global warming in transportation sector. Much research and development is under way to produce suitable liquid biofuels. These are fuels made from biomass or renewable biological resources.

Table 1: Percentage breakdown of global CO_2 emissions from fuel combustion based on sector, 2014 (IEA, 2017)

Sl. no.	*Sectors*	*CO_2 emissions in %*
1	Electricity & heat	42
2	Transport	23
3	Industry	19
4	Others	7
5	Residential	6
6	Services	3
	Total	100

Table 2: Carbon dioxide emission by fuel in million metric tons (EIA, 2017)

Sl. no.	*Sources*	*2015*	*2016*	*2017 projected*	*2018 projected*
1	Petroleum and other liquid fuels	2295	2320	2321	2349
2	Natural gas	1473	1486	1442	1500
3	Coal	1480	1353	1367	1396
4	Total fossil fuels	5259	5170	5142	5257

The first generation of biofuels are made from plant oils, sugars or starches. The sugars and starches are usually derived from feedstock crops like corn, wheat and sugar beets. These are commonly fermented by enzymes and microorganisms to produce alcohols like ethanol. Another example is pyrolysis of these plant materials to generate butanol. Alcohols like ethanol and butanol can be used as gasoline substitutes. Also a first generation biofuel, through transesterification it is generated from feedstock like animal fats and vegetable oil from oilseeds (Biofuel, 2010).

One disadvantage of first generation biofuel is that fossil fuel energy is generally required to grow, harvest and process the feedstock. Since large areas of farmland are required to produce relatively small amounts of fuel, the process may actually consume more fuel than it produces. Secondly, growing food crops for biofuel feedstock means that farmland and food must be diverted for this purpose.

The second generation biofuel involves generating alcohols or biodiesel from non-food, high cellulose materials like straw and wood. Various methods of accomplishing this are under development. One is high temperature digestion similar to that which occurs in a cow's stomach. Pyrolysis and gasification are additional techniques which can be employed to generate fluid fuel from solid biomass (Biofuel, 2010).

In this chapter the developments and advances in third generation of biofuels, more specifically the production of fuel from algae are presented. The details from a case study on the potential to produce algal biodiesel from power plant exhaust and the impact on greenhouse gas emissions.

2. OVERVIEW OF ALGAE BIOFUEL TECHNOLOGY

2.1. Algae: Characteristics

According to The American Heritage Dictionary, algae are defined as "Any of various chiefly aquatic, eukaryotic, photosynthetic organisms, ranging in size from single-celled forms to the giant kelp (The American Heritage Dictionary, 2000). In other words, commonly known as seaweed, algae are basically a large group of diverse but simple plant-like organisms which usually perform photosynthesis while growing in an aquatic environment (Graham *et al.*, 2009).

Algae occur in a wide variety of body types ranging from large to small and simple to complex. Microalgae have bodies so small that a microscope is required to see them. Many of these exist as single cells and are known as unicellular algae. Algae most commonly occur in the shape of a simple unicellular sphere. Some species arrange themselves in loosely or highly organised colonies. Algae can also occur in long, single cell strands which branch and can attach themselves to surfaces. Species that swim and float in open water are known as phytoplankton. When highly concentrated as in areas with fertilizer runoff colonies become visible and are known as algal blooms. Macroalgae like seaweed are easily visible with the unaided eye. Some consist of one very large cell, while others are formed of three dimensional arrays of cells (Graham *et al.*, 2009).

More than half of algae species perform photosynthesis. This is the process by which the organism utilizes energy from the sun to convert water and carbon dioxide into oxygen and carbohydrates. Different species have ingenious mechanisms which allow them to better perform photosynthesis in certain niche environments. Moreover, it is important to understand the manner in which algae capture energy through photosynthesis.

2.2. Algae as Fuel Feedstock

The major advantage associated with algae is that they can be grown in a wide variety of areas including sewage ponds, the ocean, on marginal landor in special reactor containers. Closed algal production systems have the advantage because these consume much less water than plant crops.

Also, algae are the fastest growing photosynthetic organisms in the world (Schneider, 2006). Research indicates that algae have the potential to produce many times more energy per unit of land area than the best fuel crops. Depending on the species and growth conditions, experts estimate algae to be anywhere from seven to several hundred times more productive per unit of land than other oil crops (Table 3) (Singh and Gu, 2010). Since they grow rapidly and can be harvested on a daily basis, algae provide a continual supply of energy (Sheehan *et al.,* 1998).

The exhaust gases from the fossil fuel power plants can be utilized to grow algae and capture a significant portion of carbon and nitrogen oxides from the exhaust gases, thus preventing them from entering the atmosphere (Demirbas and Demirbas, 2010). This has benefits in the reduction of greenhouse gas emissions and the global warming trend. Algae can also utilize another waste: contaminated water. Algae have high nutritional requirements, which they can meet by purifying sewage or waste water.

Table 3: Bio-oil yield from various renewable sources (Singh and Gu, 2010)

Source	*Oil yield (L/ha yr)*
Corn	172
Soybean	446
Peanut	1,059
Canola	1,190
Rapeseed	1,190
Jatropha	1,892
Karanj	2,590
Coconut	2,689
Oil Palm	5,950
Microalgae (30% oil by weight)	58,700
Microalgae (70% oil by weight)	136,900

2.3. Algae Production Systems

Algae require light, water, carbon dioxide and inorganic nutrients for the growth. These nutrients include nitrates, phosphates, iron and trace elements. The optimum temperature for growth tends to be in the range of 20 to 30°C (Demirbas and Demirbas, 2010). The first and most basic is simply a shallow pond. Waste water or other nutrient rich substances are added and algae are filtered out. One benefit of this system is that it works well to remove nitrogen and phosphorus from sewage ponds. The water is purified and the high level of nutrients allow for fast algal growth. Though inexpensive, it is difficult to efficiently control growth and harvest algae from these ponds (Pittman *et al.*, 2011).

The second system uses a racetrack design. It too is an open pond but it is a specifically designed circuit. Nutrients are continually added at a certain location and the algae are harvested in another one. A water wheel keeps the water circulating which also helps with nutrient mixing and prevents solids from settling (Demirbas and Demirbas, 2010).

The third system is called an algae photo-bioreactor and is a closed system, it also gaining popularity. The water and algae are circulated through an array of clear tubes exposing them to sunlight or artificial light. Nutrients are continually added including a source of concentrated carbon dioxide, example power plant exhaust gases. The growth of algae and algae's consumption reduces the power plant's greenhouse gas emissions. When growing diatom algae in a photo-bioreactor approximately 46 tons of oil can be produced in one year on one hectare of land (Demirbas, 2009).

2.4. Extraction and Processing of Algal Biomass

Once the algal biomass is produced, it must be harvested before it can be converted into fuel. This can be done by filtration, centrifugation, flotation or flocculation. Algal biomass is generally grown with a high oil percentage with the intention of using it as a feedstock for biodiesel since it is well suited for this purpose.

There are many methods for extracting the oil from the algal biomass. The first method is to press the biomass, which will extract most of the oil. The second method involves extraction by chemical solvents; hexane is a popular choice since it's inexpensive. Some researchers are exploring the use of ultrasound to generate tiny bubbles which burst and collapse algal cell walls and release the oil content (Singh and Gu, 2010). The method that currently provides the purest and concentrated oil is supercritical fluid extraction (Demirbas and Demirbas, 2010). The remnants can also be converted to bio char *via* pyrolysis. This sequesters carbon and can improve soil quality (Sohi *et al.*, 2009).

After extraction, the algal oil is in the form of triglycerides which have a very high energy density. However, the viscosity of oil is too high to be directly used in a diesel engine. A common way to convert the oil into lower viscosity biodiesel is through the chemical process of transesterification and the details are reported (Demirbas and Demirbas, 2010).

2.5. Algal Biodiesel

Algal biodiesel have some differences compared to petroleum diesel (Table 4). The algae biodiesel has slightly lower energy density. Biodiesel has lower emissions of unburned hydrocarbons and close to 100% reduction in sulphur oxide emissions. Biodiesel has much lower toxicity and much better biodegradability. The presence of polyunsaturated oils in algal biodiesel implies that it can degrade prematurely. This also gives algae biodiesel lower viscosity and therefore better cold weather performance than biodiesel from other plant sources (Zhang *et al.*, 1998; Chisti, 2007; Demirbas and Demirbas, 2010).

Table 4: Comparison of fuel characteristics for algal biodiesel and petroleum diesel (Zhang *et al.*, 1998; Chisti, 2007; Demirbas and Demirbas, 2010)

	Diesel type	
Fuel characteristic	***Biodiesel from algae***	***Diesel from petroleum***
Energy density (HHV)	~41 MJ/kg	42 MJ/kg
Emission of unburned hydrocarbons	10-33%	100%
NOx emissions	~110%	100%
SOx emissions	0%	100%
Biodegradation after 28 days	77–89%	18%
Density	~0.88 kg/L	0.83 kg/L

Although biodiesel is produced from algal biomass, methane and ethanol can also be derived from algal biomass. The former can be produced by anaerobic digestion and the latter by fermentation or gasification methods. A simple illustration of the input, output and steps involved in the production biofuels from algae are shown in Fig. 1 (Murray, 2010).

2.6. Current Status and Developments

Many countries like Israel already grow algae for food, cosmetic and medicinal purposes, and they are now developing certain algae strains for fuel. The United States though is by far the largest developer of algae fuel technology. Singh and Gu (2010) reported a detailed list of the largest algae fuel projects currently underway, and the majority are located in the USA (Singh and Gu, 2010). Some projects are focused on growing algae in large open ponds using sewage. Others are photo-bioreactors, which use power plant exhaust to supply nutrients. Genetically modified algae with superior oil production characteristics are also currently being developed (Singh and Gu, 2010).

Chen *et al.* (2012) discussed the biodiesel production from algae oil high in free fatty acids by two-step catalytic conversion. In this work they have provided the details on algae cultivation and harvesting, storage of algae paste, extraction of oil from algae paste, biodiesel production from algae oil and purification of crude algae biodiesel. They

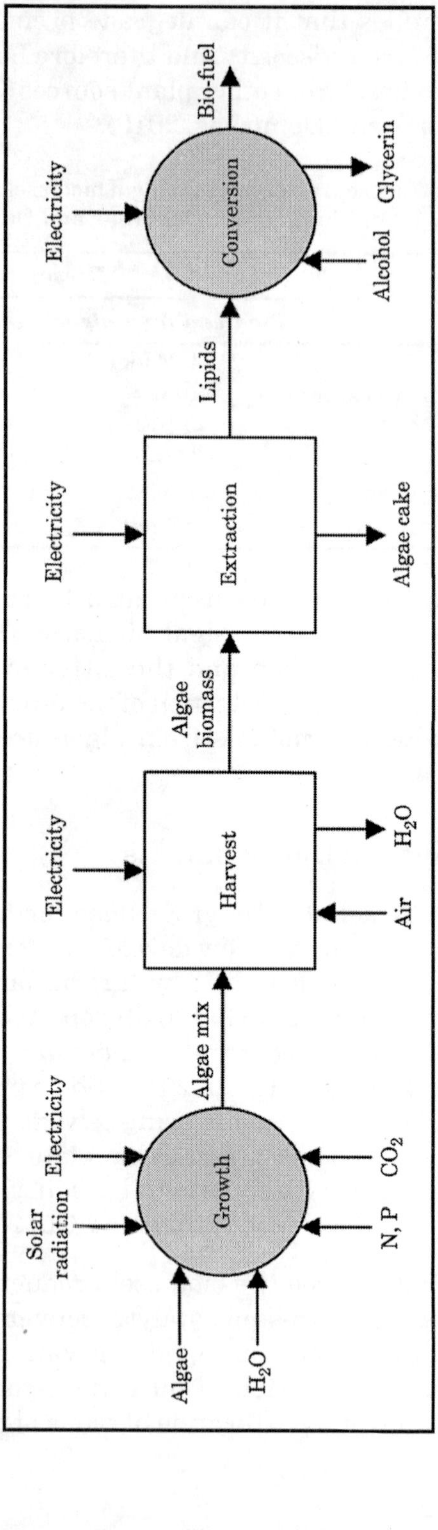

Fig. 1: A simple illustration of the inputs, outputs and steps involved from algae to biofuel conversion (Murray, 2010).

also reported the effects of storage condition on the composition of algae oil and optimization of algae biodiesel production. The proposed two-step catalytic conversion has shown good potential for production of biodiesel from algae oil high in free fatty acid. Milledge and Heaven (2013) reported a review on the harvesting of micro-algae for biofuel production. They also reported details on the comparison of microalgal harvesting methods from published literature. Nautiyal *et al.* (2014a) reported kinetic and thermodynamic studies on biodiesel production from *Spirulina platensis* algae biomass using single stage extraction-transesterification process. In the work they investigated the effects of reaction temperature, stirring intensity, catalyst concentration, algae biomass-methanol ratio and algae drying duration on biodiesel yield. Nautiyal *et al.* (2014b) studied and discussed details on the feasibility of biodiesel production from microalgae as third generation biodiesel feedstock. The investigations were conducted to evaluate the growth patterns of the algae species. From the study, it was reported that maximum biodiesel yield was obtained using simultaneous extraction and transesterification using hexane as a solvent. They also conducted systematic characterization of algae biomass, algae oil and algae biodiesel to establish the potential of microalgae for biodiesel production.

Martin and Grossmann (2014) presented design of an optimal process for enhanced production of bioethanol and biodiesel from algae oil *via* glycerol fermentation. The process details followed and the mathematical modeling of the process are reported in the work. They have optimized the integrated production of bioethanol and biodiesel from algae by using the glycerol to enhance the production of bioethanol. This process shares the dehydration technologies and results in better process, heat and product integration. The fermentation of glycerol to ethanol increases the production of bioethanol by at least 50%. Galadima and Muraza (2014) reported a critical review on the prospects of biodiesel production from algae by using heterogeneous catalysts. It covers the advantages of algae oil exploitation over edible oil feed stocks, progress made in the oil extraction, available heterogeneous catalyst systems and reaction mechanisms, optimum transesterification conditions.

Lowrey *et al.* (2015) presented a critical review on heterotrophic and mixotrophic cultivation of microalgae for biodiesel production in agricultural waste waters and the associated challenges. The review explores the possibility of leveraging the natural ability of the microorganisms to metabolize carbon heterotrophically and mixotrophically in agricultural wastewaters. This has the potential to improve the overall economics of biodiesel production. They have also identified the challenges and research gaps that should be focussed to advance this area. Fatima *et al.* (2016) reported transesterification of oil extracted from freshwater algae for biodiesel production. From the research work, it is observed that a good quantity of biodiesel can be produced from freshwater algae. Furuhashi *et al.* (2016) presented investigations and results on biodiesel and pol-unsaturated fatty acids production from algae and crop plants.

3. CASE STUDY

This case study examines the potential to produce algal biodiesel using fossil fuel power plant exhaust. More specifically, it determines how much algal biodiesel can be produced from a kilogram of exhaust CO_2, and its estimated cost.

3.1. Methodology

To estimate the amount of algal biodiesel that can be produced from power plant exhaust gases, 1 kg of CO_2 is considered for calculations. This is multiplied by the fraction of exhaust CO_2 actually utilized by the algae. More details on the case study and calculations are reported in Hundt and Reddy (2011). The amount of algal biodiesel that can be produced, from one kg of CO_2 from power plant exhaust gases is given by Hundt and Reddy (2011).

$$\begin{aligned} \alpha = {} & emissions\ (kg\ CO_2) \times fraction\ of\ plant\ exhaust\ CO_2\ utilized \\ & \times CO_2\ conversion\ factor \left(\frac{kg\ algae}{kg\ CO_2}\right) \\ & \times oil\ conversion\ factor \left(\frac{kg\ oil}{kg\ algae}\right) \\ & \times biodiesel\ conversion\ factor \left(\frac{kg\ bd}{kg\ oil}\right) \\ & \times diesel\ specific\ volume \left(\frac{L}{kg\ bd}\right) \end{aligned} \tag{1}$$

The potential global capacity to produce algal biodiesel from coal power plants exhaust gases is shown in equation 2. For natural gas power plants, the potential biodiesel production is given by equation 3. The global potential for producing algal biodiesel from power plant exhaust gases β (litres of biodiesel per year) is given by equation 4.

$$\beta c = Global\ Coal\ Electricity\ Production\ \left(\frac{MWh}{yr}\right)$$
$$\times\ Coal\ CO_2\ exhaust\ rate\ \left(\frac{kg\ CO_2}{MWh}\right) \times\ biodiesel\ production\ rate\left(\frac{L\ bd}{kg\ CO_2}\right) \quad (2)$$

$$\beta ng = Global\ NG\ Electricity\ Production\ \left(\frac{MWh}{yr}\right)$$
$$\times\ NG\ CO_2\ exhaust\ rate\ \left(\frac{kg\ CO_2}{MWh}\right) \times\ biodiesel\ production\ rate\left(\frac{L\ bd}{kg\ CO_2}\right) \quad (3)$$

$$\beta total = \beta c + \beta ng \quad (4)$$

The potential to reduce global greenhouse gas emissions γ (%) from algal biodiesel production is given by equation 5.

$$\gamma = diesel\ fuel\ demand\left(\frac{L}{yr}\right) \times\ diesel\ CO_2\ conversion\ ratio\ \left(\frac{\text{kg CO}_2}{\text{L of petrodiesel}}\right)$$
$$\times\ \text{global greenhouse gas emissions}\ \left(\frac{\text{yr}}{\text{kg CO}_2\text{ eq.}}\right) \times 100\% \quad (5)$$

3.2. Assumptions

For the case study, the following assumptions are used to generate the results.

- Algae utilize 66% of exhaust CO_2 (Based on an average of 50% on cloudy days and 82% on sunny days) (Stauffer, 2010).
- 1.83 kg of "utilized" CO_2 manufactures about 1 kg of algal biomass (Chisti, 2007).
- The algae species cultivated yield 50% oil by weight (Demirbas and Demirbas, 2010). All of which is extracted.
- 60% of the mass of algal oil is converted into fatty acids (biodiesel). The remaining 40% consists of glycerides and impurities (Levine *et al.*, 2010).
- Algal oil has a similar density as soybean oil: 0.92 kg/L (Weyer *et al.*, 2010).
- Biodiesel and petrodiesel have mass densities of 0.88 kg/L and 0.83 kg/L respectively.
- Current global petrodiesel fuel consumption is approximately 700 billion litres per year (Chisti, 2007).
- An equivalent amount of greenhouse gases is released in the production of biodiesel and petrodiesel.

- Combustion of 1 litre of petrodiesel emits 2.67 kg of CO_2 (EPA, 2005).
- Natural gas and coal power plants emit 500 kg CO_2/MWh and 950 kg CO_2/MWh respectively (Jaramillo *et al.*, 2007).
- It costs about \$1.40/L to produce algal oil (Chisti, 2007). To be competitive with petroleum diesel this needs to be reduced to about \$0.48/L. The cost of converting bio-oil into biodiesel is approximately \$0.14/L (Demirbas and Demirbas, 2010).

3.3. Results and Discussion

The rate at which algal biodiesel can be generated from power plant exhaust is shown in equation 1. The total potential global production capacity of algae biodiesel from natural gas and coal power plants exhaust gases are shown in eqns. 2 and 3 respectively. Finally, eqn. 5, was used to determine the resulting reduction in greenhouse gas emissions.

The algal biodiesel (Table 5) (Hundt and Reddy, 2011) can be produced at a rate of 0.12 L/kg of exhaust CO_2. This is equivalent to a rate of 60 to 114 L/MWh depending on the plant's fuel type. If all natural gas and coal power plants were utilized, the total global potential to produce algal biodiesel is estimated to be approximately 12 billion litres per year. Only 60% of gas and coal power plants are required in order to replace petrodiesel with algal biodiesel. The result with this approach would be a 4.7% reduction in greenhouse gas emissions.

Table 5: Case study results (Hundt and Reddy, 2011)

	Natural gas power plant	***Coal power plant***
Algal biodiesel production rate	0.12 L/kg CO_2	
	60 L/MWh^2	114 L/MWh
Algal biodiesel production cost	\$1.54/L (\$0.308/kg CO_2)	
Algal biodiesel competitive production cost	\$0.62/L (\$0.124/kg CO_2)	
Global algal biodiesel production capacity (potential)	258 billion L/yr	946 billion L/yr
Portion of coal and NG power plants requiring refit to replace petrodiesel with biodiesel	60%	
Reduction in greenhouse gas emissions if algal biodiesel replaced petrodiesel	4.7%	

From the case study, it is clear that algal biodiesel has enormous potential as a future fuel source. Algae require minimal land resources and have minimal impact on the food supply. They have a rapid growth rate, high oil content and can utilize waste streams for nutrition. If all global coal and gas power plants were harnessed to supply exhaust for photo-bioreactors, significant amount of algal biodiesel could be produced annually. Replacing petrodiesel with algal biodiesel would reduce global greenhouse gas emissions by approximately 4.7%. The case study results are in general in agreement with other studies on the potential of algal fuel technology (Singh and Gu, 2010).

4. CONCLUSIONS

Algae shows great potential as a biofuel source with several major advantages over other energy sources due to the positive characteristics of algae and an abundant supply of power plant exhaust gases, algae can be grown, harvested and converted into biodiesel and other biofuels. This offers potential to reduce global greenhouse gas emissions.

Secondly, algal biofuel production technology is still somewhat new and uneconomical. The efficiency of growth, harvesting and processing systems must be further analyzed and improved to make it more economical.

There is a need to investigate the role of carbon tax on algae biofuel and its role to focus more on fossil fuel emissions to produce algae and to reduce greenhouse gas emissions and global warming.

ACKNOWLEDGEMENTS

The authors are grateful for the financial support of the Natural Sciences and Engineering Research Council of Canada (NSERC).

REFERENCES

Biofuel, Wikipedia. 2010. Available at: *https://en.wikipedia.org/wiki/Biofuel* [Accessed 2 Nov 2010].

Chen, L., Liu, T., Zhang, W., Chen, X. and Wang, J. 2012. Biodiesel production from algae oil high in free fatty acids by two-step catalytic conversion. *Bioresource Technology,* 111(1): 208–214.

Chisti, Y. 2007. Biodiesel from microalgae. *Biotechnology Advances,* 25(3): 294–306.

Demirbas, A. and Demirbas, M.F. 2010. Algae energy: Algae as a new source of biodiesel, NY: Springer-Verlag Ltd. London, U.K.

Demirbas, A.H. 2009. Inexpensive oil and fats feed stocks for production of biodiesel. *Energy Education Science and Technology A,* 23: 1–13.

eia.gov. 2017. Environment - U.S. Energy Information Administration (EIA). [online] *https://www.eia.gov/environment/* [Accessed 1 May 2017].

EPA. 2005. Average carbon dioxide emissions resulting from gasoline and diesel fuel, electronic fact sheet. US environmental protection agency. Available at: *https://www.epa.gov/oms/climate/420f05001.htm.* [Accessed 2 Nov 2010].

Fatima, N., Mahmood, M.S., Hussain, I. and Hafeez, S. 2016. Transesterification of oil extracted from freshwater algae for biodiesel production. *Energy Sources, Part A: Recovery, Utilization, and Environmental Effects,* 38(15): 2306–2311.

Furuhashi, T., Nakamura, T., Fragner, L., Roustan, V. and Schon, V. 2016. Biodiesel and poly-unsaturated fatty acids production from algae and crop plants – A rapid and comprehensive workflow for lipid analysis. *Biotechnology Journal,* 11(1): 1262–1267.

Galadima, A. and Muraza, O. 2014. Biodiesel production from algae by using heterogeneous catalysts: A critical review. *Energy,* 78(1): 72–83.

Graham, L., Graham, J. and Wilcox, L. 2009. *Algae.* 2nd ed., Pearson, Beth Wilbur, Ed. San Francisco, CA.

Hundt, K. and Reddy, B.V. 2011. Algal biodiesel production from power plant exhaust and its potential to replace petrodiesel and reduce greenhouse gas emissions. *International Journal of Low-Carbon Technologies,* 6: 294–298.

iea.org. 2017. CO_2 emissions from fuel combustions. [online] *http://www.iea.org/publications/freepublications/publication/CO2EmissionsfromFuelCombustion_Highlights_2016.pdf* [May 2017].

Jaramillo, P., Griffin, W. and Matthews, H. 2007. Comparative life-cycle air emissions of coal, domestic natural gas, LNG, and SNG for electricity generation. *Environmental Science & Technology,* 41(17): 6290–6296.

Levine, R.B., Pinnarat, T. and Savage, P.E. 2010. Biodiesel production from wet algal biomass through *In situ* lipid hydrolysis and supercritical transesterification. *ACS Energy and Fuels,* 24(9): 5235–5243.

Lowrey, J., Brooks, M.S. and Mcginn, P.J. 2015. Heterotrophic and mixotrophic cultivation of microalgae for biodiesel production in agricultural wastewaters and associated challenges – A critical review. *Applied Phycology,* 27(1): 1485–1498.

Martín, M. and Grossmann, I.E. 2014. Design of an optimal process for enhanced production of bioethanol and biodiesel from algae oil *via* glycerol fermentation. *Applied Energy,* 135(1): 108–114.

Milledge, J.J. and Heaven, S. 2013. A review of the harvesting of micro-algae for biofuel production. *Reviews in Environmental Science and Biotechnology,* 12(1): 165–178.

Murray, M. 2010. UTSA department of geological sciences, Hydro GIS lab. Available at: *http://www.utsa.edu/hydroGIS/RenewableEnergy.html.* [Accessed 2 Nov 2010].

Nautiyal, P., Subramanian, K.A. and Dastidar, M.G. 2014a. Kinetic and thermodynamic studies on biodiesel production from *Spirulina platensis* algae biomass using single stage extraction – transesterification process. *Fuel,* 135(1): 228–234.

Nautiyal, P., Subramanian, K.A. and Dastidar, M.G. 2014b. Production and characterization of biodiesel from algae. *Fuel Processing Technology,* 120(1): 79–88.

Pittman, J.K., Dean, A.P. and Osundeko, O. 2011. The potential of sustainable algal biofuel production using wastewater resources. *Bioresource Technology,* 102(1): 17–25.

Schneider, D. Would the widespread adoption of biomass-derived transportation fuels really help the environment. *American Science,* 94(1): 408–409.

Sheehan, J., Dunahay, T., Benemann, J. and Roessler, P. 1998. A Look Back at the US Department of Energy's Aquatic Species Program—Biodiesel from Algae. *National Renewable Energy Laboratory,* Report NREL/TP-580-24190.

Singh, J. and Gu, S. 2010. Commercialization potential of microalgae for biofuels production. *Renewable and Sustainable Energy Reviews,* 14(9): 2596–2610.

Sohi, S., Loez-Capel, E., Krull, E. and Bol, R. 2009. Biochar, Climate change and soil: A review to guide future research, CSIRO Land and Water Science Report.

Stauffer, N. 2010. Algae system transforms greenhouse emissions into green fuel. MIT energy initiative. Available at: *http://web.mit.edu/erc/spotlights/alg-all.html.* [Accessed 2 Nov 2010].

The American Heritage Dictionary of the English language. 2000. 4th ed. Houghton Mifflin Company.

Weyer, K.M., Bush, D.R. and Darzins, A. 2010. Theoretical maximum algal oil production. *Bioenergy Research,* 3(2): 204–213. Available at: *http://comste.gov.ph/omages/files/theoreticalmaximumfor%20 ALGOIL%206-11-09.pdf* [Accessed 2 Nov 2010].

Zhang, X., Peterson, C., Reece, D., Haws, R. and Moller, G. 1998. Biodegradability of biodiesel in the aquatic environment. *Transactions of the ASAE,* 41(5): 1423–1430.

12

Exploring the Institutional Design of Biofuel Development Across World

Satyendra Nath Mishra[1]*

ABSTRACT

With advent of globalization, international organizations advocated for having global standards for trade to develop normative policy. The emphasis on biofuel as a commodity for global trade to overcome oil pool deficit, environmental crisis, agricultural profitability and rural development has gained momentum in last three decades. However, the scanning of policies across the world has shown that primary motive for biofuel development was not same for participating countries / regions. The biofuel programme implementation showed varied level of control by state, community organizations and private players across world, to meet the social, economic and environmental objectives. In this background, this chapter explores the institutional design of biofuel development across world – Brazil, European Union, United States, Africa and Asia.

Key words: Biofuel, Development, Institution, Policy

[1]School of Rural Management, Xavier University Bhubaneswar, India
**Corresponding author*: saty.nm@gmail.com

1. INTRODUCTION

With advent of globalization, international organizations advocated global standards for trade to develop normative policy. This dominant paradigm has been questioned in last few decades, with increasing emphasis on localization and importance of study on institutions to solve global problems (Ostrom *et al.*, 1999; Andrews, 2013). This change is advocated to address the issues like economic growth, pollution, biodiversity conservation etc., where policies are becoming more flexible and adaptable as per the regional/local requirements (Cline, 2004; Scheffer *et al.,* 2015). The emphasis on biofuel[2] as commodity for global trade to overcome oil pool deficit, decreasing greenhouse gases (GHGs) emission, employment generation and agricultural profitability has gained momentum in last three decades (Gonsalves, 2006; Staley and Bradley, 2008). In this background the dilemma of biofuel development across the world can be put in following words 'is the [bio]fuel driving the economy or the economy driving the [bio]fuel'. Exploring the possible linkages between economy and [bio]fuel sources requires logical scanning of political economy of country/region. This is pertinent as world faces various environmental issues and uncertainty of fossil fuel availability due to geo-political reasons (Amirahmadi, 1997; Munk, 2005; Nathan, Kulkarni and Ahuja, 2013). Climate change, intermittent oil crisis due to varied reasons and regional natural resource management strategies are playing pivotal role in acceptance of renewable energy options and developing appropriate technologies. This has led to the scaling-up, investment and advancement in all fronts of renewable energy options and its uses. It is important to note that magnitude of energy consumption is one of the important (measurable) variables of human development (Goldemberg *et al.*, 1985, 1988; Cabraal *et al.*, 2005), in which issues like type of energy sources; location of production; mode of production and type of users and uses, should be included for better assessment.

The production and use of biofuel have been controversial for policy makers and scholars across domain and disciplines (Kher, 2005; Maheshwari, 2008). This is due to the allocation of resources, like [waste]land and water, against food production (Braun, 2008; Mitchell, 2008). Studies have shown that biofuels produced on wasteland, marginal land and on low input grassland could reduce GHG emissions,

2 Biofuel is used for bioethanol and biodiesel in recent literatures, excluding the biomass based biofuel mostly used in rural areas of developing countries.

can be economically and ecologically viable in long run (Hill *et al.*, 2006; Tilman, Hill and Lehman, 2006). Brazil and United States (US) account for about 87 percent of world bioethanol production by using sugarcane and corn respectively. US, European Union (EU) – dominated by Germany and France – and Brazil accounts for about 80 percent of biodiesel production of world with the use of soyabeans, used rapeseed oil, cooking oil and sweet sorghum (Deurwaarder, 2005; REN, 2016). The biofuel development in countries/regions like, Brazil, US, EU and Asian countries got thrust either from one or/and more of these factors, *viz.*, i) the oil crisis of 1970s, ii) the need to develop new avenues and markets for agricultural outputs and iii) environmental concerns (Demeritt, 2001; Rajagopal and Zilberman, 2007). However, the scanning of policies showed that primary motive for biofuel development was not same for all these countries/regions. In this background, the chapter attempts to understand the institutional design of biofuel development across world – Brazil, US, EU, Asia, Africa and India – by exploring the political economy of country/region.

2. BRAZIL

In 1975, after the petroleum oil shock, the Brazilian government launched a programme named 'Pro-Alcohol', to promote the sugarcane based ethanol. This was aimed at restricting the high oil import bill, negative growth and the high rate of inflation. The programme gave push to rural employment by revitalizing the sugar industry. PETROBRAS, the state-owned petroleum oil company played a proactive role in developing the supply chain of ethanol from farm to the pump. Small farmers who produced about 30 percent of total sugarcane produce were given small-on-farm ethanol distilleries and linked with industries for ethanol procurement. Favorable agro-climatic condition of Sau Paulo for sugarcane cultivation proved boon for the Brazil. The Sau Paulo region accounts for more than 70 percent of sugarcane and ethanol production (Abramovay, 2008). By 2016, sugarcane based ethanol constituted 36 percent of Brazil's total energy matrix, mostly fueling transport and industrial sector (Freitas, 2016). The success of Brazil is an exemplary case for the world in general and in particular for developing countries. Staley and Bradley (2008) stated that the success of the ethanol programme in Brazil can be attributed to the adjustment between the state, industries and technology development as per country's own energy requirements. This was in synergy with sugar industry, rather than being pushed by the international development organizations. During the initial phase of the Pro-Alcohol programme, the Brazilian government gave tax breaks and social labels to sugarcane industry for the purchase of raw material from small farms (Abramovay and Magalhaes, 2007). This led to more inclusive development of sugarcane cultivation and self-sustaining energy matrix of Brazil. By the end of 2016, Brazil was expected to produce about 30.68 billion litre of ethanol. The ethanol production was supported by about 370 sugar industries, with the average production of 1.2 million tons per year. During 2015 the ethanol blend in gasoline was in the range of 25 to 27 percent (Barros, 2015). However, this blending percent varied based on the price of gasoline, yearly sugarcane crop cultivation and ethanol production. It is to be noted that the ethanol programme in Brazil is not out of scientific investigation and political debate, as it has raised the problem of increasing land concentration, issue of monoculture and huge subsidy to industry by State (Rothkopf, 2007).

3. EUROPEAN UNION (EU)

The EU initiative on biofuel production came in 1992, with reform in the 'Common Agricultural Policy' (CAP). Under CAP farmers could grow non-food crops on land set aside for income generation which was not under agricultural use (Henniges and Zeddies, 2006; Kojima, Mitchell and Ward, 2007). In May 2003, the European Parliament and the Council adopted the 'directive on the promotion of the use of biofuels or other renewable fuels for transport'. The directive stated that the member states should replace minimum of 5.75 and 10 percent of the transport fuel requirement with biofuel by 2010 and 2020 respectively (Deurwaarder, 2005; Laak *et al.*, 2007). The objective for biofuel promotion was to i) sustain the farming activities, ii) reduce the cost of farm support policies iii) diversify energy supplies and iv) create a value-added product and employment in countryside. The issues of reducing the GHG emissions, net production cost and energy efficiency of the biofuel crops remained in controversies. In spite of that, biofuel production gained momentum in Europe as farmers were benefited, *via*, i) revised CAP allowed farmers to grow energy crops on set aside land, which otherwise would have been fallow, with ensured earning of about €100 to €500 per hectare ii) the dung from high density animal husbandry activities was profitable now, as the dung was used as manure for biofuel cultivation, iii) a premium €45 was paid per hectare for assisting energy crop cultivation and iv) biofuel production led to higher prices for agricultural commodities due to requirement of feedstock (Henniges and Zeddies, 2006; Kojima *et al.*, 2007). However, in 2007 the EU revised the CAP guidelines for subsidy on 'set aside' agriculture land in wake of increasing food prices (Waterfield, 2007). The increase in the global food price in 2012 was attributed to the increasing demand of biofuel from EU (Kelly, 2012). Due to increasing food prices and environmental concerns, the EU reduced the target of mixing 10 percent biofuel by 2020 for transport purpose to 5 percent, from raw material like wheat, rapeseed or soya (Bauwens, 2012)[3]. The guidelines were again revised by the European Parliament in April 2015. The revised Renewable Energy Directive (RED) stated that by 2020 up to 7 percent of final energy consumption should be replaced by first generation biofuel, *viz.*, biofuel crop grown in agriculture land (Flach *et al.*, 2015).

4. UNITED STATES (US)

In US the ethanol industry is protected from the foreign competition (especially Brazil's ethanol) since 1980 under Ethanol Import Tariff. In addition, US government promoted use of ethanol, *via*, tax credit under Energy Tax Act 1978 until 2011. About \$0.54 per gallon of tariff is levied on imported ethanol. The issues of subsidies, tax credit, vehicle norms and regulation about the US biofuel programme are more complex than other countries. This was due to the nature of raw material – corn – used for ethanol production. Corn is an agriculture crop used as staple food. The post production tax credits and relaxed fuel efficiency for flexed vehicle is more favorable in US than in other countries (Staley and Bradley, 2008). About 95 percent of ethanol was produced from corn which was heavily subsided. From 1995 to 2010 corn cultivation received subsidies of around \$89 billion, against \$38 and \$34 billion for wheat and cotton respectively[4]. US government policies support the ethanol industry from both supply and demand side. On supply side the agricultural subsidies led to low cost for feedstock. On demand side, the mandatory blending and tax incentives on ethanol consumption enhances the sales in open market. The US Energy Policy Act of 2005 gave direct support to ethanol programme through Volumetric Ethanol Excise Tax Credit (VEETC) and tax credit of \$0.51 per gallon of ethanol

---

[3] Accessed from <http://blogs.nature.com/news/2013/09/european-parliament-votes-to-limit-crop-based-biofuels.html> on 06th October 2015.

[4] Accessed from <http://farm.ewg.org/progdetail.php?fips=00000&progcode=corn> on 12th December 2014.

mixed with gasoline (Koplow, 2007; Staley and Bradley, 2008). US Farm Bill 2008 increased the support for biofuel programme from $2.5 billion in 2002 to $7.14 billion in 2008. Section IX on Energy of US Farm Bill 2008 supported 'Biomass Crop Assistance Program', which provided 75 percent support for establishing an eligible bioenergy crop. Support was extended (for two years for collection, harvest, storage, and transportation of raw material to a biomass conversion facility) in terms of annual payments for production and the matching payments of up to $45 per ton[5]. The 2007 Energy Independence and Security Act passed by US Congress increased the investment for ethanol production. By 2015 under the Renewable Fuel Standard (RFS) the US was required to produce 57 billion litres of ethanol for transportation fleet (Carter and Schaefer, 2015; Elliott, 2015). Concerns were raised about the relation between use of corn as feedstock and rising food prices in the international market, which hit the poor and the poorest of world most (Babcock, 2008; Braun, 2008). On environment side the potential of corn based ethanol to reduce the GHG emission has been in debate and questioned (Tilman *et al.*, 2006; Searchinger *et al.*, 2008). This protectionist policy of US government and trade options for biofuel industry indicates the mandate for making the agriculture and related bioenergy production more viable and financially profitable, irrespective of environmental concerns.

5. AFRICA AND ASIA

The role of developed countries has been controversial when it comes to extracting energy resources from developing countries (Margonelli, 2008). The development of biofuel programme in African and Asian countries have been no different. There are controversies over large scale *versus* small scale biofuel programme design to fulfill the requirement of industries *versus* local economy respectively (Eijck and Romijn, 2008). The structure of financial support from the international organization for biofuel programme provide less scope to design processes as per local requirement. The biofuel story had mixed outcome with less of success and more of failure and controversies in this region. Few small-scale projects in African countries, like in Mali, Mozambique, Burkina Faso, Ghana, Guinea and Senegal, ran successfully due to following reasons, *viz.*, multipurpose approach of biofuel project for

---

[5] Accessed from <http://www.ers.usda.gov/FarmBill/2008/Titles/TitleIXEnergy.htm#biobased> on 11th March 2015.

energification[6] rather than electrification only, the participation of local communities, women's participation, small land requirement of about 10 to 150 ha with catchment of 20 to 50 square kilometres, decentralized approach and multiple stakeholders (Henning, 2000; Jongh, 2006). Small scale success story on local innovation and the use of biodiesel from *Jatropha* and Karanja was also observed in India like, Powerguda and Chaloadi villages in Adilabad district of Andhra Pradesh, Gardih village in Bokaro district of Jharkhand, Gudiyatham in Vellore district of Tamil Nadu and Ranidahara village in Chhattisgarh (WII, 2009, 2010). The commonalities of these success stories have been fulfilling requirement of local economy with aid of supportive technology and people's participation. The implementation of large scale projects in Africa and Asia was marked with i) targeted production of biofuel for transport fuel ii) coercive policy of the state and industries for land consolidation and other resource mobilization and iii) large scale displacement of people (Cotula *et al.*, 2008). Biofuel project in Indonesia and Colombia faced the problem of land consolidation, land allocation, land tenures in addition to social acceptance by the local populace (Peskett *et al.*, 2007; Rothkopf, 2007). The conflict over land rights of local people and the access to resources are important questions for biofuel programme implementation in Africa and Asia (Cotula *et al.*, 2008).

5.1. Case of India

In 2003 the Government of India (GoI) published the 'Report of the Committee on Development of Biofuel', to promote biofuel (ethanol production from sugarcane and biodiesel from *Jatropha* and Karanja) on wasteland. The biodiesel programme caught the attention of stakeholders due to the requirement of resources from rural India, *viz.*, wasteland, manpower for employment and other related input, with emphasis on fulfilling the fuel requirement of transport sector (Planning Commission, 2003). The rationale given for programme implementation was biofuel's potential role in providing an alternative to the petroleum fuel, checking GHG emissions and generating employment in rural areas (Kher, 2005; Bhojvaid 2007). The *Jatropha* based biofuel programme was implemented in two-phases. First phase involved demonstration project (from 2003 to 2007) for plantation of *Jatropha* on wasteland, through Joint Forest Management (JFM) and Non-JFM approach, across

[6] The focus is on providing energy for cooking, irrigation, lighting, heating etc, unlike just focusing on lighting.

eight compact areas[7]. The second phase (from 2007 to 2012) envisioned that villagers would take up the plantation work at community level, with institutional and financial support of GoI and state government[8]. To tap the resources (fund, land, administrative support etc.) from GoI, state government and other national/international donor organizations many states (like Chhattisgarh, Madhya Pradesh, Rajasthan, and Uttarakhand) constituted separate biofuel development board[9].

In 2008 the GoI approved the National Policy on Biofuel to expedite the programme implementation. The National Policy on Biofuel targeted blending of biofuel (ethanol and bio-diesel) up to 20 percent by 2017. The Minimum Purchase Price (MPP) of bio-ethanol was based on the actual cost of production and import price of the bio-ethanol. For biodiesel, the MPP was linked with the prevailing market price of diesel. The policy envisaged that biofuel may be brought under the ambit of 'Declared Goods', by the government to ensure unrestricted movement within and outside different states of India[10]. The MPP for biofuel was a welcome step for biofuel crop grower for diversifying the revenue generation options. On other hand the unrestricted movement of biofuel may open it to be influenced by market fluctuation and defy the welfare goal as stated in policy for rural employment, agricultural productivity and energification in the long run.

---

[7] Each compact area is a group of districts covering 50,000 to 60,000 hectares of wasteland for Jatropha plantation and establishing forward and backward linkage for biodiesel production and use. The four JFM based compact areas were in the states of Gujarat, Chhattisgarh, Tamil Nadu and Tripura. The four Non-JFM based compact areas were in the state of Uttar Pradesh, Madhya Pradesh, Maharashtra and Andhra Pradesh (Planning Commission 2003).

[8] In 2010 The Energy and Resources Institute carried out details study regarding present status of Jatropha plantation in nine states of India (Andhra Pradesh, Chhattisgarh, Karnataka, Madhya Pradesh, Orissa, Rajasthan, Tamil Nadu, Uttarakhand and Uttar Pradesh). The executive summary of the report can be accessed from *<http:/ /www.teriin. org/index.php?option=com_completed&task =details&pcode=2009CM04>*. Based on the report, Ministry of Rural Development, GoI, had put on hold the Jatropha plantation programme. Accessed from <http://dailypioneer.com/nation/66165-jairam-has-no-faith-in-jatropha-biofuel.html> on 04th December 2013.

[9] For further details on policy implementation across different State of India read chapter title 'Biofuel Policy in India: Explicating Policies and Strategies across India States', by me Mishra (2018).

[10] Accessed from <http://pib.nic.in/release/release.asp?relid=42733> on 12th September 2008.

The programme received mixed reaction for selection of feedstock and promotion of plantation on the wastelands. It is to be noted that planning commission report emphasized that local institutions would be given responsibility for the resources planning and development through i) the involvement of Gram Panchayat (GP) ii) providing right of resources to the local user and iii) priority for local energy use and self-sufficiency (Planning Commission, 2003). However, the report remained silent on how to mobilize and use the local/regional resources for the local/regional energification process. In contract, the financial and environmental standards stated in the report were for transport sector, showing scalar approach of the state resource mobilization for Research and Development (R&D) and biodiesel production. Biodiesel Purchase Policy (MoP and NG, 2005) gave rights to the state-owned oil marketing company (OMC) for purchase of B100 (with effect from 2006) from local producers and entrepreneurs at INR 25 per litre. The policy document raised concerns about large scale land ownerships by big OMC and private industries. However, no measures were stipulated to check the land consolidation and to empower local institutions (like GP, community group etc.) or Rural Business Hubs. The report can be criticized for overlooking the R&D required for identifying the suitability of bioenergy crop varieties for different agro-climatic regions of India. The target of cultivating 65 million hectares wasteland with biofuel crop was criticized on the ground of being an encroachment to the commons of rural communities, who depend on it for livelihood and for various daily requirements (SPWD, 2007).

6. CONCLUSIONS: INSTITUTIONAL DESIGN OF BIOFUEL DEVELOPMENT

Institutions are humanly devised informal and formal constraint that structures the social, political and economic interaction, to create order and reduce uncertainty in exchange (North, 1991). The biofuel programme implementation showed varied level of control by state, community organizations and private players across world, to meet the social, economic and environmental objectives. Within the above defined boundary this section explored the design aspect of institutions across major biofuel producing countries/regions of world (Table 1).

The biofuel programme of EU and US emphasized on i) centralized production approach ii) use of large productive agriculture land and ii) huge [non]farm subsidies by government. This approach indicates

Table 1: The institutional design of biofuel programme across the world

Country/ Region	***Politico-economic condition***	***Policy design***	***Major focus of programme***	***Ongoing (possible) implications***	
				Pros	***Cons***
Brazil	Petroleum oil shock, high import bill, negative growth, high inflation	Sugarcane cultivation by farmers and industries, institutional support for small farmers, farm-industry linkage.	Sugarcane based ethanol production, cutting import bill, rural employment.	Modernization of sugar industry, skilled job, environmentally and economically compatible ethanol production, self-sufficiency in vehicular fuel need.	Land concentration by industries, issue of monoculture, subsidy to industry.
EU	Reducing dependence on petroleum fuel, reducing GHGs emission, agricultural profitability.	Reform in CAP in 2002, promotion of biofuel based agriculture on set aside agriculture land.	Use of productive agriculture land, agricultural profitability, reducing cost of farm support fund, diversify energy supplies, value added [by]product and employment generation	Increased farm income, production of biodiesel.	Question on achieving set environmental standard, possible implication on higher price of food produce.
US	Petroleum oil shock, increasing transport fuel demand, decreasing dependence on petroleum import.	Corn based ethanol production, support to big farms and subsidies, incentive for using blended ethanol in	Bulk ethanol production, protection to domestic ethanol industry from Brazil.	Ethanol as substitute for gasoline, higher corn prices for farmers.	Question on environmental and economic viability of ethanol production from corn, possible implication on

Table 1: (*Contd...*)

Table 1: (*Contd...*)

Country/ Region	*Politico-economic condition*	*Policy design*	*Major focus of programme*	*Ongoing (possible) implications* Pros	Cons
		vehicle.			world food (corn) price
African continent	Reducing oil dependency, rural energy option, rural employment and income generation, aid based programme design.	Privately acquired farms by multinational companies, community farming at experimental level(for diversified energy use at local level)	Biofuel production by multi-national companies for export, experimental programme, to fulfill rural energy need.	Employment generation, experimental success at small level.	Clash of interest between local people and industries on local resource use like land, water etc.
India (*Jatropha* based biodiesel programme)	Reducing petroleum oil dependency, rural development and employment.	Plantation on wasteland, field bunds, road side plantation; creation of rural business hub, OMC support for biofuel procurement for transport use.	Right of resource use to local institutions, fulfilling rural energy need, employment generation, biodiesel mix for use in vehicle.	Employment generation, experimental success at small level for local energy use.	Encroachment on commons, lack of support to local institutions, lack of relevant R&D.

Source: Compiled by author

market oriented protectionist trade policy, with profit motive for [large] farmers/industries and fulfilling the fuel requirement of domestic transport sector. Brazil adapted a mixed approach with decentralize way of acquiring raw material and centralized way of final processing to produce ethanol. In Brazil 70 percent of land under sugarcane cultivation was owned by the 340 industries, while rest of the 30 percent owned by about 60,000 farmers having average land holding of about 27.5 ha (Cotula *et al.,* 2008). Brazilian government support in form of subsidies to sugar industry was gradually reduced and now removed. Over the year, Brazil's ethanol programme led to two-pronged social change. First, the technological innovations led to the expansion in skilled job opportunities in sugar industries, with positive spin off across regions and many sectors. Second, it led to the concentration of land and wealth in hands of few industries and corporations (Abramovay, 2008). To overcome this design flaw of ethanol programme, the National Programme for the Production and Use of Biodiesel (PNPB) of Brazil charted out steps for the inclusion of small and low income farmers. This was done by setting up of favorable institutional arrangements, subsidies and capacity building, in the supply chain of biodiesel. This indicates how biofuel programme matured, learned and unlearned in Brazil, for more inclusive growth to suit domestic socio-technical and economic environment. In African and Asian context, the approach of small verses large scale production of biofuel is becoming a cause of rift between the local people, industries and the State. This is due to skewed land to people ratio, agriculture been labour intensive and dependence of people on commons for not only income, but also for food, fuel and fodder to larger extent. The issues of undefined property rights of land (Gopalakrishnan, 2012) are only going to aggravate this problem.

The promise for pro-poor development depends on how the [local] resources and the actors, who control and access it get involved in the programme implementation (Brittaine and Lutaladio, 2010). Due to this, the potential implication of biofuel programme would be influenced by contextual factors. This includes role of actors (international donor and environmental agencies, national government, role and right of local government and agencies), the allocation of input resources (type of land, structure of fund allocation, labor force, water etc.), approach of use of [by]products (for local, regional, or for international trade), issues of ownership of resources, control of resources production, processing and uses (by state, private agencies, community or hybrid of different design). The diversified institutional design based on control of state/market/community (or different combination of it) would have differential influence and implication on achieving the goal of [rural] development particularly in African and Asian context and environmental issues across world. In this context, we must be conscious of how in future the policy will change over period, factors influencing the programme implementation and its potential implication. Above all, it is the policy design of respective governments that will have deciding influence on scope and scale of biofuel programme for the development of respective country/region.

ACKNOWLEDGEMENTS

The manuscript is largely based on the Fellow Programme in Rural Management (Equivalent to PhD) thesis work done by author at Institute of Rural Management Anand (IRMA), Gujarat, India. The author is thankful to the Reserve Bank of India for funding his study at IRMA. An early version of this chapter was published in Journal of Rural Development.

REFERENCES

Abramovay, R. 2008. A political-cultural approach to the biofuels market in Brazil." São Paulo: University of Sao Paulo (2008). Accessed from <*http://www.econ.fea.usp.br/abramovay/artigos_jornal/2008/Abramovay_Ethanol_biodiesel_Brazial_political_cultural_approach.pdf*> on 10th March 2009.

Abramovay, R. and Magalhães, R. 2007. The access of family farmers to biodiesel markets: partnerships between big companies and social movements, Department of Economics and Environmental Program of University of São Paulo: Sao Paulo. Accessed from <*http://pubs.iied.org/pdfs/G03239.pdf*> on 27th March 2011.

Amirahmadi, H. 1997. World oil and geopolitics to the year 2010. *In*: *Fuel and Energy Abstracts,* 1(38): 1–12.

Andrews, M. 2013. The limits of institutional reform in development: Changing rules for realistic solutions. Cambridge University Press, p. 254.

Babcock, B.A. 2008. Breaking the link between food and biofuels. Center for Agricultural and Rural Development: Iowa State University, pp. 1–11.

Barros, S. 2016. Brazil: Biofuel Annual. GAIN Report Number: BR15006. Accessed from <*https://gain.fas.usda.gov/Recent%20GAIN%20Publications/Biofuels%20Annual_Sao%20Paulo%20ATO_Brazil_8-4-2015.pdf*> dated 13 December 2016.

Bauwens, D. 2012. EU Cap 'Only Boosts Biofuels, Accessed from < *http://www.ipsnews.net/2012/09/eu-cap-only-boosts-biofuels/*> on 12th December 2012.

Bhojvaid, P.P. 2006. Biofuels: Towards a greener and secure energy future. New Delhi: TERI Press, p. 281.

Braun, J.V. 2008. Biofuels, international food prices, and the poor. International Food Policy Research Institute: Rome.

Brittaine, R. and Lutaladio, N. 2010. *Jatropha*: *A small holder bioenergy crop. The potential for pro-poor development*. Integrated Crop Management, Vol. 8. Rome: FAO.

Cabraal, R.A., Barnes, D.F. and Agrawal, S.G. 2005. Productive uses of energy for rural development. *Annual Review of Environmental Resources,* 30: 117–144.

Carter, C.A. and Schaefer, K.A. 2015. US biofuels policy, global food prices and international trade obligations. Economic Perspective: American Enterprise

Institute. Accessed from <*https://www.aei.org/wp-content/uploads/2015/05/US-biofuels-policy.pdf*> dated 20 November 2016.

Cline, W. 2004. Trade policy and global poverty. Washington, DC: Center for Global Development and Institute for International Economics.

Cotula, L., Dyer, N. and Vermeulen, S. 2008. Fuelling Exclusion? The Biofuels Boom and Poor People's Access to Land. London: IIED. p. 72.

Demeritt, D. 2001. The construction of global warming and the politics of science. *Annals of the Association of American Geographers,* 91(2): 307–337.

Deurwaarder, E.P. 2005. Overview and analysis of national reports on the EU biofuel Directive: Prospects and barriers for 2005. Energieonderzoek Centrum Nederland.

Eijck, J.V. and Romijn, H. 2008. Prospects for *Jatropha* biofuels in Tanzania: An analysis with strategic niche management. *Energy Policy,* 36(1): 311–325.

Elliott, K.N. 2015. The time to reform US biofuel policy is now. Center for Global Government: Washington. Accessed from < *http://www.cgdev.org/sites/default/files/Biofuels-Final.pdf*> 16th November 2016.

Flach *et al.*, 2015. EU: Biofuel Annual 2015. GAIN Report Number: NL5028. Accessed from < *https://gain.fas.usda.gov/Recent%20GAIN%20Publications/Biofuels%20Annual_The%20Hague_EU-28_7-15-2015.pdf*> 10th November 2016.

Freitas, G. 2016. Sugarcane fuels win in Brazil as cheap ethanol beats gasoline. Accessed from < *https://www.bloomberg.com/news/articles/2016-03-01/sugar-cane-fuel-wins-in-brazil-as-cheap-ethanol-beats-gasoline*>, 13th October 2016.

Goldemberg, J., Johansson, T.B., Reddy, A.K.N. and Williams, R.H. 1985. Basic needs and much more with one kilowatt per capita. *Ambio.,* 14(4–5): 190–200.

Goldemberg, J., Johansson, T.B., Reddy, A.K.N. and Williams, R.H. 1988. Energy for Development. Washington: World Resource Institute.

Gonsalves, J.B. 2006. An assessment of the biofuels industry in India. Geneva: United Nation Conference on Trade and Development.

Gopalakrishnan, S. 2012. Undemocratic and Arbitrary: Control, Regulation and Expropriation of India's Forest and Common Lands. New Delhi: SPWD.

Henniges, O., and Zeddies, J. 2006. Bioenergy in Europe: experiences and prospects'. *In*: Hazell, P. and Pachauri, R.K. (*eds.*), Bioenergy and agriculture: Promises and challenges. International Food Policy Research Institute: Washigton DC.

Henning, R.K. 2000. Use of *Jatropha curcas* oil as raw material and fuel: An integrated approach to create income and supply energy for rural development (Experience of *Jatropha* project in Mali, West Africa). Presented at the international meeting on "Renewable energy – A vehicle for local development-II" at Denmark in August 2000. Accessed from <*http://www.angelfire.com/ks3/go_diesel/files042803/jathropa2.pdf*> on 3rd December 2012.

Hill, J., Nelson, E., Tilman, D., Polasky, S. and Tiffany, D. 2006. "Environmental, economic and energetic costs and benefits of biodiesel and ethanol biofuels. *Proceeding of National Academy of Science,* 103(30): 11206–11210.

Jongh, D.J. 2006. State of art *Jatropha* Development Mozambique, Arrakis, Veldhoven, Netherlands. Accessed from <*http://www.arrakis.nl/reports/State%20of%20Art%20Jatropha%20Development%20Mozambique.pdf*> on 10th February 2009.

Kelly, R. 2012. The hunger grains: The fight is on, time to scrap EU biofuel mandates. Oxford: Oxfam Briefing Paper. p. 161.

Kher, R. 2005. Biofuels: The Way Ahead. *Economic and Political Weekly,* 40(51): 5376–5378.

Kojima, M., Mitchell, D. and Ward, W. 2007. Considering trade policies for liquid biofuels, Energy Sector Management Assistance Program. Washington DC: World Bank.

Koplow, D.N. 2007. Biofuels, at what Cost? Government Support for Ethanol and Biodiesel in the United States: 2007, Canada: International Institute for Sustainable Development.

Laak Van der, W.W.M., Raven, R.P.J.M. and Verbong, G.P.J. 2007. Strategic niche management for biofuels: Analysing past experiments for developing new biofuel policies. *Energy Policy,* 35(6): 3213–3225.

Maheshwari, R. 2008. The science behind the biofuel controversy. *Current Science,* 95(5): 594–602.

Margonelli, L. 2008. Oil on the Brain: Petroleum's long, strange trip to your tank. New York: Broadway Books. p. 352.

Mishra, S.N. 2018. Biofuel Policy in India: Explicating Policies and Strategies Across India States. In: Paul, A.K., Bhattacharyya, D.K. and Anand, S. (*eds.*), Handbook of Research on Strategic Green Business Practices', Pennsylvania: IGI Global.

Mitchell, D. 2008. A note on rising food prices. Washington DC: World Bank Policy Research Working Paper Series.

MoPNG. 2006. Bio-diesel Purchase Policy. New Delhi: Government of India.

Munk, B.E. 2005. The End of Cheap Oil, Once Again: Geopolitics or Global Economics? Accessed from <*http://www.ecomentary.com/Newsletters/The%20End%20of%20Cheap%20Oil%20Once%20Again.pdf*> on 10 October 2016.

Nathan, H.S.K., Kulkarni, S.S. and Ahuja, D.R. 2013. Pipeline politics – A study of India's proposed cross border gas projects. *Energy Policy,* 62: 145–156.

North, D.C. 1991. Institutions. *The Journal of Economic Perspectives,* 5(1): 97–112.

Ostrom *et al.* 1999. Revisiting the Commons: Local Lessons, Global Challenges. *Science*, 284(5412): 278–282.

Peskett, L., Slater, R., Stevens, C. and Dufey, A. 2007. Biofuels, agriculture and poverty reduction. *Natural Resource Perspective,* 107: 1–6.

Planning Commission 2003. Report of the Committee on Development of Biofuel. New Delhi: GoI.

Rajagopal, D. and Zilberman, D. 2007. Review of Environmental, Economic and Policy Aspects of Biofuels, Washington DC: World Bank Policy Research Working Paper 4341.

REN. 2016. Renewables-Global Status Report 2016. Paris: Renewable Energy Policy Network for the 21 Century.

Rothkopf, G. 2007. A blueprint for green energy in the Americas: Strategic analysis of opportunities for Brazil and the Hemisphere. Washington DC: Inter-American Development Bank.

Scheffer *et al.* 2015. Creating a safe operating space for iconic ecosystems. *Science,* 347(6228): 1317–1319.

Searchinger *et al.* 2008. Use of US croplands for biofuels increases greenhouse gases through emissions from land-use change. *Science*, 319(5867): 1238–1240.

SPWD (Society for Promotion of Wastelands Development) 2007. Allocation of wastelands for bio-fuel plantation in Rajasthan. Jaipur: Institute of Development Study. p. 36.

Staley, B.C. and Bradley, R. 2008. Plants at the Pump: Reviewing Biofuels' Impacts and Policy Recommendations. Washtington DC: World Resources Institute. p. 52.

Tilman, D., Hill, J. and Lehman, C. 2006. Carbon-negative biofuels from low-input high-diversity grassland biomass. *Science*, 314(5805): 1598–1600.

Waterfield, B. 2007. Set aside subsidy halted to cut grain price. Accessed from < *http://www.telegraph.co.uk/news/uknews/1564327/Set-aside-subsidy-halted-to-cut-grain-prices.html*> on 12th December 2012.

WII-Winrock International India. 2009. Sixth International Conference on Biofuels: Conference Proceedings. New Delhi: WII.

WII-Winrock International India. 2010. Seventh International Conference on Biofuels: Conference Proceedings. New Delhi: WII.

13

Human Rights Implications of Biofuel Production

Tarini Mehta[1*]

ABSTRACT

Biofuel production is being promoted as a viable alternative to fossil fuels that can bring about rural empowerment and economic development for developing countries. However, this one-sided representation ignores important facts pertaining to human rights violations caused by the industry. This chapter focuses primarily on violations of the human rights to food, land and water, with the aim of bringing attention to the social and environmental implications of the biofuels industry.

Key words: Biofuels, Right to food, Food insecurity, Right to water, Water pollution, Depletion of natural resources, Land rights, Land grabbing, Forced evictions, Participation, Consultation, Human rights, Women's rights, Indigenous rights.

[1]Advocate, Environmental and Human Rights Law
Corresponding author: mehtatarini@yahoo.com

1. INTRODUCTION

Nearly a century ago Henry Ford predicted that biofuels would be *"the fuel of the future"*.[1] Global needs and demands have proved him right. The use of biofuels as a replacement for fossil fuels is fast moving into the mainstream and has become an important part of agricultural production in several countries, aided by supportive government policies and an increase in market demands. Projections suggest that demand for biofuels will increase over the coming decades, especially in the context of their possible role in addressing energy and climate concerns. Access to affordable energy can also play an important role in the eradication of poverty and hunger, improvement of basic health-care services and enhancement of the rights to education, livelihood, standard of living and food. While biofuel production has the potential to contribute to rural development and the Millennium Development Goals, its adverse impacts cannot be ignored. Various forms of environmental damage, social injustice and human rights violations have been identified that give rise to degradation of natural resources and pose risks to biodiversity, food security, land rights and availability and access to water. The direction that the industry will take depends on legal safeguards put in place, nationally and internationally, to ensure that biofuel production remains sustainable and socially beneficial. This chapter presents the less discussed impacts of biofuel production on land, food and water resources, that impact the poor, marginalized and under-represented the worst; people whose voices are often drowned in the clamour for more energy to fuel the engine of economic development. Taking international human rights law as a framework for analysis, it focuses on the rights to food, land and water, which at times offer some form of protection and at others leave people open to exploitation.

2. FOOD VERSUS FUEL

Rapid increase in global population is resulting in greater demand for food, while the need for alternative fuel sources is giving rise to increasing demand for biofuels. This results in conflicts and shortages in both areas. Nearly 60% of the world's population suffers from malnourishment. The Food and Agriculture Organisation of the United

---

1 'Ford Predicts Fuel from Vegetation'. 19 September, 1925. New York Times, Boston. p. 24.

Nations (FAO) estimates that in 2014–2016 little over one out of nine persons suffered from hunger and undernourishment.[2] Understandably, growing fuel crops on land that could be used for food production is seen as a waste by many, especially since land resources are limited. These concerns are particularly pertinent in the case of commercial large-scale biofuel production, which is often done on land that is suitable for growing food crops. This adversely impacts access to food especially for the poor and the vulnerable. As David Pimentel *et al.* state, it 'squanders land, water and energy resources vital for the production of food for human consumption.'[3] Several studies have also predicted an increase in world food prices owing to the demand for biofuel feedstocks.

99.7% of food for human consumption is land-based and the majority of land is already in use –11% for growing crops, 9% as urban land, 27% as pasture land and 32% as forests. The remaining 21% is not suitable for agriculture, pasture or forestry; the soil either lacks fertility or the area is plagued by adverse climatic conditions. To add to this, global population growth has resulted in the average available cropland per capita to diminish to less than 0.22 hectares. The amount required for the production of a nutritious, diverse diet of both plant and animal-based products is 0.5 hectares. This was last available in 1960 when human population was a mere 3 billion.[4] The use by the biofuel industry of fertile land that can be used for growing food crops conflicts with food production requirements. Yet, biofuel production is being encouraged and supported through policy and law. Developing countries in Latin America, Eastern Europe, Asia and Africa are keen to further the growth of this industry, motivated by the need for economic progress, rural development and poverty eradication. Endowed with favourable climatic conditions and an abundance of land resources, they are ideal hosts for the biofuels industry.

---

[2] Food and Agriculture Organisation of the United Nations. 2016. State of Food Insecurity in the World. Food and Agriculture Organisation, Rome. p. 2. <*http://www.fao.org/3/a-i4671e.pdf* >.

[3] Pimentel, D. *et al.* 2008. Biofuel Impacts on World Food Supply: Use of fossil fuel, land and water resources, *Energies* 41(1) *http://www.mdpi.com/1996-1073/ 1/2/41/htm#B23-energies-01-00041*

[4] Giampietro, M. and Pimentel, D. 1994. Energy Utilization. *In*: Arntzen, C.J. and Ritter, E.M. (*eds.*), Encyclopedia of Agricultural Science Vol. 2. Academic Press, USA. pp. 63–76; Pimentel, D. *et al.* (See above n 6).

2.1. Hunger and Malnutrition

Over one billion people suffer from under-nourishment, a state where a person is unable to acquire sufficient food to meet the daily minimum energy requirements for a period of one year or more. Hunger is seen as being synonymous with chronic under-nourishment.[5] 90% of deaths due to hunger are a result of chronic, long-term lack of access to adequate food, and the majority of those facing such hunger and malnutrition are landless people, small landholders, and to a large extent girls and women in rural areas. Malnutrition also leads to the deaths of nearly six million children each year, amounting to 50% of preventable deaths.[6]

Recognition of the urgency and seriousness of the situation has resulted in the right to food and its various elements being widely acknowledged. In the Rome Declaration on World Food Security, adopted in 1996 at the World Food Summit, States declared that,

> 'We consider it intolerable that more than 800 million people throughout the world, and particularly in developing countries, do not have enough food to meet their basic nutritional needs. This situation is unacceptable. Food supplies have increased substantially, but constraints on access to food and continuing inadequacy of household and national incomes to purchase food, instability of supply and demand, as well as natural and man-made disasters, prevent basic food needs from being fulfilled. The problems of hunger and food insecurity have global dimensions and are likely to persist, and even increase dramatically in some regions, unless urgent, determined and concerted action is taken, given the anticipated increase in the world's population and the stress on natural resources'.[7]

---

[5] Food and Agricultural Organisation. 2015. State of Food Insecurity in the World 2015: The FAO Hunger Map 2015. *<http://www.fao.org/hunger/en/>*.

[6] Office of the UN Commissioner for Human Rights. 2010. The Right to Adequate Food: Fact Sheet No. 34. *<http://www.ohchr.org/Documents/Publications/FactSheet34en.pdf>*.

[7] Food and Agriculture Organisation. 1996. Rome Declaration on World Food Security. Adopted by the World Food Summit at Rome, Italy, 13-17 November 1996. *http://www.fao.org/WFS/*

The Declaration aimed to 'eradicate hunger in all countries' and achieve 'food security for all',[8] through the adoption of a Plan of Action, envisaged as,

> 'An ongoing effort to eradicate hunger in all countries, with an immediate view to reducing the number of undernourished people to half their present level no later than 2015, and a mid-term review to ascertain whether it is possible to achieve this target by 2010'.[9]

This was, however, not the first time that countries had come together to further food security. The Universal Declaration on the Eradication of Hunger and Malnutrition, adopted in 1974 during the World Food Conference, declared that 'within a decade no child will go to bed hungry, that no family will fear for its next day's bread, and that no human being's future and capacities will be stunted by malnutrition'.[10] Needless to say, this goal was not achieved. Several declarations stressing the urgency and fundamental nature of the right to be free from hunger and to have access to food have been made, such as the World Declaration on Nutrition (adopted at the 1992 International Conference on Nutrition), the 1994 International Conference on Population and Development, as well as the 1995 Copenhagen Declaration on Social Development.

At the time of the 1996 World Food Summit there were 823 million undernourished people in developing countries. If the commitments made during the 1996 Summit had been followed assiduously Eide conjectures that there should have been a reduction in that number by around 20 million every year, which over a decade should have led to a reduction of at least 583 million. However, the number of undernourished people has not decreased at that rate.[11] The fact that

---

[8] Rome Declaration (See above n 7).

[9] Food and Agriculture Organisation. 13-17 November 1996. World Food Summit Plan of Action. Point 7. <*http://www.fao.org/docrep/003/w3613e/ w3613e00.htm#PoA*>.

[10] Universal Declaration on the Eradication of Hunger and Malnutrition, Adopted on 16 November 1974 by the World Food Conference convened under General Assembly Resolution 3180 (XXVIII) of 17 December 1973 and endorsed by General Assembly Resolution 3348 (XXIX) of 17 December 1974.

[11] Eide, A. 2008. The right to food and the impact of liquid biofuels (agrofuels), Food and Agriculture Organization of the United Nations <*http://publish.uwo.ca/~dgrafton/righttofood.pdf*>

these targets have not been met has been acknowledged by the Human Rights Council. In its words, 'the promises made at the World Food Summit in 1996 to halve the number of persons who are undernourished are not being fulfilled'.[12]

In 2000 the world community once again came together to re-commit to working on this issue, this time through the United Nations Millennium Declaration, adopted by the UN General Assembly. States promised to free people from deprivation, based on the dignity and equality of all persons and their right to a basic standard of living, which necessitates the right to freedom from hunger. This resulted in the promulgation of the Millennium Development Goals, with indicators and targets to measure progress at the international and national levels over a period of 25 years, from 1990 to 2015. Eradication of hunger and poverty is the first Millennium Development Goal (MDG 1). It targeted a reduction of hunger by half by 2015.[13] Whether this has been achieved is questionable. The UN has announced that 'projections indicate a drop of almost half in the proportion of undernourished people in the developing regions, from 23.3% in 1990–1992 to 12.9% in 2014–2016. This is very close to the MDG hunger target'.[14] However, there is some debate over the correct estimate of people still suffering from hunger and malnourishment. The FAO tells us that currently there are 795 million people (one in every nine persons) who lack adequate food and are undernourished. Of them 780 million are in developing countries, particularly in Asia and Africa, which are incidentally also the regions exploring the development of the biofuels industry to meet energy, sustainability and rural development targets.[15] The FAO in its report on the progress of MDG 1 explains that the progress that has been

---

[12] Human Rights Council. 27 March 2008. Resolution 7/14 on the Right to Food, Para 19. <*http://ap.ohchr.org/documents/E/HRC/resolutions/A_HRC_RES_7_14.pdf*>.

[13] UN Millennium Development Goals, 'Goal 1: Eradicate Extreme Hunger and Poverty', <*http://www.unmillenniumproject.org/goals/gti.htm#goal1*>.

[14] *Quoted in* Institute for Food and Development Policy. 2016. *Food First Backgrounder*, 22(2A): 1–4. <*https://foodfirst.org/wp-content/uploads/2016/06/Summer2016Backgrounder.pdf*>

[15] World Food Programme. 2016. Hunger Statistics. <*https://www.wfp.org/hunger/stats*>; Food and Agriculture Organisation of the United Nations. 2015. The State of Food Insecurity in the World: Meeting the 2015 International Hunger Targets: Taking Stock of Uneven Progress. <*http://www.fao.org/3/a-i4646e.pdf*> p. 4.

made towards food security has been uneven, with some regions progressing rapidly such as Central and Eastern Asia, Caucasus, Northern Africa and Latin America, and others moving at a slower rate, such as Western and Southern Asia, Caribbean, Oceania and Sub-Saharan Africa. Furthermore, in some places there has been a reduction of hunger, but not of malnutrition caused by lack of access to adequately nutritious food - also an element of the right to food.[16]

2.2. The Right to Food

The right to food is more than a mere moral duty. Article 25 (1) of the Universal Declaration of Human Rights affirms that, 'everyone has the right to a standard of living adequate for the health and well being of himself and his family, including food, clothing, housing'.[17] Article 11 of the International Covenant on Economic Social and Cultural Rights (ICESCR) explicitly recognizes the right to freedom from hunger and lays down measures for the achievement of this, which include 'developing or reforming agrarian systems in such a way as to achieve the most efficient development and utilization of natural resources'.[18] The UN General Assembly, the Commission on Human Rights and the Economic and Social Council (ECOSOC) have also passed resolutions and declarations affirming the right to food.[19] Other international instruments and declarations that include the right to food are the Convention on the Elimination of All Forms of Discrimination against Women, the Convention on the Rights of the Child and the 2006 Convention on the Rights of Persons with Disabilities. Regional instruments have also recognised the right, such as the Additional

[16] Committee on Economic Social and Cultural Rights. 12 May 1999. General Comment No. 12: The Right to Adequate Food (Article 11). Adopted at the Twentieth Session of the Committee on Economic, Social and Cultural Rights. E/C.12/1999/5, para 8 <*http://www.refworld.org/docid /4538838c11.html*>.

[17] Universal Declaration of Human Rights. Adopted by the United Nations General Assembly on 10 December 1948. UNGA Res 217 A(III).

[18] International Covenant on Economic, Social and Cultural Rights. 16 December 1966. Adopted and opened for signature, ratification and accession by General Assembly Resolution 2200A (XXI), entered into force 3 January 1976, Article 11(2)(a).

[19] United Nations General Assembly. Resolution on the Right to Food Adopted on 18 December 2009. A/RES/64/159 <*http://repository.un.org/handle/11176/147271*>; Human Rights Council. 27 March 2008. Resolution 7/14 on the Right to Food. (See above n 12).

Protocol to the 1988 American Convention on Human Rights in the Area of Economic, Social and Cultural Rights (the Protocol of San Salvador), the 1990 African Charter on the Rights and Welfare of the Child and the Protocol to the African Charter on Human and Peoples' Rights on the Rights of Women in Africa.

The right is also recognised implicitly through other human rights, such as the right to life, for the protection of which states must take steps to eliminate malnutrition. General Comment 6 of the Human Rights Committee states that,

> 'The expression "inherent right to life" cannot properly be understood in a restrictive manner, and the protection of this right requires that States adopt positive measures. In this connection, the Committee considers that it would be desirable for States parties to take all possible measures to reduce infant mortality and to increase life expectancy, especially in adopting measures to eliminate malnutrition and epidemics'.[20]

The African Commission on Human and Peoples' Rights has made a similar interpretation, stating that the right to food is implicitly protected through the right to life, as well as through the rights health and to economic, social and cultural development under the African Charter on Human and Peoples' Rights.

The human right to food is now accepted as a basic norm of international human rights, through treaty law, customary international law, as well as state practice. Several countries have also embraced it as a legal obligation within their domestic legal systems.

Normative content of the right to food

The Committee on Economic, Social and Cultural Rights states that the right to adequate food is realised when each and every individual and community has access to adequate food or has the means to procure it.[21] The core content of the right entails,

---

[20] UN Human Rights Committee. 30 April 1982. General Comment No. 6: Article 6 (Right to Life). Adopted at the Sixteenth Session of the Human Rights Committee. para 5.

[21] The committee expanded upon the normative content of the right in General Comment No. 12.

> The availability of food in a quantity and quality sufficient to satisfy the dietary needs of individuals, free from adverse substances, and acceptable within a given culture; the accessibility of such food in ways that are sustainable and that do not interfere with the enjoyment of other human rights'.[22]

A state's obligations towards human rights have been defined in various ways, with Shue characterizing them as an obligation to 'forbear, protect and aid' and Van Hoof promulgating the fourfold characterization of- 'respect, protect, ensure and promote'.[23] The approach adopted by the Committee on Economic, Social and Cultural Right is to 'respect, protect and fulfill' human rights. In relation to the right to food the obligation to respect requires that states not take steps that prevent the availability and access to adequate food'.[24] The obligation to fulfill has been subdivided into a duty to fulfill by facilitating, providing[25] and promoting.[26] It requires that the State 'proactively engage in activities intended to strengthen people's access to and utilization of resources and means to ensure their livelihood, including food security'.[27] States are also obliged to provide the right to those who are unable to gain access to adequate food, due to reasons out of their control, such as natural disasters; an obligation that is of a more immediate nature, while other aspects require that achievement of the full realization of the right be done through the implementation of progressive measures. The core obligation is to alleviate hunger, as stated in Article 11(2) of ICESCR,[28] and at the very least to ensure

---

[22] UN Committee on Economic Social and Cultural Rights, General Comment No. 12. para 8. (See above n 16).

[23] Dowell-Jones M. 2004. Contextualising the International Covenant on Economic, Social and Cultural Rights: Assessing the Economic Deficit. Martinus Nijhoff Publishers, Holland. p. 29.

[24] UN Committee on Economic Social and Cultural Rights. General Comment No. 12. Para 15. (See above n 16).

[25] UN Committee on Economic Social and Cultural Rights. General Comment No. 12. (See above n 16).

[26] UN Committee on Economic Social and Cultural Rights. 11 August 2000. General Comment 14 - The Right to the Highest Attainable Standard of Health (Article 12). Adopted at the Twenty-second Session of the Committee on Economic, Social and Cultural Rights, on 11 August 2000. U.N. Doc. E/C.12/2000/4.

[27] UN Committee on Economic Social and Cultural Rights. General Comment No. 12. Para 15. (See above n 16).

[28] UN Committee on Economic Social and Cultural Rights. General Comment No. 12. Para 6. (See above n 16).

satisfaction of the 'minimum essential level required to be free from hunger'.[29] Failure to achieve this amounts to a violation of the Covenant. A determination of the unwillingness or inability of a state to comply also factors into a determination of whether its' actions amount to a violation. It must be noted, however, that the right to food does not imply a right to be fed. Governments are only required to provide food in situations such as natural disasters, armed conflicts and so forth. What states must do is provide an environment in which people can procure or produce adequate food for themselves and their families. To do this certain resources are necessary – land, water, seeds, money, access to markets and so forth.[30]

With regards to biofuel production, violation of the right would arise if a state introduces policies that encourage biofuel production at the cost of the right to food, or if a State fails to regulate such initiatives by non-state entities. The Committee on Economic, Social and Cultural Rights has stated that although the obligation to implement the right is primarily upon the State, non-state actors also have a responsibility to respect the right and ensure that there is no infringement.[31] The right to food in relation to biofuels also engages the concept of sustainability, which is linked to adequate food being available for present as well as future generations.[32]

Connection with other human rights

The enjoyment of one human right depends upon and is intertwined with the enjoyment of others. Human rights are indeed *"universal, inalienable, interrelated and interdependent"*.[33] Food is the basis for survival, making the right to food a fundamental right, the violation of

---

[29] UN Committee on Economic Social and Cultural Rights. General Comment No. 12. (See above n 16).

[30] Office of the UN Commissioner for Human Rights. 2010. The Right to Adequate Food: Fact Sheet No. 34. (See above n 6) 3-4; Committee on Economic Social and Cultural Rights. General Comment No. 12. (See above n 16).

[31] UN Committee on Economic Social and Cultural Rights. General Comment No. 12, para 20. (See above n 16).

[32] UN Committee on Economic Social and Cultural Rights. General Comment No. 12, para 7. (See above n 16).

[33] Vienna Declaration and Programme of Action. 25 June 1993. Adopted by the World Conference on Human Rights in Vienna, UN Doc. A/CONF.157/23. part I, para. 5.

which impacts several areas of human life, including livelihood, standard of living and health. ICESCR incorporates this view, linking the right to adequate food with the right to an adequate standard of living. Article 11(1) of the covenant affirms 'the right of everyone to an adequate standard of living for himself and his family, including adequate food, clothing and housing, and to the continuous improvement of living conditions'.[34] The Committee on Economic, Social and Cultural Rights expands upon this position in General Comment 12:

> 'The Committee affirms that the right to adequate food is indivisibly linked to the inherent dignity of the human person and is indispensable for the fulfillment of other human rights enshrined in the International Bill of Human Rights. It is also inseparable from social justice, requiring the adoption of appropriate economic, environmental and social policies, at both the national and international levels, oriented to the eradication of poverty and the fulfillment of all human rights for all'.[35]

Moreover, the connection of the right to food with other rights is not linear. A variety of rights impact its actualization as well, such as to water and land. These have a profound impact on the right to food. Therefore, an approach that views the right to food in isolation from other rights is not appropriate, neither on the ground, nor theoretically.

2.3. The Right to Food and Biofuels

Several African countries are attracting foreign investment for biofuel production. The South Korean firm Daewoo's attempt to get a 99 year lease over 1.3 million hectares of agricultural land in Madagascar received a lot of media attention. The deal fell through, but if it had been successful the company would have gained control over half of the country's arable land. While this may seem shocking, it is in fact just one of the many mega land deals in Africa. 1.5 million hectares of fertile farm land in Sudan has been leased to Egypt, South Korea and the Gulf states for 99 years. Sun biofuels – a UK Company – has acquired 80,000 hectares of land in Ethiopia, 8000 hectares in Tanzania and

[34] International Covenant on Economic, Social and Cultural Rights. Art. 11(1). (See above n 18).

[35] UN Committee on Economic Social and Cultural Rights. General Comment No. 12. Para 4. (See above n 16).

5000 hectares in Mozambique for jatropha plantations. The CAMS group, another UK company, acquired 45,000 hectares in Tanzania to grow sweet sorghum for ethanol. Brazilian, Swedish, Dutch, Norwegian, British and German companies are vying for land to grow biofuel crops in Ghana; 136,000 acres of Ghanaian land is already being farmed by seven private companies. German companies have expanded their business ventures into a number of countries. Acazis leased 56,000 hectares in Tanzania for a period of 50 years, with a further concession of 200,000 hectares, to grow castor. In 2009, it acquired another 13,000 hectares for biofuel crop cultivation in Ethiopia's Oromia State. The Prokon Group, also a German company, initiated a 10,000 hectare jatropha plantation programme in Tanzania's Mpanda district.[36] Swedish companies have acquired 250,000 acres in Mozambique. China, another major player in Africa, has been investing in the region since 2000. It has embarked on several large-scale biofuel projects, such as in Congo where 7 million acres of land have been acquired for palm oil plantations, as well as in Zambia where 1.8 million hectares of land will be cultivated by Kaidi Biomass Zambia Ltd. - a Chinese-Zambian joint venture. This will produce biodiesel from soybean, jatropha, castor and palm oil, and bioethanol from cassava, sweet sorghum, sisal and sugarcane. In fact, it is estimated that in 2004–2009 nearly 2.5 million hectares of agricultural land in Africa[37]

Such large-scale land acquisition for biofuel crop production has resulted in a decline in food production. In Southern Africa it 'left millions facing starvation, with rural populations being the most affected'.[38] Large-scale production of biofuel crops adversely impacts food security in a number of ways. Firstly, the use of highly fertile soil for growing energy crops instead of food crops raises concerns, particularly for developing countries that engage in biofuel production despite the fact that they have to import food grains for domestic consumption. Secondly, changes in livelihood patterns contribute to food

---

[36] Herre, R. (*ed.*), 2010. German Investment Funds involved in Land Grabbing. FIAN Deutschland e.V., Köln, Germany. p. 18.

[37] Kugelman, M. and Levenstein, S.L. (*eds.*), Land Grab? The Race for the World's Farmland. 2009. Woodrow Wilson International Center for Scholars, Washington D.C.; Cotula, L. *et al.*, 2009. Land grab or development opportunity? Agricultural investment and international land deals in Africa. FAO, IIED and IFAD, London/Rome.

[38] Bamikole, A. *et al.,* 2010. Biofuels and Sustainability in Africa. *Renewable and Sustainable Energy Reviews,* 1–13.

insecurity. Large-scale monoculture plantations, a characteristic of biofuel production, adversely affect small-scale farmers, as well as those who depend on the land for customary uses that are vital for their livelihoods and survival. Thirdly, an increase in biofuel production has an impact on food prices. The rise in grain prices by about 30% between 2000-2007 was caused by an increase in the demand for biofuels.[39] The OECD estimates that Canadian, EU and US biofuel blending mandates caused a 20% increase in the price of vegetable oil in 2014. The International Food Policy Research Institute suggests that by 2020 the prices of certain crops could rise by up to 70%. A number of other studies have also calculated impacts, albeit not to the extent predicted by the International Food Policy Research Institute. A rise in food prices owing to biofuel production would contribute in a major way to the world food crisis.[40] As pointed out by the former Director-General of the FAO, Jacques Doiuf, there is 'a very serious risk that fewer people will be able to get food', particularly in the developing world.[41] The greatest sufferers will be the marginalized and the poor. Considering these impacts, it is no wonder that the former UN Special Rapporteur on the Right to Food, Jean Ziegler, called biofuel production a 'crime against humanity'.[42]

In fact, concerns regarding food security, farmers' rights and environmental degradation, have led some African countries to put the brakes on biofuel production. In Tanzania, pressure from environmental organisations and farmers groups led in 2009 to the suspension of biofuel investments and land acquisition for biofuel production. South Africa

---

[39] Bamikole, A. *et al.* (See above n 38); Elobeid, A. and Hart, C. 2007. Ethanol Expansion in the Food versus Fuel Debate: How Will Developing Countries Fare? *Journal of Agricultural & Food Industrial Organization,* 5(6): 1–21.

[40] Ahmed, A.U. *et al.* 2007. The world's most deprived: Characteristics and Causes of Extreme Poverty and Hunger 2020, *International Food Policy Research Institute. Discussion Paper,* 43: 1–96. Elobeid, A. and Hart, C. 2007. Ethanol Expansion in the Food versus Fuel Debate: How Will Developing Countries Fare? *Journal of Agricultural & Food Industrial Organization,* 5(2): 1–21. p. 1. Chakravorty, U., Hubert, M., Moreaux, M. and Nøstbakken, L. 2015. The Long-Run Impact of Biofuels on Food Prices, *Resources for the Future*; Ghana goes biofuel, despite global food crisis. 2008. *Ghana Business News.*

[41] Rosenthal, E. 2007. World food stocks dwindling rapidly, UN warns. *New York Times*

[42] *Quoted in* Ferrett G. 2007. Biofuels 'crime against humanity'. *BBC News.* This sentiment is echoed by the IMF, which pointed out that increasing reliance on biofuels could have serious consequences for the poor.

has excluded maize and potatoes from biofuel production, in view of the fact that they are the staple foods of the country. Mozambique imposed a moratorium on biofuel projects from 2009–2011, to give time for national land use planning to be conducted. In 2014 the Government started drafting the biofuels sustainability criteria framework document, which asserts that investments should not have an adverse effect on food security.[43]

3. BIOFUELS AND LAND RIGHTS

3.1. Introduction

Millions of hectares of land have already been acquired for biofuel production, and this is only going to increase as the growth of the biofuels industry is encouraged through government policies that support investments and land acquisition. The term 'land grabbing' – used first by agrarian and environmental rights movements – has been used to describe the acquisition of the commons and the dispossession of indigenous and peasant communities in the Global South by foreign companies and governments. Attempts are made to re-frame projects that result in land grabbing – biofuels being a case in point – as pro-poor, sustainable development initiatives that give rise to improved livelihood opportunities for poor farmers and rural communities. Several institutions, including the World Bank and the Food and Agriculture Organisation, support such initiatives as being beneficial to both parties, claiming that they promote broad-based, sustainable economic development in developing nations; a win-win situation for the foreign corporation or government, as well as the developing country and its local communities. However, this position is questionable in view of research highlighting the adverse impacts of land grabbing on local communities. Furthermore, the clearing of grasslands and forests to make room for biofuel crops negatively impacts biodiversity, as does the conversion of mixed farm land into monoculture plantations. Without the introduction of sustainable agricultural practices this can lead to soil erosion and a reduction in soil fertility.

---

[43] Government of South Africa Department of Minerals and Energy. December 2007. Biofuels Industrial Strategy of the Republic of South Africa. p. 3; South African Government. 2017. Energy; Fundira, T. and Henley, G. 2017. Biofuels in Southern Africa: Political economy, trade, and policy environment, *WIDER Working Paper 2017/48, United Nations University*.

3.2. Land Acquisition for Biofuels

The acquisition of large tracts of land by biofuel companies has become a cause of landlessness and hunger in developing countries. Land diverted for biofuel production is often land that was previously used by local communities for a variety of purposes; it is productive for them, but not for the government, which sees it instead as 'marginal' – land that is lying unused and can, therefore, be diverted for commercial use. Since these lands are 'unused' it can be proclaimed that their use for biofuel production does not conflict with food production or livelihood generation. In this way, the project would 'avoid using lands used for food production. During the launch of the Central African Mining and Exploration Company's sugarcane ethanol project (the 'ProCana' Project), when the London-based company announced its intention to invest US$510 million in 30,000 hectares of land in the Gaza province, the Mozambican President, Armando Guebuza, declared that, 'biofuel development will not dislodge Mozambican farmers from their lands', as the land that would be allocated comes within the category of under-utilised or empty lands. In this way, the programmer would 'avoid using lands used for food production'.[44] Out of the 36 million hectares of land available in Mozambique, the Government reported that only 9% is in use, owing to which it felt that it would be able to bring 41.2 million hectares of marginal land into use. Mozambique's rate of land concessions is, in fact, one of the highest in Africa. Just in 2004–2009 more than 2.7 million hectares of land was granted by the Mozambican government to 405 large-scale investment projects, amounting to 7% of the country's arable land.[45]

Classification of land that is not a part of commercial production regimes as marginal or waste land is an artificial construct that ignores the many forms of local and traditional land use, and fails to take into account local communities dependent on these lands for their livelihoods. A variety of socio-economic factors make people use land in different

[44] M. Borras Jr *et al.*, 2011. The politics of agro-fuels and mega-land and water deals: insights from the ProCana case, Mozambique. *Review of African Political Economy,* 38: 216–217.

[45] The Oakland Institute. 2011. Understanding Land investment Deals in Africa: Country Report Mozambique. The Oakland Institute, Oakland, CA.; Dietz, K. *et al.*, 2014. The Political Ecology of Agrofuels. Routledge, USA. 98; World Bank Group. 2011. Rising Global Interest in Farmland: Can it Yield Sustainable and Equitable Benefits? World Bank, Washington DC.

ways at different times, including customary uses that also have cultural significance. Indigenous communities are known to have deep cultural and historical ties to their traditional lands.

Losing access to these lands leads to vulnerability and marginalization on many levels. A farmer thrown off her land in an operation in which several thousand farmers were dispossessed of their lands by Addax Energy (a Swiss Company that acquired land in Sierra Leone to produce ethanol from sugarcane) described her plight,

> 'Now I don't have a farm. Starvation is killing people. We have to buy rice to survive because we don't grow our own now'.[46]

Madagascar, Ethiopia, Ghana, Mali and Guinea are some of the countries where fertile land has been given away to biofuel companies; the land acquired has better access to irrigation, higher rainfall and greater market proximity. All recorded land deals in Mali are concentrated in parts of the country that are more agriculturally productive. Acheampong and Campion conducted interviews with 109 Ghanaians to understand the amount of land lost by them to jatropha plantations. They found that 69.7% had lost up to ten acres of land and 7.3% more than hundred acres. One person had lost thousand acres. The majority of farmers in Agogo, Ghana reported that land granted to ScanFarm Ghana Ltd. (formerly ScanFuel Ltd.) had been used by them for growing crops such as maize, yam, plantain and cocoa. As a result of this they were compelled to move to infertile lands, not suitable for agriculture.[47] Similarly, 700,000 hectares of land was allocated to an Italian company in Guinea for a jatropha plantation, displacing farmers dependant on the land. The Massingir district of Mozambique, where ProCana had been allocated land, also benefits from higher rainfall than other areas of the Gaza province and is close to the Rio do Elefantes River, making it suitable for agriculture. Land allocated to ProCana had been extensively used by local communities for cattle grazing, subsistence farming and charcoal production. Its importance for their

---

[46] GRAIN. 2013. Land Grabbing For Biofuels Must Stop: EU Biofuel Policies Are Displacing Communities and Starving The Planet. GRAIN, Barcelona, Spain. p. 2.

[47] Acheampong, E. and Campion, B.B. 2014. The Effects of Biofuel Feedstock Production on Farmers' Livelihoods in Ghana: The Case of Jatropha curcas. *Sustainability,* (6): 4587–4607. p. 4592.

livelihoods and survival should have been quite apparent. There were also villages, schools and health centers present. Yet, the Director of District Service for Economic Activities in Massingir stated that the area was uninhabited and not being used by local communities. There is also no record of ProCana being granted any sort of EIA approval. Although, the project was eventually cancelled, the language used to validate it by the company and the government highlights the problem of devaluing local and customary uses of land through the promulgation of mainstream capitalist definitions of land use. In many countries, such as Ethiopia, Tanzania, India and China, land classified as waste or marginal is used by local communities for a variety of activities, such as grazing, collection of medicinal plants, fodder, fuel and so forth. In Ethiopia grazing and shifting cultivation takes place on such lands. In Tanzania too, dry pastureland that is classified as 'waste' or 'marginal' is commonly used for grazing by pastoralists. In India, so-called waste lands are often used for fodder, fuel wood, food and thatch collection. Jatropha plantations in China are found on lands classified by the state as 'barren', even though village collectives own and use the land.

Local customs play a key role in delineating the rights of individuals and communities. It is estimated that 65% of the world's land is held under customary and community-based land tenure systems, particularly in Africa, Asia and South America. According to the World Bank only 2–10% of land in Africa is held under formal land tenure systems.[48] Yet, these systems of land use and ownership are generally not protected or legally recognised. Lack of recognition of customary land rights creates an environment in which land grabbing and eviction of people can occur with impunity.

Some countries have tried to give legal recognition to customary rights, such as India, Philippines, Indonesia, Burkina Faso, Botswana, Ghana, Lesotho, Malawi, Mali, Mozambique, Niger, Namibia, Swaziland, South Africa, Tanzania and Uganda. Various strategies have been adopted to achieve this, including giving customary land rights formal protection under the law, managing customary rights through the creation of administrative bodies and incorporating customary dispute resolution bodies into the framework of institutions dispensing justice, with the possibility of their decisions being appealed at the

---

[48] Rights and Resources Initiative. 2015. Who Owns the World's Land? A global baseline of formally recognized indigenous and community land rights. 1-17; Deininger, K. Land Policies and Land Reform. *World Bank*, 1–30. p. 8.

highest court of law. Nonetheless, customary rights still have a lower level of protection than formal legal rights.

Customary land rights in Botswana

The Tribal Land Act introduces three kinds of land rights – customary, common law and freehold. Customary rights are not ownership rights, but rights to use land within tribal areas, acquired through custom. Common law land rights, on the other hand, are granted for residential, commercial or industrial land use, including commercial farming initiatives to national investors and foreigners (Art. 23, 24).

Customary land rights can only be cancelled in a limited number of cases [listed in Art. 15(1) (a-d) of the Act] that include failure to observe restrictions related to the use of the land and use of the land for a purpose not authorized by customary law or that which is contrary to customary law. In 1993 an amendment was made to incorporate a provision that allows the repossession of unused land, permitting the grant of a right to be cancelled and repossessed if 'without sufficient excuse, the land has not been cultivated, used or developed to the satisfaction of the land board....in accordance with the purpose for which the grant was made'.[49] This provision gives land boards excessive authority in decisions regarding the status of land. It also diverts from customary norms that do not assume that unused land is abandoned land, and rarely permit the termination of land rights.

The Tribal Land Act does contain provisions that formalize existing customary land rights and eliminate the need for complex registration procedures to legitimise such claims. However, along with this it still allows customary land to be given away for commercial purposes, particularly in the case of land allocated as per customary norms before the passing of the Act. Customary rights before the passing of the Act do not have to be registered, which works against them. Upon

[49] Tribal Land Act (Chapter 32:02), Date of original text: 1968. Date of consolidation/ reprint: 31 December 2008.

receiving an application for a common law land grant, the Land Board is supposed to investigate the registry to ensure that the land is not held under any customary rights. If not, an agreement is drafted and sent to the minister for final authorization [Art. 24 and 31, regulations 20 (1–3)]. Apart from checking the registry, no other process of investigation is stipulated. Neither is a public hearing required. While, consent of the land holder is required for transfer of land to a common law grant [Art. 24, regulations 20 (3–4)], what constitutes 'consent' is not clear. The Act does not include measures to protect holders of customary rights from forced consent, nor to ensure that they are fully aware of the fact that they have the right to withhold consent. Sufficient protection is not given to prevent the acquisition of customary land by powerful companies through common law grants.

Customary land rights in Tanzania

The framework for land rights is laid down in the Land Act of 1999 and the Village Land Act of 1999. Under the Land Act all land belongs to the State, while 'radical title' lies with the president. Usufruct rights, therefore, define land tenure in Tanzania. The Act lays down three categories of land – general, village and reserved land. Village land is that which is held under customary rights of occupancy. Only general land is supposed to be allocated to biofuel companies. However, the definition of general land is problematic. The Land Act defines general land as 'all public land which is not reserved land or village land and includes unoccupied or unused village land'.[50] The Village Land Act omits the classification of 'unoccupied' or 'unused village land' from the definition, simply stating that general land is 'all public land which is not reserved land or village land'.[51] This variance in definitions is of significance to those with customary rights to use or occupy land and has caused a great deal of conflict, as it has made it possible for village land to be given away for

[50] Land Act 1999. Act No. 4 of 1999. Government of Tanzania. Section 2.
[51] Village Land Act 1999. Act No. 5 of 1999. Government of Tanzania. Section 2.

commercial enterprises. Land that seems to be unoccupied can still be important for the livelihoods of local communities, being used for a variety of purposes, such as gathering forest produce, grazing livestock and so forth. In fact, a large amount of land that is identified for biofuel production is village land that is used or occupied by local communities.

Once village land has been converted into general land it can be transferred to foreign investors. This can be done by the president for public interest reasons, which include 'investments of national interest'. The procedures that have to be followed for this are given in the Village Land Act. Importantly, the transfer cannot be vetoed by the village. Once the amount, type, timing and method of payment of compensation is agreed upon by the Commissioner of Lands and the Village Council, the transfer is complete, or in the alternate once the High Court decides upon an interim measure.

The compensation process laid down in the village land act is not satisfactory. Payments given are not sufficient to enable people to effectively engage in alternative livelihoods. To make matters worse local communities do not have a clear understanding of their rights. Owing to this companies make promises that are not included in the written agreements, thereby they are able to avoid having to keep them.

Of course, before the Village Land Act the rights of the local community had largely been neglected. By attempting to give equal status to customary and formal land rights, and creating a hybrid of customary and codified law that makes the village the centre of local land administration, the Village Land Act has indeed reformed previous land law. Nevertheless, Tanzania's land tenure system is still plagued by a lack of respect for customary rights.

Mozambique's Lei de Terras (1997)

Title to all land in Mozambique is held by the state. Land allocation is a lease agreement between the state and the applicant investor, whether a citizen, national corporation, foreign individual or foreign corporation.

Customary rights are protected and any individual living or working on the land for a period of at least ten years in 'good faith' or 'through customary norms and practices', has an automatic de jure 'right of use and benefit' over the land (Article 12). Their rights are secure and can be transferred through inheritance or to third parties through a consultative process. Community consultations are mandatory to ensure that land rights of local communities are not set aside by the government or by investors. If a community so wishes it is also possible to conduct a 'community land delimitation process' to map customary claims and facilitate their inclusion in the national cadastre. The problem is that there are no legal mechanisms to check the fairness or inclusivity of the consultation proceedings, nor any accountability mechanisms to ensure that benefits promised to communities by investors are actually delivered.[52]

Discriminatory treatment of certain groups also contributes to the creation of an environment in which land rights can be violated. A case in point is Botswana where the Tribal Land Act does not provide for the formalization of customary rights to non-Tswana ethnic groups such as the Basarwa (San) people, a traditional hunter-gatherer community. Customary rights to hunt and gather have been denied to them, despite the fact that the court held that they have the right to live, hunt and gather on their traditional lands. The Government has taken a number of steps to weaken these legal guarantees, including preventing the Basarwa from using water boreholes on their lands and forcing them off their lands, at times with violence. Having lost their traditional lifestyles the Basarwa have succumbed to depression, alcoholism, and diseases such as HIV/AIDS and tuberculosis.

Botswana's Tribal Grazing Lands Policy (TGLP) and National Policy on Agricultural Development (NPAD) contribute to the privatization of large tracts of communal land, allowing the wealthy and influential to claim extensive areas purportedly to facilitate 'full productive use' of the land. As Adams writes, 'one of the most frequent complaints against

[52] Lei de Terras (1997), Lei No. 19/1997. Government of Mozambique.

the land boards is that they allocate land inequitably, that they favour those with influence and many cattle, and ignore the land claims of those who are politically inarticulate and have few animals'.[53] Owing to this, a small number of wealthy families control vast areas of Botswana's grazing and arable land, substantially reducing the amount of grazing land available to small-scale, subsistence tribal herders. A consequence of this is an increase in rural landlessness.

There have been a number of cases of land grabbing and forced evictions in Africa, Asia and South America resulting in the illegal eviction of people, aided by inadequate legal protection of customary rights and a lack of full and effective consultation and consent from local communities. They also do not receive complete information on proposed projects and terms given in land deals. Neither is consultation conducted in a satisfactory manner, often merely a one-time event. Communities are not always given the opportunity to approve the minutes of the consultations before they are sent to other agencies, making it easier to not incorporate agreements related with benefit-sharing, access to resources and so forth into legally binding documents and contracts. Instead of attempting to make procedures more just and transparent, governments tend to support and enforce measures that reduce barriers to accessing land for commercial purposes. For instance, in 2002 the Mozambican government introduced a 90 day time limit to process land applications, including the community consultation process. This is not enough time for a meaningful consultation process in the context of large communities and the power disparity between companies and local people. Designing a successful consultation strategy is a complex process; several elements need to be kept in mind, such as the amount of time given, method of consultation, context and so forth. Community perception of the consultative process as fair and honest is also imperative for its success. Without this the process cannot be considered a 'genuine consultation', as required by the Committee on Economic, Social and Cultural Rights, as a part of the procedural protections that apply in forced evictions.[54]

---

[53] *Quoted in* Knight, R.S. 2010. Statutory recognition of customary land rights in Africa an investigation into best practices for lawmaking and implementation. Food and Agriculture Organization, Rome. p. 89.

[54] UN Committee on Economic, Social and Cultural Rights (CESCR). General Comment 7: The right to adequate housing (Art.11.1): forced evictions. Adopted at the Sixteenth Session of the Committee on Economic, Social and Cultural Rights on 20 May 1997. UN Doc. E/1998/22. para 15.

There have been several cases of illegal evictions by the biofuels industry. Over 20,000 people were evicted from their lands in Dakatcha, Kenya, for the development of a 50,000 hectare jatropha plantation, without any respect for their right to consultation and consent. In Sierra Leone, thousands of subsistence farmers lost their lands so that socfin-SL, a subsidiary of socfin, could set up a palm oil plantation, again without consultation and consent. The company signed a 50-year lease agreement with the Government of Sierra Leone for 6500 hectares of agricultural land in Pujehun District, with the option of increasing this by an additional 5000 hectares. The land acquired by them was being used by over 9000 farmers and around 120 families who were extremely critical of the manner in which the lease agreement was finalized with the help of corrupt local chiefs. In the words of a local farmer,

> "We landowners were not consulted on the project and therefore do not know the content of the agreement. So we all rejected the agreement as landowners. With one voice we said no! We expressed our refusal to give up our land. But the chief told us that he is the sole repository of land and everything he says is final".[55]

Eddy Kamara, of Sanh Village, Sierra Leone, speaking about the palm oil plantation deal by Socfin Agricultural Company SL Ltd that acquired large tracts of their land

"The police arrested and beat many of us... about 38 at this time. Later in the night they came to the village, knocking on the door and forcefully took people from their homes. Some of us fled to the river. The company executives designated people one by one to the police who then stopped before taking them to Pujehun. Once in custody, we were told that we would all rot in prison... Fifteen were left in the cell and charged. The lawyer who represented us was not allowed to post bail and we spent the night in jail. Then we were transferred to Pujehun. We stayed in a cell for eight days. We were not given any food".

[55] Land Justice 4 West Africa. June 3, 2014. When law fails to protect small farmers against land grabbers in Sierra Leone. *https://landjustice4wa.org/2014/06/03/when-law-fails-to-protect-small-farmers-against-land-grabbers-in-sierraleone/*

> Further protests against the project in October 2011, led to the arrest of 40 people. They were protesting the lack of adequate consultation with local communities, the company's lack of transparency, as well as the paucity of information regarding resettlement plans. They were also upset by the corruption, poor working conditions and pressure put on village chiefs and land owners to give away their lands.[56]

A Norwegian Company – Biofuel Solar – took advantage of the traditional system of communal land ownership in Ghana to take ownership of 38,000 hectares of land in Kusawgu from an illiterate chief who was deceived into the agreement. As elaborated by Bakari Nyari, Vice Chairman of the Regional Advisory and Information Network Systems (RAINS) and a member of the Ghana and African Biodiversity Network Steering Committee, 'bypassing official development authorization and using methods that hark back to the darkest days of colonialism, this investor claimed legal ownership of these lands by deceiving an illiterate chief to sign away 38,000 hectares of land with his thumbprint'.[57] By the time the community realized that they were being cheated 2,600 hectares of land had already been deforested for jatropha plantations, resulting in the loss of dawadawa and shea trees – valuable sources of food. It is reported that in 2007–2010 769,000 hectares of land were acquired by companies, such as Jatropha Africa (UK), ScanFuel and Biofuel Africa (Norway), Kimminic Corporation (Canada), Galten Global Alternative Energy (Israel), Agroils

[56] Land Justice 4 West Africa. June 3, 2014. When law fails to protect small farmers against land grabbers in Sierra Leone. *https://landjustice4wa.org/2014/06/03/when-law-fails-to-protect-small-farmers-against-land-grabbers-in-sierraleone/*

[57] *Quoted in* Acheampong, E. and Campion, B.B. p. 4602. (See above n 47).

(Italy) and Gold Star Farms (Ghana).[58] There have been similar cases in South America and Asia as well.[59]

3.3. Forced Relocation and International Law

Certain basic needs, such as for food, shelter, freedom from violent assault and threats, are necessary for survival, as well as for living a productive and dignified life. These are some of the needs guaranteed and protected by human rights. Needs, therefore, form the basis for human rights; a view affirmed by Green who considers it justifiable to *'regard instinctoid basic needs... as rights'*.[60] Access to land can be counted as a basic human need, for it is a basis for the achievement of other needs that are vital for the survival of individuals and communities, such as shelter and livelihood. Loss of access to land inevitably impacts food security, access to shelter and so forth, increasing the vulnerability of affected persons. This is not all, being a foundation for the development of cultures and traditions, land plays a role in the formation of identity. The protection of the right to land is, therefore, not merely about survival, but also about the dignity and development of a person. As Donelly, a scholar who goes beyond the needs-based assessment argues, the source of human nature is not just needs, but also man's moral nature. Where a scientist's analysis of human nature would conclude that 'beyond this we cannot go', the moral nature that grounds human beings and therefore human rights proclaims that there

[58] Friends of the Earth. 2010. Africa up for grabs: The scale and impact of land grabbing for agro-fuels. Friends of the Earth Europe, Brussels, Belgium.

[59] Parakuiyo Pastoralists Indigenous Community Development Organization (PAICODEO). 2013. Report on The State of Pastoralists' Human Rights in Tanzania: Survey of Ten Districts of Tanzania Mainland 2010/2011. PAICODEO, Tanzania; Rist, L. *et al.*, 2009. Biofuels: Social Benefits. *Science,* 326(5958): 1344; Land Justice 4 West Africa. April 1, 2014. Land Grabbing in Northern Ghana: How Fair were the Deals Signed? *<https:// landjustice4wa.org/2014/04/01/ land-grabbing-in-northern-ghana-how-fairwere- the-deals-signed/>*; Kennedy, J. 2013. Sierra Leone Farmers Evicted for Sugarcane Biofuel Plantations. *Business and Human Rights Resource Centre. <https://business-humanrights.org/en/sierra-leone-farmers-evicted-forsugarcane-biofuel-plantations>*; See also Vermeulen, S. and Cotula, L. 2010. Over the heads of local people: consultation, consent and recompense in largescale land deals for biofuels projects in Africa. The *Journal of Peasant Studies,* 37(4): 899–916.

[60] *Quoted in* Donnelly, J. 2003. Universal Human Rights in Theory and Practice, 2nd edn. Cornell University Press, USA. p. 13.

is a level beneath which men and women must not descend. Human rights, thus, protect that which is imperative for a life worthy of a human being.[61] The preamble of the Universal Declaration of Human Rights (UDHR) expresses this through an affirmation of the importance of the 'diginty and worth of the human person', proclaiming it to be the basis of human rights and declaring in its very first article that every person is born 'free and equal in dignity and rights'.[62]

Forced evictions without adequate compensation, protection and assistance cause homelessness, poverty and starvation, demeaning a person to the extent that it may be said that he or she is reduced to living like an animal – a life with no dignity that is vulnerable to human rights abuses. As stated by the Office of the High Commissioner for Human Rights (OHCHR), forceful eviction of people from their lands is one of the 'most supreme injustices any individual, family, household or community can face'.[63] Land grabbing and forced evictions are facilitated by the lack of a legal and policy environment for the protection of customary land rights and the promotion of foreign investments in large-scale development projects at the cost of human rights. This has become a worldwide problem, to the extent that this century has been called "the century of displacement".[64]

Forced evictions have been defined by the Committee on Economic Social and Cultural Rights, as '...*The permanent or temporary removal against their will of individuals, families and / or communities from the homes and / or land which they occupy, without the provision of, and access to, appropriate forms of legal or other protection*'.[65] The violation of the right to adequate housing is the most obvious impact of this, however, several other rights are also affected, directly or indirectly, such as to an adequate standard of living, information, participation,

---

[61] Bay believes that *'needs establish human rights'*. Donnelly, J. 2003. Universal Human Rights in Theory and Practice, 2nd edn. Cornell University Press, USA. pp. 13–14.

[62] Universal Declaration of Human Rights. Article 1. (See above n 17).

[63] Office of the High Commissioner for Human Rights. 1996. Fact Sheet No. 25, Forced Evictions and Human Rights. No. 25. p. 2.

[64] Office of the High Commissioner for Human Rights. 1996. Fact Sheet No. 25. p. 2.

[65] UN Committee on Economic, Social and Cultural Rights. 1997. General Comment 7 on 'The Right to Adequate Housing' (Article 11(1)): Forced Evictions. (See above n 54).

freedom of movement and residence, expression, food, as well as health, recognised in a number of international instruments and national legislations. Tribal and indigenous people, nomadic herders, pastoralists, squatters and other marginalised communities are some of the most affected, owing to their particular vulnerability in terms of lack of secure land tenure. For groups with unique cultures and lifestyles that are intimately linked with traditional lands this also implies the destruction of traditional knowledge and belief systems that have been passed down through generations.

Rights to life and security of person are entailed when evictions are conducted in a violent manner. Land acquisition is often accompanied by coercive measures and violence, as the State and the company attempt to enforce their authority (as has happened in Colombia and Ghana) often leading to protests. An Agogo farmer (Ghana) narrates his experience,

> 'I resisted the Company taking up my land with a gun. They used to operate at night. I struggled with them for about 3 years. I threatened them with death and they stopped. What I don't like is that everything on the land is cleared. When the Company leaves, there will be chaos as all boundary markers are gone. No landowner will be able to identify his or her land'.[66]

In today's day and age, disregard for human rights for the sake of development is too easily accepted, owing to the belief that development serves the 'public interest' and the 'greater good'. From this perspective forced evictions are seen as just a 'side effect'[67] of development projects, rather than as a human rights issue. Nor is it the case that all forced evictions are considered illegal. As the Committee on Economic, Social and Cultural Rights asserts, although 'instances of forced eviction are prima facie incompatible with the requirements of the Covenant' they can be justified 'in the most exceptional circumstances and in accordance with the relevant principles of international law'.[68] Governments can

---

[66] Acheampong, E. and Campion, B.B. p. 4592. (See above n 47).

[67] Office of the High Commissioner for Human Rights. 1996. Fact Sheet No. 25. p. 2. (See above n 63).

[68] UN Committee on Economic Social and Cultural Rights. 1991. General Comment 4: The Right to Adequate Housing: (Art. 11 (1) of the Covenant). Adopted at the Sixth Session of the Committee on Economic, Social and Cultural Rights, on 13 December 1991. U.N. Doc. E/1992/23. para 18.

evict people from their lands, but can only do so if it is absolutely necessary, and conducted in a manner that is in conformity with the law and with provisions of international human rights treaties.[69] In this regard attention has to be paid to the scale of displacement, the extent of constructive dialogue and participation with affected people, as well as the development of comprehensive solutions that minimize negative impacts and protect fundamental rights.[70]

Nonetheless, in view of the increase in forced evictions the Committee on Economic, Social and Cultural Rights is of the view that states 'must refrain from forced evictions', calling for strict legal measures to prevent forced evictions whether undertaken by state agents or third parties as,

> '...legislation against forced evictions is an essential basis upon which to build a system of effective protection. Such legislation should include measures which (a) provide the greatest possible security of tenure to occupiers of houses and land, (b) conform to the Covenant and (c) are designed to control strictly the circumstances under which evictions may be carried out'.[71]

Agenda 21 also supports the protection of people against forced evictions, asserting that 'people should be protected by law against unfair eviction from their homes or land'.[72] The Commission on Human Rights has adopted a Resolution seeking enforcement of measures to prohibit forced evictions. It urges,

> 'Governments to undertake immediately measures, at all levels, aimed at eliminating the practice of forced eviction by, inter alia, repealing existing plans involving forced evictions as well as any legislation allowing for forced

---

[69] UN Committee on Economic, Social and Cultural Rights. 20 May 1997. General Comment 7 on 'The Right to Adequate Housing' (Article 11(1)): Forced Evictions. Para 3. (See above n 54).

[70] Office of the High Commissioner for Human Rights. May 1996. Fact Sheet No. 25. p. 4. (See above n 63).

[71] UN Committee on Economic, Social and Cultural Rights. General Comment 7. para 9. (See above n 63).

[72] Agenda 21. 1992. Adopted at the United Nations Conference on Environment & Development Rio de Janerio, Brazil, 3 to 14 June 1992. A/CONF.151/26/ Rev.1 (vol. I), annex II, Agenda 21. Chapter 7.9 (b)).

evictions, and by adopting and implementing legislation ensuring the right to security of tenure for all residents'.[73]

Indigenous people are particularly affected by displacement. The U.N. Special Rapporteur on the situation of human rights and fundamental freedoms of indigenous people has observed that:

> 'Wherever [large-scale projects] occur in areas occupied by indigenous peoples it is likely that their communities will undergo profound social and economic changes that are frequently not well understood, much less foreseen, by the authorities in charge of promoting them...The principal human rights effects of these projects for indigenous peoples relate to loss of traditional territories and land, eviction, migration and eventual resettlement, depletion of resources necessary for physical and cultural survival, destruction and pollution of the traditional environment, social and community disorganization, long-term negative health and nutritional impacts as well as, in some cases, harassment and violence'.[74]

Evictions from ancestral lands impact a range of human rights, well illustrated by the example of pastoralists, one of the many communities that face displacement due to large-scale production of biofuel and other cash crops. Many of the so-called marginal or waste lands that have been converted into biofuel plantations had been used by them. The resultant loss of diverse vegetative cover and biodiversity has impacted food security and disrupted traditional lifestyles, now recognised as having value for biodiversity and wildlife conservation, prevention of desertification and sequestration of carbon. It has also rendered pasture land useless, since crops such as jatropha curcas are toxic to humans, as well as to livestock. Neither the bi-products of the production process

---

[73] Commission on Human Rights. 16 April 2004. Resolution: 2004/28 on Prohibition of Forced Evictions. CHR Res 2004/28,E/CN.4/RES/2004/28. para 2.

[74] UN Commission on Human Rights. 2003. Report of the Special Rapporteur on the situation of human rights and fundamental freedoms of indigenous people, Rodolfo Stavenhagen, submitted in accordance with Commission resolution 2001/65 (Fifty-ninth session). U.N. Doc. E/CN.4/2003/90. para 21; The Inter- American Court of Human Rights relied upon this analysis by the UN Commission on Human Rights in *Saramaka People v. Suriname* - Preliminary Objections, Merits, Reparations and Cost (Judgement of 28 November 2007)[Ser. C, No. 174] para 135.

nor the crop residue can be used as fodder to feed livestock. What is more, jatropha curcas plants can survive for up to 50 years, during which time the land cannot be used for anything else. Its return to pastoral use thereafter is also doubtful. Some argue that the seed cake of the varieties grown in Africa could be used as fertilizer. However, to do so its toxicity has to be removed through a process that is expensive and complex. The production of second generation biofuels also poses a threat to pastoralists, as it entails the harvesting and removal of large amounts of biomass from savannah and grassland areas, which can impact soil nutrient levels. The consequence of replacing so-called 'marginal' or 'waste' lands with biofuel plantations – economically, environmentally, as well as in terms of the lifestyle and culture of traditional and indigenous communities – needs to be factored into an impact assessment of biofuel projects and taken into account by policy makers. Emphasis ought to be put on models of development that are pro-poor and beneficial for local communities.

3.4. A Human Rights Based Approach to Development

Development projects are in no way outside the scope of the international human rights framework, and human rights obligations have to be incorporated into every stage of the development process. This requires that consideration be given to civil, cultural, economic and social rights in the planning and implementation of projects. However, as Cernea points out, examination of these risks is not a component of conventional project analysis, increasing the risk of impoverishment due to loss of livelihood opportunities and sources of income from arable, grazing and forest lands. This in turn causes a decline in standards of living and adversely affects food security and health, leading to further marginalization of those impacted by such policies.[75]

A range of human rights based approaches have been developed depending on the sector, as well as the social and political context. All these approaches are united by the principles of non-discrimination, equality, participation, and accountability. These have been affirmed by the UN General Assembly in its resolution on the right to

---

[75] Cernea, M.M. 2004. Impoverishment Risks, Risk Management, and Reconstruction: A Model of Population Displacement and Resettlement. Paper presented at the UN Symposium on Hydropower and Sustainable Development, Beijing: 41.

development. Twomey adds to this the principle of prioritization of vulnerable groups.[76]

The principle of equality includes not only formal equality but also substantive equality, which requires that states address the causes of discrimination and work towards their elimination. This necessitates prioritization of the protection of vulnerable groups. Policies should be based on empowering people so that they have the ability and capacity to affect positive change in their communities and lives.

Accountability, a core aspect of the human rights based approach, enables translation of universal standards into national and local benchmarks, in order to measure progress, find solutions to policy failures and develop effective policies, laws and mechanisms for redress. Another key element of good development practice is the participation of local communities in decisions related with project planning and implementation. To be effective this needs to be active, free, meaningful and inclusive of all stakeholders. Mere formal consultation is not sufficient; processes of participation need to be put in place that empower people. Of course, this does not mean participation in the technical aspects and details of a project, but as stated by OHCHR, it does mean that 'when alternative policy options are being explored by experts, the implications of these options for the interests of various population groups must be made transparent and presented in an understandable manner to the general public, including the poor, so that they can have an opportunity to argue for the options that serve their interests best'.[77] This gives a voice to vulnerable groups and people affected by development projects, enabling them to advocate measures and policy formulations that are of benefit to them.

There is urgent need for comprehensive solutions that strike a balance between the rights and needs of affected people and the nations' development goals. However, the role of states in land grabbing makes

---

[76] UN General Assembly. Resolution adopted by the General Assembly on 18 December 2009 (Sixty-fourth session). A/RES/64/172; Twomey P. 2007. Human Rights-Based Approaches to Development: Towards Accountability. *In:* Baderin, M.A. and McCorquodale, R. (*eds.*), Economic, Social and Cultural Rights in Action. OUP, UK. pp. 45–69.

[77] Office of the United Nations High Commissioner for Human Rights. Principles and Guidelines for a Human Rights Approach to Poverty Reduction Strategies. HR/PUB/06/12. para 68.

the implementation of a human rights based approach a challenge. Owing to the legal principles of eminent domain, public interest and sovereignty, imposing regulations in this area, particularly at the international level, becomes difficult.

Effective participation

The fact that sustainable development requires effective participation by local communities was acknowledged by the Rio Declaration of 1992, and the important role that local communities play in development because of their 'knowledge and traditional practices' has been highlighted in Principle 22. It asserts that states should support their 'identity, culture and interests and enable their effective participation in the achievement of sustainable development'.[78] Agenda 21, which further expands upon the concept of sustainable development, reaffirms the significance of public participation by individuals, groups and organisations in decisions that affect their communities.[79]

Mere formal consultation does not amount to effective participation. As observed by the Inter-American Commission in *Maya Indigenous Communities of the Toledo District v. Belize*, 'fully informed consent' requires 'at a minimum, that all of the members of the community are fully and accurately informed of the nature and consequences of the process and provided with an effective opportunity to participate individually or as collectives'.[80] It is, thus, better defined as a process through which people's concerns and needs in the shaping of the project are heard and acknowledged, thereby enabling them to protect their human rights. Participation in this sense is interwoven with the basic right to freedom of expression and participation enshrined in Articles 19 and 25 of the International Covenant on Civil and Political Rights (ICCPR). For, there can be no meaningful participation if people are unable to express their opinions on matters that affect their lives, knowing that their views will be respected by those in power. The

[78] Rio Declaration on Environment and Development. 1992. Adopted at the United Nations Conference on Environment & Development Rio de Janerio, Brazil, 1992. A/CONF.151/26. Principle 22.

[79] Agenda 21. (See above n 72).

[80] *Maya Indigenous Communities of the Toledo District v. Belize* (2004) Case 12.053 Report No. 40/04, Inter-American Commission on Human Rights, OEA/Ser.L/V/II.122 Doc. 5 rev. 1 at 727. Para 142.

importance of the right to participation is evidenced by the fact that it has been addressed in several international instruments, including the Universal Declaration of Human Rights, International Covenant on Economic, Social and Cultural Rights, Convention on the Elimination of Discrimination against Women, Convention on the Rights of the Child and the Declaration on the Right to Development. A survey of general comments, recommendations and concluding observations of treaty-based bodies shows a degree of consensus among treaty supervisory bodies with regards to the right.

Special importance has been given to the right to participation of minority and indigenous communities. Article 27 of ICCPR imposes a positive duty on States to 'ensure the effective participation of members of minority communities in decisions which affect them'.[81] The Committee on Economic, Social and Cultural Rights (CESCR) has also expanded upon the right of indigenous peoples to prior, free and informed consent in General Comment 21,[82] on the basis of Article 15 of the International Covenant on Economic, Social and Cultural Rights (ICESCR), which affirms the right of participation in cultural life. CESCR interprets this to include the right to restitution or return of lands and resources that have been traditionally used, if they were taken from them without prior and informed consent.[83] It calls upon States Parties to the ICESCR to 'respect the principle of free, prior, and informed consent of indigenous peoples in all matters covered by their specific rights'.[84] The Human Rights Committee has reiterated in General Comment 23 that protection of the traditional way of life of indigenous people includes control over and use of resources and lands and is connected with the right to culture.

---

[81] UN Human Rights Committee. 1994. CCPR General Comment 23: Article 27 (Rights of Minorities). CCPR/C/21/Rev.1/Add.5, para 3.2.

[82] UN Committee on Economic, Social and Cultural Rights. 2009. General Comment no. 21, Right of everyone to take part in cultural life (art. 15(1)(a) of the International Covenant on Economic, Social and Cultural Rights). Adopted at the Forty-third Session of the Committee on Economic, Social and Cultural Rights, on 2009. U.N. Doc. E/C.12/GC/21.

[83] UN Committee on Economic, Social and Cultural Rights. General Comment 21. p. 36 (See above n 82).

[84] UN Committee on Economic, Social and Cultural Rights, General Comment 21. p. 37. (See above n 82).

> 'With regard to the exercise of the cultural rights protected under article 27 [of ICCPR], the Committee observes that culture manifests itself in many forms, including a particular way of life associated with the use of land resources, especially in the case of indigenous peoples. That right may include such traditional activities as fishing or hunting and the right to live in reserves protected by law. The enjoyment of those rights may require positive legal measures of protection and measures to ensure the effective participation of members of minority communities in decisions which affect them'.[85]

The Committee on the Elimination of Racial Discrimination (CERD) also calls upon '[S]tates parties to recognize and protect the rights of indigenous peoples to own, develop, control and use their communal lands, territories and resources'.[86] It mandates consultation, qualifying this with the requirement that consent should be informed.[87] In General Recommendation 23 CERD urges States Parties to 'ensure that members of indigenous peoples have equal rights in respect of effective participation in public life and that no decisions directly relating to their rights and interests are taken without their informed consent'.[88] In cases where they have been deprived of their lands without prior and informed consent the committee stipulates restitution.[89] The Human Rights Committee also urges States Parties to respect their duty to conduct consultations with indigenous people before allowing

---

[85] UN Human Rights Committee. 1994. CCPR General Comment 23 (See above n 81).

[86] U.N. Commission on the Elimination of Racial Discrimination. 1997. Report of the Committee on the Elimination of Racial Discrimination: Annex V - General Recommendations XXIII. U.N. Doc. Supplement No. 18 (A/52/18), para 5.

[87] International Convention on the Elimination of All Forms of Racial Discrimination. 1965. 660 U.N.T.S. 195. Entered into force on 4 January 1969.

[88] U.N. Commission on the Elimination of Racial Discrimination. Report of the Committee on the Elimination of Racial Discrimination: Annex V - General Recommendations XXIII. Para 4(d). (See above n 86).

[89] U.N. Commission on the Elimination of Racial Discrimination. Report of the Committee on the Elimination of Racial Discrimination: Annex V - General Recommendations XXIII. p. 5. (See above n 86).

any resource concession or development project on their traditional lands.[90]

The UN Declaration on the Rights of Indigenous Peoples (henceforth 'UN Declaration') calls for the prior, free and informed consent of indigenous people when a State adopts administrative or legislative measures that have an impact on them or gives approval to any project 'affecting their lands or territories and other resources, particularly in connection with the development, utilization or exploitation of mineral, water or other resources'.[91] In cases of relocation as well, prior, free and informed consent is required (Article 10). There are varying views regarding what the right to prior, free and informed consent entails. Some scholars see it as giving indigenous people the right to veto, while others are of the opinion that it merely gives them the right to meaningful participation in decisions that affect their lands and resources. The Special Rapporteur on the Situation of Human Rights and Fundamental Freedoms of Indigenous Peoples, James Anaya, clarified that the Declaration aims to ensure that the right to self-determination of indigenous peoples is respected through the requirement of consultation which is aimed at prior, free and informed consent. This, however, does not amount to veto power, but rather indicates that the purpose of consultations is consent, based on a mutually acceptable agreement.[92] The UN Declaration is a soft law instrument and not legally binding, but it is frequently referred to by international tribunals, such as the Inter-American Court of Human Rights (IACHR), which has a substantive body of jurisprudence on

---

[90] U.N. High Commissioner for Human Rights. 2007. Concluding Observations of the Human Rights Committee: Chile. U.N. Doc. CCPR/C/CHL/CO/5, para 19; U.N. High Commissioner for Human Rights. 2008. Concluding Observations of the Human Rights Committee: Panama. U.N. Doc. CCPR/C/PAN/CO/3. para 21; U.N. High Commissioner for Human Rights. 2008. Concluding observations of the Human Rights Committee: Nicaragua. U.N. Doc. CCPR/C/NIC/CO/3. para 21; Darrow, M. 2003. Between Light and Shadow: The World Bank, The International Monetary Fund, and International Human Rights Law. Hart Publishing, UK. pp. 255–258.

[91] United Nations Declaration on the Rights of Indigenous Peoples. 2007. Adopted by the General Assembly by resolution No. 61/295. Art. 32. See also Article 19.

[92] UN Human Rights Council. 2009. Report of the Special Rapporteur on the Situation of Human Rights and Fundamental Freedoms of Indigenous People. U.N. Doc. A/HRC/12/34. Paras 41, 44, 46–48.

indigenous rights. There is, in fact, some debate regarding its customary law status.[93]

IACHR has affirmed the duty of states to consult with indigenous communities and gain free, prior and informed consent from them when planning large development projects on their traditional lands. In *Saramaka Peoples v. Suriname*, the IACHR stated that, 'regarding large-scale development or investment projects that would have a major impact within Saramaka territory, the State has a duty, not only to consult with the Saramakas, but also to obtain their free, prior, and informed consent, according to their customs and traditions'.[94] Not doing so would amount to a violation of their rights, despite the fact that under the law their rights to land are customary in nature and ownership of all land and natural resources vests in the State. The IACHR has also recognised indigenous peoples' collective rights to land and traditionally used natural resources. The decisions of the court have been grounded on the rights to culture, property and judicial protection, as outlined in the American Declaration on the Rights and Duties of Man (a source of legal obligations for Member States of the Organisation of American States, though not legally binding) and the American Convention on Human Rights, a legally binding treaty

---

[93] See Anaya, S.J. 2009. The Right of Indigenous Peoples to Self-Determination in the Post-Declaration Era. *In:* Chartres, C. and Stavenhagen, R. (*eds.*), Making the Declaration Work: The United Nations Declaration on the Rights of Indigenous Peoples. International Work Group for Indigenous Affairs, Denmark. p. 184; Clavero B. 2005. The Indigenous Rights of Participation and International Development Policies. *Arizona Journal of International and Comparative Law,* 22(1): 41–51. p. 42; Xanthaki, A. 2009. Indigenous Rights in International Law over the Last 10 Years and Future Developments. *Melbourne Journal of International Law,* 10(1): 36; Davis, M. 2008. Indigenous Struggles in Standard-Setting: The United Nations Declaration on the Rights of Indigenous Peoples. *Melbourne Journal of International Law,* 9(2): 458– 59, 465.

[94] *Saramaka People v. Suriname.* Judgment of 28 November 2007. Preliminary Objections, Merits, Reparations and Cost, Inter-American Court of Human Rights, [Series C, No. 174] para 131, 133 and 136.

(ACHR). The decisions of the Court are binding on States Parties who have recognised its jurisdiction.[95]

The requirement of consent with regards to projects taking place on traditional lands has been affirmed through a series of cases. In *Mary and Carrie Dann v. United States*[96], brought before the Inter- American Commission of Human Rights, on the basis of the IACHR's decision in *Mayagna (Sumo) Awas Tingni Community v. Nicaragua*, the Commission recognised that any determination related to indigenous land rights must 'be based on fully informed consent of the whole community, meaning that all members be fully informed and have the chance to participate'.[97] Furthermore, States must ensure that measures exist to facilitate the recognition of indigenous peoples' collective rights over their lands and resources.[98]

In the case of *Mayagna (Sumo) Awas Tingni Community v. Nicaragua*, the IACHR affirmed that the right to property enshrined in the ACHR protects the right to communal property of indigenous people as well. The State does not have the right to grant concessions to third parties on such lands. The Court also held that the State would have to create an effective mechanism for the delimitation, demarcation and titling of community lands of indigenous people, in accordance with customary norms.[99]

In *Maya Communities of the Toledo District v. Belize* the Commission found that the State had violated the property rights of Maya Communities by granting concessions on their lands 'without effective

[95] American Declaration of the Rights and Duties of Man. *Adopted at the Ninth International Conference of American States, Bogota, Colombia, 1948.* O.A.S. Res. XXX [hereinafter American Declaration]; *American Convention on Human Rights. Adopted at an Inter-American Specialized Conference on Human Rights, San José, Costa Rica, 1969.* O.A.S. T.S. No. 36 [hereafter "American Convention"]; Page A. 2004. Indigenous Peoples' Free Prior and Informed Consent in the Inter-American Human Rights System. *Sustainable Development Law and Policy,* 4(2): 16–20.

[96] *Mary and Carrie Dann v. U.S.* Inter-American Commission of Human Rights, Case 11.140, (2002) Report No. 75/02, OEA/Ser.L/V/II.117, doc. 5. Rev, para 1.

[97] *Mary & Carrie Dann v. U.S.* p. 18. (See above n 96).

[98] *Mary & Carrie Dann v. U.S.* p. 15. (See above n 96).

[99] *Mayagna (Sumo) Awas Tingni Community v. Nicaragua.* Merits, Reparations, and Costs, Inter-American Court of Human Rights (Judgment of 31 August 2001) [Series C, No. 79].(See above n 80).

consultations with and the informed consent of the Maya people'.[100] For, 'the duty to consult is a fundamental component of the States obligations in giving effect to the communal property rights of the Maya people in the lands that they have traditionally used and occupied'.[101] It also found the lack of full and effective demarcation and titling of communal lands traditionally used and occupied by them to be a violation of their property rights.[102]

Both, the Inter-American Commission for Human Rights and the Inter-American Court of Human Rights have asserted that the land rights of indigenous people cannot be taken away if doing so would result in a violation of their right to survival.[103] The African Commission on Human and People's Rights (African Commission) and the African Court on Human and People's Rights (African Court) have been influenced by the decisions of the Inter-American Court and Commission. The African Commission has affirmed the rights of traditional communities over their natural resources, as laid down in Article 21 of the African Charter on Human and People's Rights (African Charter),[104] stating that 'the protection of rights to land and natural resources is fundamental for the survival of indigenous communities in Africa and such protection relates both (sic) to Articles 20, 21, 22 and 24 of the African Charter.'[105]

---

[100] *Maya Indigenous Communities of the Toledo Dist. v. Belize*, Inter-American Commission for Human Rights. para 153. (See above n. 80).

[101] *Maya Indigenous Communities of the Toledo Dist. v. Belize*. para 155. (See above n 80).

[102] *Maya Indigenous Communities of the Toledo Dist. v. Belize.* para 194. (See above n 80).

[103] *Saramaka People v. Suriname.* (See above n 94); *Case of the Indigenous Community Yakye Axa v. Paraguay*. Merits, Reparations and Costs. Judgment of June 17, 2005 Series C No. 125; *Case of the Indigenous Community Sawhoyamaxa v. Paraguay*. Merits, Reparations and Costs. Judgment of March 29, 2006. Series C No. 146.

[104] Article 21(1): All peoples shall freely dispose of their wealth and natural resources. This right shall be exercised in the exclusive interest of the people. In no case shall a people be deprived of it. Article 21 (2): In case of spoilation, the dispossessed people shall have the right to the lawful recovery of its property as well as to an adequate compensation. African Charter on Human and People's Rights. Adopted 27 June 1981, entered into force 21 October 1986. OAU Doc. CAB/LEG/67/3 rev. 5, 21 I.L.M. 58 (1982).

[105] African Commission on Human and Peoples' Rights. 2003. Report of the African Commission's Working Group of Experts on Indigenous Populations/ Communities. Adopted by the African Commission at its 34th Ordinary Session on November 2003. p. 21.

In the Ogoniland case brought before the African Commission, the Government of Nigeria was engaged in oil exploration on lands belonging to the Ogoni people, adversely impacting their environment, as well as their health. Protests by the Ogoni were quelled through violent means, at times even leading to fatalities. The African Commission decreed that the activities of the government had amounted to a violation of the right to health of the Ogoni people,[106] as well as the right to 'a general satisfactory environment favourable to their development'.[107] It also held that their right to 'freely dispose of their wealth and natural resources'[108] had been infringed, owing to degradation of their lands, lack of participation in decisions related with the development of Ogoniland, as well as the fact that they had not received any material benefits from the project.

> '[I]n all their dealings with the oil consortiums, the government did not involve the Ogoni communities in the decisions that affected the development of Ogoniland. The destructive and selfish role played by oil development in Ogoniland, closely tied with repressive tactics of the Nigerian Government, and the lack of material benefits accruing to the local population, may well be said to constitute a violation of Article 21.'[109]

The Commission held that degradation of the environment caused by the project endangered the traditional livelihoods and survival of the Ogoni, giving rise to an infringement of the right to life.[110] It also recognised rights not explicitly mentioned in the African Charter – such as the right to be protected against forced evictions – through a construction of the right as a part of the right to housing. Forced evictions have been defined by the African Commission in accordance with the definition given by the Committee on Economic Social and Cultural Rights.[111]

---

[106] African Charter on Human and Peoples' Rights. (See above n 104). Article 16.
[107] African Charter on Human and Peoples' Rights. (See above n 104). Article 24.
[108] African Charter on Human and Peoples' Rights. (See above n 104). Article 21.
[109] *The Social and Economic Rights Action Center for Economic and Social Rights v. Nigeria*. 2001. Comm. No. 155/96. para 55.
[110] African Charter on Human and Peoples' Rights. (See above n 104). Article 4.
[111] *The Social and Economic Rights Action Center for Economic and Social Rights v. Nigeria*. 2001. Comm. No. 155/96. para 63.

In the Endorois case,[112] the African Commission once again asserted the importance of the rights of indigenous people over their traditional lands. This case dealt with the eviction of the Endorois – a pastoralist community in Kenya – from their traditional lands, for the establishment of a game reserve. Private concessions had also been granted over the land to a mining company. Acknowledging the importance of traditional lands for indigenous people it stated that,

> '[T]he African Commission is of the view that Endorois culture, religion, and traditional way of life are intimately intertwined with their ancestral lands - Lake Bogoria and the surrounding area. It agrees that Lake Bogoria and the Mochongoi Forest are central to the Endorois' way of life and without access to their ancestral land, the Endorois are unable to fully exercise their cultural and religious rights, and feel disconnected from their land and ancestors.'
>
> The Commission went on
>
> '[T]he first step in the protection of traditional African communities is the acknowledgment that the rights, interests and benefits of such communities in their traditional lands constitute "property" under the Charter and that special measures may have to be taken to secure such "property rights'.[113]

The African Commission found that Kenya had violated Article 1 (adopting legislative measures to give effect to the rights in the charter), Article 8 (freedom of conscience and religion), Article 14 (right to property), Article 17 (cultural rights), Article 21 (right to dispose off their wealth and natural resources) and Article 22 (right to economic, social and cultural development) of the African Charter. The violations committed were considered to be 'those that go to the heart of indigenous rights – the right to preserve one's identity through identification with

---

112 *Centre for Minority Rights Development (Kenya) and Minority Rights Group International on behalf of Endorois Welfare Council v. Kenya*. 2009. Comm. No. 276/03.

113 *Centre for Minority Rights Development (Kenya) and Minority Rights Group International on behalf of Endorois Welfare Council v. Kenya*. paras 155, 187. (See above n 112).

ancestral lands, cultural patterns, social institutions and religious systems.' [114]

This decision aligns the African Commission with decisions of the Human Rights Committee on Article 27 of ICCPR. Jurisprudence from the Inter-American system was also a source of inspiration, and the judgments given in *Saramaka v. Suriname* and *Moiwana v. Suriname* were discussed at length. Unfortunately, the implementation of the decision has been weak. The Commission had recommended that Kenya recognise the ownership rights of the Endorois, restitute their traditional lands, give them access to key surrounding sites for grazing, as well as religious and cultural purposes, pay adequate compensation as well as royalties to the Endorois and engage in constructive dialogue with them to implement the recommendations of the commission. Kenya failed to comply, owing to which a resolution was issued by the commission in November 2013 urging Kenya to implement the recommendations. This did not make a difference either.

This explains why the later case regarding the Ogiek people was transferred by the African Commission to the African Court of Human and People's Rights. The Court has stronger enforcement powers than the Commission. The Ogiek people are forest-dwelling hunter-gatherer communities, who live in the Mau Forest and in forests around Mt. Elgon in Kenya, and are dependent on these lands for their survival and cultural identity. The African Court found that the nature of the violations had led to a 'situation of extreme gravity and urgency, as well as a risk of irreparable harm to the Ogiek of the Mau Forest'. It held that the eviction of the Ogiek people without consultation had led to a violation of several rights contained in the ACHPR, including the right to non-discrimination (Art. 2), freedom of religion (Art. 8), property (Art. 14), culture and traditional values [Art. 17(2) and (3)], free disposal of wealth and natural resources (Art. 21) and economic, social and cultural development (Art. 22).[115]

---

[114] *Centre for Minority Rights Development (Kenya) and Minority Rights Group International on behalf of Endorois Welfare Council v. Kenya*. para 157. (See above n 112).

[115] *The Matter of African Commission on Human and Peoples Rights v Republic of Kenya* [2013] eKLR. Application NO.006/2012. Order dated 15 March 2013. para 25. Final judgment delivered on 26 May 2017.

The Inter-American and African systems attempt to deal with the problem of competing interests of indigenous people and the State, to ensure that indigenous rights are not set aside in the name of development. Nonetheless, restrictions can still be put on the land rights of indigenous people, if considered necessary to achieve 'a legitimate objective in a democratic society'.[116] This is also the position under international law, as understood through treaty provisions, general comments and case law. As stated by the Committee on Economic, Social and Cultural Rights, forced evictions can be justified, but only in the 'most exceptional circumstances, and in accordance with the relevant principles of International law'.[117] The fact that the possibility of the abrogation of the land rights of indigenous people is allowed under the current legal environment is a matter of concern, since there is a history of 'legitimate' objectives being given to justify undermining indigenous rights over their lands and resources, a practice that continues to this day.

Traditional communities and sustainable development

Having rights over land encourages sustainable management, as it gives rise to an interest in the maintenance of the land and its resources, ensuring its continued productivity. Recognising customary rights over traditional land has this benefit. Legal and policy arrangements that result in the commercialization of customary lands, result in the moving away from traditional natural resource management systems, harming the environment, as well as the well-being and standard of living of people. The Food and Agriculture Organisation of the United Nations (FAO) states in its report that 'progressive fragmentation of previously common rangelands into private parcels by a growing number of agrarian capitalists, elites, state agents and multinational investors is...likely to have profound impacts on the ability of households to get out of, or keep out of, poverty'.[118] This is well illustrated by policies such as Botswana's Tribal Land Grazing Policy and National Policy on

---

[116] The Court stated that "a State may restrict the use and enjoyment of the right to property where the restrictions are: a) previously established by law; b) necessary; c) proportional, and d) with the aim of achieving a legitimate objective in a democratic society." - *Saramaka People v. Suriname* (See above n 94).

[117] UN Committee on Economic Social and Cultural Rights, *General Comment 4: The Right to Adequate Housing: (Article 11(1))* General Comment No. 4 (13/12/91). para 18.

[118] Knight, R.S. p. 91. (See above n 53).

Agricultural Development. These policies facilitate the allocation of large tracts of communal land to cattle ranchers, giving rise to degradation due to over-use and overcrowding of areas that had been conserved and sustainably used for centuries through the implementation of customary norms.

Case study: Botswana

Nearly half of Botswana's land has been communally used and occupied. Concerns regarding over-grazing of communal land resulted in the adoption in 1975 of the Tribal Land Grazing Policy (henceforth 'TLGP'), based on the supposition of there being vast tracts of unused land that can be handed over to large-scale commercial ranch owners, purportedly to take the pressure off over-grazed communal lands. Despite protests the government re-zoned tribal land to give exclusive rights for commercial ranching to certain groups and individuals. Large tracts of communal grazing land were allocated to wealthy cattle owners through 50 year leases,without adequate consultation with the hunter-gatherer and pastoralist communities that were dependant on them, and despite the fact that under Article 17(4) of the Tribal Land Act land boards are meant to make such zoning decisions only after consultation with local authorities. The Land Policy Review explains the situation well:

> 'The Tribal Grazing Land Policy brought to the surface the insecurity of tenure faced by minority ethnic groups...[in that it] assumed that there were extensive areas of tribal land that could be zoned and fenced for commercial ranching, but it emerged that the veld was used and occupied by... Basarwa. In some cases, land boards... drew up appendices to the lease to be given to prospective ranchers, spelling out in detail the rights of [Basarwa] ranch residents to continue to hunt, gather, graze livestock, cultivate fields and even send their children to school. However, the Attorney General ruled these appendices illegal. Ranch lessees were given exclusive rights to the land'.[119]

[119] Government of Botswana (January 2003), *Quoted in* Knight, R.S. p. 83. (See above n 53).

> Matters were made worse by the 1991 National Policy on Agricultural Development (NPAD), which gave borehole owners 50-year exclusive rights to water and to the natural resources surrounding their borehole wells, allowing them to cordon off grazing lands around their boreholes, amounting to quite large areas – at times 3600–6400 hectares. By 2005, NPAD had enabled the enclosure of over 2 million hectares, amounting to 8% of Botswana's land area. This has had a detrimental effect on sustainable management of communal areas and customary land rights of tribal groups such as the Basarwa (San). It has also weakened the decision-making powers of land boards in relation to land and natural resources, as set out in the Tribal Land Act.
>
> A small number of wealthy families have now gained control over extensive areas of grazing and arable land, reducing the amount of such land available to small-scale and subsistence pastoralists. As a result a landless rural underclass has emerged. In addition, despite assertions to the contrary, these policies have increased degradation of communal grazing areas. Owners of ranches have in effect been given dual grazing rights – to communal areas during the monsoons and to their exclusively controlled ranches during the dry summer months and periods of drought. If their commercial ranches get over-grazed they simply move their cattle onto communal areas. What is more, the supposedly 'unused' land allocated for commercial ranches had been used by local communities as a safety net during the dry season. Since entry into these areas has been prohibited, communal grazing lands have become critically overgrazed and degraded.[120]

Access to land, sustainable management of the environment and eradication of hunger and poverty are inter-connected. Poverty and hunger can only be addressed through an approach that acknowledges the relationship between natural resources, the local communities dependent on them and the role of these communities in conservation. This was acknowledged during the 1996 Rome Declaration of the World Food Summit. The World Food Summit's Plan of Action, therefore, urges states to, 'establish legal and other mechanisms, as appropriate, that

[120] *Quoted in* Knight, R. S. p. 89. (See above n 53).

advance land reform, recognize and protect property, water, and user rights, and enhance access for the poor and women to resources. Such mechanisms should also promote conservation and sustainable use of natural resources (such as land, water and forests), lower risks, and encourage investment.'[121] Chapter 3 of Agenda 21, which is on poverty, also affirms that policies dealing with environment and conservation need to take cognizance of people dependent on natural resources, warning of adverse impacts on conservation if this is not done.[122]

Chapter 10 of Agenda 21 outlines an integrated approach to the management and planning of land resources, recognising the importance of 'private property rights, the rights of indigenous people and their communities and other local communities' to the achievement of sustainable management, and to avoid 'competition and conflicts[...] resulting in suboptimal use of both land and land resources'.[123] Governments are encouraged to enable equitable access to forest, water and land resources for indigenous people, small farmers, rural communities, women and the landless, as well as to implement policies that provide secure land tenure to vulnerable groups.[124] Chapter 14 of Agenda 21 urges governments to 'review and refocus existing measures to achieve wider access to land [and] assign clear titles, rights and responsibilities for land and for individuals or communities', because 'the greater the degree of community control over the resources on which it relies, the greater will be the incentive for economic and human resources development'.[125] Procedures are also required through which groups that have hitherto been excluded, such as indigenous people and women, are encouraged to participate in decision-making and implementation processes related to land resources and development.[126]

---

[121] World Food Summit Plan of Action (See above n 9).

[122] Agenda 21. Chapter 3. (See above n 72). - 'While managing resources sustainably, an environmental policy that focuses mainly on the conservation and protection of resources must take due account of those who depend on the resources for their livelihoods. Otherwise it could have an adverse impact both on poverty and on chances for long-term success in resource and environmental conservation. Equally, a development policy that focuses mainly on increasing the production of goods without addressing the sustainability of the resources on which production is based will sooner or later run into declining productivity, which could also have an adverse impact on poverty.'

[123] Agenda 21. Chapter 10, paras 10.1, 10.5 (See above n 72).

[124] Agenda 21. Chapter 14, para 14.17 (b) (See above n 72).

[125] Agenda 21. Chapter 10, para 14.16 (See above n 72).

[126] Agenda 21. Chapter 14, para 10.10 (See above n 72).

The Convention on Biological Diversity also affirms that protection and encouragement of 'customary use of biological resources in accordance with traditional cultural practices' is an element of the 'sustainable use of components of biological diversity'. States are obliged to adhere to this, but 'as far as possible and as appropriate,' giving them the discretion to design policies that they consider suitable for their country and people.[127]

Traditional agricultural and conservation practices of indigenous communities are being threatened by large-scale commercial biofuel production. This poses a risk to the survival of these communities, to their cultural integrity, as well as to the natural resources upon which they are so dependent. The environmental concerns arising out of this are but obvious. It is imperative that laws and policies are put in place that ensure that customary rights are not violated by biofuel companies, and that local communities are embraced as key stakeholders in sustainable development and conservation.

4. THE IMPACT OF BIOFUELS ON WOMEN

Every day, women spend hours gathering fire wood, collecting water, working in the fields. They are, in fact, the primary food producers in many countries, undertaking 60-80% of food production in developing countries.[128] Electricity and mechanization would enable them to have the time to engage in education, self-development activities and additional income generation, thereby raising their standards of living and that of their families. Growing biofuel crops can also help generate a better income. Value addition to the lives of women has, in fact, been cited as one of the reasons for engaging in biofuel production by several governments.

However, this viewpoint ignores some key issues. Firstly, biofuel production impacts women differently and disproportionately, owing to the inequalities that exist between men and women. Gender dynamics

---

[127] Convention on Biological Diversity. 1992. 1760 UNTS 79. Entered into force on 29 December 1993. Article 10 (c).

[128] Karlsson, G. and Banda, K. (*eds*). 2009. Biofuels for Sustainable Rural Development and Empowerment of Women: Case studies from Africa and Asia. Energia, Netherlands; Secretariat of the Convention on Biological Diversity. September 2011. Gender and Biodiversity: Living in Harmony with Nature. Secretariat of the Convention on Biological Diversity, Quebec, Canada.

and culturally influenced roles within the household and society result in lower levels of participation by women in decisions regarding natural resource allocation and use. Lack of ownership of land, gender discrimination within institutions and poor flow of information contribute to this. Secondly, in a number of countries women are excluded from benefit-sharing arrangements arising out of biofuel projects, as it is more common for farming contracts to be signed with male heads of households. Finally, their dependence upon the land means that they are particularly vulnerable to changes that occur when cash crop and biofuel plantations encroach upon land used by them for fodder and fuel collection, cattle grazing, farming and so forth. Availability of food, as well as income and employment prospects are some of the areas of women's lives that are affected.[129]

The importance of access to land and security of tenure for women has been highlighted by several international instruments. The Universal Declaration of Human Rights affirms the right of all people to own and administer property without discrimination (articles 2 and 17). The Convention on the Elimination of all Forms of Discrimination against Women (CEDAW) [articles 14(2) (g) and 15 (2)] asserts the right of women to equal treatment and rights in land and property matters.[130] Among soft law instruments, Resolution 15 of the UN Sub-Commission on Prevention of Discrimination and Protection of Minorities states that discrimination against women with regards to securing and acquiring land is a violation of human rights law, particularly the rights to protection against discrimination, equality, adequate housing and standard of living. It urges governments to amend laws and policies that infringe upon women's rights to equal access to land and security of tenure, as well as to work towards the transformation of discriminatory traditions and customs.[131] Women's empowerment and gender equality has also been laid down as an objective of the World Food Summit Plan of Action,[132] which envisions the introduction and

---

[129] This is the case across Africa and Asia.

[130] The Convention on the Elimination of All Forms of Discrimination against Women. Adopted and opened for signature, ratification and accession by General Assembly resolution 34/180 of 18 December 1979. 1249 U.N.T.S. 13. Art. 14(2)(g) and 15(2).

[131] Sub-Commission on Prevention of Discrimination and Protection of Minorities. 1998. Resolution 1998/15 on 'Women and the Right to Land, Property and Adequate Housing. paras 1 and 3.

[132] World Food Summit Plan of Action, Objective 1.3. (See above n 9).

enforcement of legislation that ensures women's access to and control over natural resources, such as water and land.[133] Chapter 24 of Agenda 21, which specifically focuses on women and 'sustainable and equitable development', has made the participation of women in ecosystem management a priority area of focus. Governments are to 'formulate and implement clear governmental policies and national guidelines, strategies and plans for the achievement of equality in all aspects of society, including...participation in key decision-making positions and in management of the environment, particularly as it pertains to their access to resources, [by] taking measures towards ensuring women's access to property rights.'[134] The Convention on Biological Diversity also acknowledges 'the vital role that women play in the conservation and sustainable use of biological diversity,'[135] affirming the 'need for the full participation of women at all levels of policy-making and implementation for biological diversity conservation'.[136] Similarly, the 1994 Convention to Combat Desertification asserts the importance of women's participation in order to manage and overcome desertification.[137] Article 8 of the Regional Implementation Annex for Africa also requires the participation of women in national action programmes.

Although the important role of women in natural resource management, as managers and guardians, has been acknowledged in a number of instruments, their participation in decisions related to natural resource management is still limited. The Second National Report of the Convention on Biological Diversity (2011) highlighted that only 25% of the States Parties had fully incorporated women and women's organisation in activities undertaken under the CBD. The Fifth National Report (2014) does not indicate much progress in this area either. Only 19% of the States Parties had recognised the different roles of men and women and the importance of women as managers of natural resources. Merely 5% had acknowledged the traditional knowledge of women, 28% had accepted that women are key stakeholders and only 8% included women's organisations in the implementation process. This

---

[133] World Food Summit Plan of Action, para. 16(b). (See above n 9).

[134] Agenda 21. Chapter 24. (See above n 72).

[135] Convention on Biological Diversity. Preamble. (See above n 127).

[136] Convention on Biological Diversity. Preamble. (See above n 127).

[137] United Nations Convention to Combat Desertification in Countries Experiencing Serious Drought and/or Desertification, Particularly in Africa. 1994. Adopted on 17 June 1994, entered into force 26 December 1996. 1954 UNTS 3. Articles 5, 10 and 19.

hinders the design and implementation of policies that could improve the position of women. Under-representation in biodiversity management means that their knowledge of biodiversity is not acknowledged and the impacts on them of decisions pertaining to natural resources are not sufficiently taken into account.[138]

5. BIOFUELS AND WATER

Biofuel crops are heavy on water, requiring it for the growth as well as production stage. On an average the amount of water needed for the production of 1 liter of biofuels is 850 liters.[139] This varies considerably across regions, depending on the method of irrigation, availability of water and type of crop. Water use in biofuel production can, therefore, lead to competition for water and impact the right to access water. For instance, sugarcane and maize crops require large quantities of water, problematic where water scarcity is an issue, such as in parts of Asia and Sub-Saharan Africa. The average consumptive water use per gallon of maize-based ethanol in China is almost 1.5 times that in the United States. In China, 2400 liters of water is required for the production of 1 liter of ethanol from maize.[140] Plans to boost domestic production of biofuels raises serious concerns regarding water sustainability and the distribution of limited water resources between food and biofuel crops. Biofuel production can lead to an increase in water scarcity, especially in water scarce areas. The problem is different in rain fed areas, here biofuel crops use 'green water', *i.e.*, water stored in the soil, which if over-exploited could lead to a depletion of groundwater, along with the impact it has on rivers. Thus, biofuel production causes competition

---

[138] IUCN Global Gender Office. 2017. Advancing gender equality in biodiversity conservation: A review of the fifth national reports to the Convention on Biological Diversity (CBD). IUCN Global Gender Office, Washington D.C.; IUCN Global Gender Office. 2016. Inclusion and characterization of women and gender equality considerations in the fifth national reports to the CBD. IUCN Global Gender Office, Washington D.C.

[139] León-Moreta, M. 2011. Biofuels - A Threat to the Environment and Human Rights? An Analysis of the impact of the production of feedstock for agro-fuels on the rights to water, land and food. *European Journal of Legal Studies,* 4(1): 102–120.

[140] de Fraiture, C. *et al.,* 2008. Biofuels and Implications for Agricultural Water Use: Blue Impacts of Green Energy. *Water Policy,* (10)1: 67–81. p. 79; Varghese, S. 2007. *Biofuels and Global Water Challenges.* Institute for Agriculture and Trade Policy, Minnesota, USA. p. 4.

over water resources and infringes the right to water if the state is not able to reconcile opposing interests.

Currently, the role of biofuels in energy production is limited, but this sector is expected to grow to the extent that, as de Fraiture estimates, globally 30 million hectares of agricultural land will be required, as well as an additional withdrawal of irrigation water amounting to 180 km^3. This may not add up to much globally, but at the level of individual countries can take a heavy toll, depending on the availability of water resources. It is, in fact, thought that developing countries such as India and China will not be able to 'meet future food, feed and biofuel demand without substantially aggravating already existing water scarcity problems, or importing grain.'[141] Assessments and projections regarding water management in the agricultural sector indicate that if adequate policy measures are not implemented water crises will only increase.

Case study: Ethiopia

The growth of the biofuels industry in Ethiopia has had an impact on water use, rural livelihoods and the price of food. One of the impacts of increasing the area under biofuel cultivation has been a reduction in the cultivation of cotton, which impacts the textile industry and thereby employment. Pastoralists who graze their animals on cotton stubble are also affected. Reduction of water available for the irrigation of food crops is another consequence, resulting in a rise in food prices.

Biofuel production also leads to water pollution caused by pesticide and fertilizer run-off. The 'dead zone' in the Gulf of Mexico, where no marine life can survive, was a result of the washing of nitrogen into the Mississippi River. Similarly, the cultivation of sugarcane in Brazil has been a cause of water pollution. The Ipojuca River in Northeastern Brazil, has suffered an increase in turbidity, nitrate leaching and acidification, as well as an imbalance of oxygen. Water and soil contamination also occurs through a practice known as 'ferti-gation' (ferti-irrigation), practiced commonly in Sao Paulo. This is a method for disposing vinasse – a highly polluting effluent created by the production of sugarcane based ethanol – into sugarcane fields. Vinasse harms aquatic life by causing a reduction in oxygen levels and an increase in water and soil

[141] de Fraiture, C. *et al.* p. 78. (See above n 140).

acidification. It also leads to nutrient run-off and sediment contamination due to soil erosion. Each year, soil erosion in sugarcane fields is as high as 30 tonnes per hectare. Yet, the sugarcane industry continues to support the use of vinasse, claiming that it helps soil nutrient enrichment and moisture replacement. Adverse impacts on water quality have also been reported in ethanol producing countries such as South Africa and India.[142] Thus, although biofuel crops can help boost farmers' incomes, as well as energy production, the industry needs to be managed properly to avoid negative environmental and social consequences.

5.1. Right to Water and Biofuels

The human right to water has been recognised, explicitly or implicitly, in several international instruments, including declarations and treaties. In 2010, the UN General Assembly passed a resolution declaring that clean drinking water is 'essential to the full enjoyment of life and all other human rights' and that every person has the right to access sufficient quantities of water for domestic and personal use.[143]

> 'The Assembly recognized the right of every human being to have access to sufficient water for personal and domestic uses (between 50 and 100 liters of water per person per day), which must be safe, acceptable and affordable (water costs should not exceed 3 per cent of household income), and physically accessible (the water source has to be within 1,000 meters of the home and collection time should not exceed 30 minutes).'[144]

---

[142] Charlotte de Fraiture and Göran Berndes. 2009. Biofuels and Water. *In:* Howarth, R.W. and Bringezu, S. (*eds.*), Biofuels: Environmental Consequences and Implications of Changing Land Use. Proceedings of the Scientific Committee on Problems of the Environment (SCOPE) International Biofuels Project Rapid Assessment, 22-25 September 2008, Gummersbach Germany; The National Academy of Sciences. 2007. Water Implications of Biofuels Production in the United States. National Academies Press, Washington D.C.; Varghese, S. 2007. *Biofuels and Global Water Challenges*. Institute for Agriculture and Trade Policy, Minnesota, USA.

[143] This led to several states stating that the right to water is a part of their treaty obligations under the ICESCR even though the treaty only accepts the right implicitly, such as Germany, United Kingdom and the Netherlands.

[144] UN General Assembly Resolution No. 10967 on the Human Right to Water and Sanitation. July 2010. A/64/L.63/Rev.1 + Add.1. The General Assembly adopted the resolution recognizing access to clean water and sanitation as a human right by a vote of 122 in favour and 41 abstentions. None voted against the resolution.

The UN General Assembly also urged States, as well as international organisations, to 'provide financial resources, build capacity and transfer technology, particularly to developing countries, in scaling up efforts to provide safe, clean, accessible and affordable drinking water and sanitation for all.'[145] Although non-binding, the resolution strengthens the affirmation of the right to water at the international level.

The Committee on Economic, Social and Cultural Rights (CESCR) echoes the view of the General Assembly in its description of water as a 'limited natural resource and a public good fundamental for life and health'.[146]

The content and scope of the right to water has been outlined by the CESCR in General Comment 15, which explains that 'the human right to water entitles everyone to sufficient, safe, acceptable, physically accessible and affordable water for personal and domestic uses.'[147] Safety refers to water sources being free from pollutants, while accessibility connotes both availability and accessibility of water for domestic and personal needs, including both physical and economic accessibility. Pollution and depletion of water sources violate the right to safe drinking water.

The significance of the right to water for the fulfillment of other human rights, such as the right to life and health, cannot be emphasised enough and is reflected in its expression in various international instruments. The 1989 Convention on the Rights of the Child (CRC), the 1979 Convention on the Elimination of Discrimination against Women (CEDAW) and the 2006 Convention on the Rights of Persons with Disabilities (CRPD) are some of the treaties that recognize the importance of this right. Article 24 of CRC recognizes the importance of the right to water for the fulfillment of the right to health. Article 14 of CEDAW connects the right to water with the right to adequate living conditions in relation to problems faced by women in rural areas.[148] CRPD affirms the right of disabled persons to access clean water.[149]

---

[145] UN General Assembly Resolution No. 10967. (See above n 144).

[146] UN Committee on Economic, Social and Cultural Rights. General Comment 15 on the Right to Water (Adopted at the Twenty-ninth session of the Committee on Economic, Social and Cultural Rights, 2003). U.N. Doc. E/C.12/2002/11. para 1.

[147] General Comment 15, para 2. (See above n 146).

[148] Convention on the Elimination of All Forms of Discrimination against Women. Article 14(2)(h). (See above n 130).

[149] The Convention on the Rights of Persons with Disabilities. Adopted by the General Assembly on 24 January 2007. A/RES/61/106. Article 28.

Regional instruments have affirmed the right as well, such as the African Charter on the Rights and Welfare of the Child, which views it in relation to the right to health (article 14) and the Protocol to the African Charter on Human and People's Rights on the Rights of Women in Africa, which brings it within the right to food security [article 15(a)]. Several countries have also taken steps to protect the right within their domestic legal systems.[150]

Access to clean water for domestic and personal use continues to be impacted by increasing demands from industry, agriculture and energy generation projects. Growth in demand for biofuels will only give rise to more problems. If effective management practices are not put in place to ensure that biofuel production does not adversely impact the environment, contamination and depletion of water sources will increase. Furthermore, rapidly growing economies like India and China are not likely to be able to meet biofuel and food demands without giving rise to water scarcity. In China, 550 of its largest cities are already facing water shortage and its cities and industries compete with farmers for access to water. Both China and India have growing middle class populations, aspiring for a lifestyle comparable to that of the United States. These competing interests and needs will only aggravate the water crisis.[151]

There are a variety of agricultural practices and technologies that can increase crop yields and reduce adverse impacts on water resources, such as methods of irrigation that conserve water, water recycling, techniques that prevent erosion and those that promote fertilizer efficiency. Such practices have the potential to impact the environment

---

[150] *Subhash Kumar v. State of Bihar* (1991) 1 SCC 598 (Supreme Court of India); *Delhi Water Supply and Sewage Disposal Undertaking v. State of Haryana* (1996) 2 SCC 572 (Supreme Court of India); *Matsipane Mosetlhanyane and Ors v. The Attorney General* (2011) Civil Appeal No. CACLB-074-10 (Botswana); *Jorge Hernán Gómez Ángel v. Alcalde Municipal de Versalles–Valle del Cauca y el Gerente de la Empresa de Servicios Públicos de Versalles* (2003) C.C. T-410/03 (Colombia).

[151] Stanford University Woods Institute for the Environment. The Impacts of Large-Scale Biofuel Use on Water Resources. <*https://woods.stanford.edu/sites/default/files/files/Biofuels-Policy-Brief-4-20061205.pdf*>; Varghese, S. 2007. *Biofuels and Global Water Challenges*. Institute for Agriculture and Trade Policy, Minnesota, USA; International Water Management Institute. Water Policy Brief: Water Implications of Biofuel Crops: Understanding Tradeoffs and Identifying Options. Issue 30.

positively and should be encouraged through policy and law, as should policies that encourage farmers to produce biofuels from cellulosic alternatives (second generation biofuels) that have less impact on water quality per unit of energy gained.

6. CONCLUSION

Biofuels are seen as an answer to the energy requirements of nations, bringing about greater security through reduced dependence on fossil fuels. It is also claimed that they are a solution to climate change since they can help decrease greenhouse gas emissions from automobiles; the validity of this is, however, questioned by several studies. If the biofuels industry is to achieve the purposes for which it is purportedly encouraged – rural empowerment, economic development, reduction of GHG emissions and so forth – the host of problems caused by their production – environmental and social – need to be addressed. Human rights and environmental standards need to be incorporated into each stage of biofuel production and a thorough impact assessment conducted before they are sanctioned. Where rights exist under international law, such as to consultation, participation and consent, they should be implemented in the planning and implementation process; where there are gaps – customary land rights being a case in point – they should be addressed through domestic, regional and international laws.

In the absence of an international instrument governing biofuel production, regional and national legislations play a key role in filling legal and policy gaps. A variety of binding and non-binding instruments have been adopted to regulate the industry. An example of non-binding guidelines being the RSB Principles, developed by the Roundtable on Sustainable Biofuels (RSB), that incorporate safeguards related with human rights and the environment. Principle 2 addresses the requirement of consultation and participation, stating that:

> 'Sustainable biofuel operations shall be planned, implemented, and continuously improved through an open, transparent, and consultative impact assessment and management process...'[152]

[152] RSB Principles and Criteria for Sustainable Biofuel Production. 2010. RSBSTD-01-001. Principle 2.

Principle 12 deals with land rights, both formal and customary, Principle 9 with water rights, including respecting customary water rights[153] and Principle 4 with human and labour rights.[154] The RSB principles also focus on ensuring that biofuel production does not compromise food security.[155]

The European Union, which had a target of 10% as a 'binding minimum target to be achieved by all Members States for the share of biofuels in overall EU transport petrol and diesel consumption by 2020',[156] has introduced laws to regulate the industry and ensure that it respects human rights and environmental standards. The EU amended its 10% target in 2015 to reduce indirect land use change, and ensure food security and sustainability. While the overall target has remained 10%, it was decided that the share towards the 2020 targets of biofuels from crops grown on agricultural land would be limited to 7%. New rules were formulated in 2015 to amend the Renewable Energy Directive and the Fuel Quality Directive accordingly. In addition, it is mandated that the supply of biofuels from non-EU countries should meet sustainability criteria.[157]

On the other hand, the regulatory landscape for biofuel production in Africa, South America and Asia is still developing, with some countries having stronger laws than others. Codification of principles of environmental and human rights law in the form of a treaty governing

---

153 *Biofuel operations shall maintain or enhance the quality and quantity of surface and ground water resources, and respect prior formal or customary water rights* - RSB Principles & Criteria for Sustainable Biofuel Production. (See above n 153).

154 *Biofuel operations shall not violate human rights or labor rights, and shall promote decent work and the well-being of workers* - RSB Principles & Criteria for Sustainable Biofuel Production. (See above n 152).

155 Principle 6: *Biofuel operations shall ensure the human right to adequate food security in food insecure regions*- RSB Principles & Criteria for Sustainable Biofuel Production. (See above n 152).

156 Council of the European Union. 2007. Presidency Conclusions of the Brussels European Council (8/9 March 2007). 7224/1/10 Rev 1 Annex I. para 7.

157 Commission of the European Communities. 2008. Proposal for a Directive of the European Parliament and of the Council on the promotion of the use of energy from renewable sources. COM (2008) 19 final 4-8; Council Directive 2009/28/EC of 23 April 2009 promoting the use of energy from renewable sources and amending and subsequently repealing Directives 2001/77/EC and 2003/30/EC (2009) OJ L140/16. paras 6, 65-66.

biofuel production could bring much-needed harmonization and standardization in the application of these principles, encouraging their application across the board.

Striking a balance between economic, social and environmental interests in the planning and implementation of biofuel projects would be beneficial for the sustainability of businesses as well, as adverse environmental and human rights consequences reduce the economic viability of a business in the long-run. It would also help to ensure that biofuel projects actually minimise GHG emissions, and bring about rural empowerment and economic development that is equitable and sustainable.

Finally, some hard questions about the growth of the biofuels industry. Should it be encouraged at all considering its environmental and social costs? For the sake of following the mainstream conception of development, in itself not sustainable, are we willing to continue the destruction of forests and grasslands, and cause the large-scale displacement of people? When a large percentage of the global population is facing hunger and poverty, how much space should be given to this fast-growing industry?

BIBLIOGRAPHY

Books

Alston P. and Tomasevski, K. (*eds.*), 1984. The Right to Food. Martinus Nijhoff Publishers, Netherlands.

Anaya, S.J. 2009. The Right of Indigenous Peoples to Self-Determination in the Post–Declaration Era. *In*: Chartres, C. and Stavenhagen, R. (*eds.*), Making the Declaration Work: The United Nations Declaration on the Rights of Indigenous Peoples. International Work Group for Indigenous Affairs, Denmark.

Bouthillier, Yves Le, Cowie, A., Martin, P., McLeod-Kilmurray, H. The Law and Policy of Biofuels. Edward Elgar Publishing, UK.

Darrow, M. 2003. Between Light and Shadow: The World Bank, The International Monetary Fund, and International Human Rights Law. Hart Publishing, UK.

De Fraiture, C. and Berndes, G. 2009. Biofuels and Water. *In*: Howarth, R.W. and Bringezu, S. (*eds.*), Biofuels: Environmental Consequences and Implications of Changing Land Use. Scope, <*http://np-net.pbworks.com/f/SCOPE%20(2009)%20Biofuels%20%26%20Water.pdf*>.

Donnelly, J. 2003. Universal Human Rights in Theory and Practice, 2nd edn. Cornell University Press, USA.

Dowell-Jones, M. 2004. Contextualising the International Covenant on Economic, Social and Cultural Rights: Assessing the Economic Deficit. Martinus Nijhoff Publishers, Holland.

Eduardo Chiziane *et al.* 'Legal frameworks enabling sustainable land-use investment in Mozambique: Current strengths and opportunities for improvement' (Center for International Forestry Research, 2015).

León Moreta, M.A. 2015. The Human Rights Fundaments of Conservation in the Context of the Extraction of Energy Resources. V&R Unipress, Germany.

Moeckli, D. *et al.*, 2013. International Human Rights Law. Oxford University Press, UK.

Parliament of Great Britain, House of Commons, Environmental Audit Committee. Are Biofuels Sustainable? First Report of Session 2007-08: Volume II. January 15, 2008. (The Stationary Office, London).

Salomon, M.E. 2007. Global Responsibility for Human Rights: World Poverty and the Development of International Law. OUP, UK.

Twomey, P. 2007. Human Rights-Based Approaches to Development: Towards Accountability. *In:* Baderin, M.A. and McCorquodale, R. (*eds.*), Economic, Social and Cultural Rights in Action. OUP, UK.

Journal Articles

Acheampong, E. and Campion, B.B. 2014. The effects of biofuel feedstock production on farmers' livelihoods in Ghana: The case of *Jatropha curcas. Sustainability,* (6): 4587–4607.

Bamikole, A. *et al.*, 2010. Biofuels and sustainability in Africa. *Renewable and Sustainable Energy Reviews*, pp. 1–13.

Bastos Lima, M.G. 2009. Biofuel governance and international legal principles: Is it equitable and sustainable? *Melbourne Journal of International Law,* 10(1): 1–22.

Biswas, W.K. *et al.* 2001. Model for empowering rural areas through renewable energy technologies in Bangladesh. *Environmental Sciences & Policy,* (4): 333–344.

Borras, S.M. Jr. *et al.*, 2011. The politics of agro-fuels and mega-land and water deals: Insights from the ProCana case, Mozambique. *Review of African Political Economy,* 28(128): 215–234.

Borras, S.M. Jr. *et al.*, 2012. Land grabbing in latin America and the Caribbean. *Journal of Peasant Studies,* 39(3–4): 845–872. pp. 857-858.

Clavero, B. 2005. The indigenous rights of participation and international development policies. *Arizona Journal of International and Comparative Law,* 22(1): 41–51.

Davis, M. 2008. Indigenous struggles in standard-setting: The United Nations Declaration on the Rights of Indigenous Peoples. *Melbourne Journal of International Law,* 9(2): 458–59.

De Fraiture, C. *et al.* 2008. Biofuels and implications for agricultural water use: Blue impacts of green energy. *Water Policy,* (10)1: 67–81.

Deda, P. and Rubian, R. 2004. Women and biodiversity: The long journey from users to policy-makers. *Natural Resources Forum,* 28(3): 201–204.

Elobeid, A. and Hart, C. 2007. Ethanol expansion in the food versus fuel debate: How will developing countries fare? *Journal of Agricultural & Food Industrial Organization,* 5(6): 1–21.

International Water Management Institute. Water Policy Brief: Water Implications of Biofuel Crops: Understanding Tradeoffs and Identifying Options. Issue 30.

Kachika, T. Land Grabbing in Africa: A Review of the Impacts and the Possible Policy Responses. Oxfam International.

León-Moreta, M. 2011. Biofuels - A threat to the environment and human rights? An analysis of the impact of the production of feedstock for agro-fuels on the rights to water, land and food. *European Journal of Legal Studies,* 4(1): 102–120.

Madhu Ramnath. 2008. Surviving the Forest Rights Act: Between Scylla and Chaybdis. *Economic and Political Weekly,* 43(1): 37–42.

McGahey, D.J. 2008. Bioenergy – a new opportunity or a major threat to pastoralism? *SOS Sahel International UK. Haramata,* 54: 14–16.

McGahey, D.J. 2008. Bioenergy and Pastoralism: Challenging the Wastelands Myth. *SOS Sahel International UK,* pp. 1–28.

Nalepa, R.A. 2011. A question of scale: The construction of marginal lands and the limitations of global land classifications. Paper presented at the International Conference on Global Land Grabbing.

Nuffield Council on Bioethics. 2011. Biofuels: Ethical Issues. Nuffield Press. 4.12.

O'Flaherty, M. 2006. The Concluding Observations of United Nations Human Rights Treaty Bodies. *Human Rights Law Review,* 6(1): 27–52.

Olanya, D.R. 2012. From global land grabbing for biofuels to acquisitions of African water for commercial agriculture. *Current African Issues,* 50: 4–31.

Page, A. 2004. Indigenous peoples free prior and informed consent in the interamerican human rights system. *Sustainable Development Law and Policy,* 4(2): 16–20.

Phalan, B. 2009. The social and environmental impacts of biofuels in Asia: An overview. *Applied Energy,* 86: 21–29.

Rachel, A. Nalepa 2011. 'A question of scale: The construction of marginal lands and the limitations of global land classifications'. Paper presented at the International Conference on Global Land Grabbing.

Rist, L. *et al.* 2009. Biofuels: Social Benefits. *Science,* 326(5958): 1344.

Saturnino, M., Borras, Jr. *et al.* 2011. The politics of agrofuels and mega-land and water deals: insights from the ProCana case, Mozambique. *Review of African Political Economy*, 38: 216–217.

Tenenbaum, D.J. 2008. Food *vs.* Fuel: Diversion of crops could cause more hunger. *Environmental Health Perspectives*, 116(6).

The National Academy of Sciences. 2007. Water Implications of Biofuels Production in the United States.

Tramontana, E. 2010. The contribution of the Inter-American human rights bodies to evolving international law on indigenous rights over lands and natural resources. *International Journal of Minor Group Rights,* 17(2): 241–263.

Veit, P. 2010. Biofuels and Land Use in Tanzania. *World Resources Institute, Landesa.*

Vermeulen, S. and Cotula, L. 2010. Over the heads of local people: consultation, consent and recompense in large-scale land deals for biofuels projects in Africa. *The Journal of Peasant Studies,* 37(4): 899–916.

Ward, T. 2011. The right to free, prior and informed consent: Indigenous peoples' participation rights within international law. *Northwestern Journal of International Human Rights,* 10(2): 54–84.

Xanthaki, A. 2009. Indigenous rights in international law over the last 10 years and future developments. *Melbourne Journal of International Law,* 10(1): 36.

Research Papers

Ahmed, A.U. *et al.* 2007. The world's most deprived: Characteristics and causes of extreme poverty and hunger 2020. *International Food Policy Research Institute, Discussion Paper,* 43: 1–96.

Cernea, M.M. 2004. Impoverishment risks, risk management and reconstruction: A model of population displacement and resettlement. Paper presented at the UN Symposium on Hydropower and Sustainable Development, Beijing.

Chakravorty, U., Hubert, M., Moreaux, M. and Nøstbakken, L. 2015. The long-run impact of biofuels on food prices. *Resources for the Future.*

Cotula, L. *et al.* 2008. Fuelling exclusion? The biofuels boom and poor people's. Food and Agriculture Organization of the United Nations, Rome.

Cotula, L. *et al.* 2009. Land grab or development opportunity? Agricultural investment and international land deals in Africa. FAO, IIED and IFAD, London/Rome. p. 71.

Deininger, K. Land Policies and Land Reform. *World Bank,* pp. 1–30.

Dietz, K. *et al.* 2014. The Political Ecology of Agrofuels. Routledge, USA. 98; World Bank Group. 2011. Rising Global Interest in Farmland: Can it Yield Sustainable and Equitable Benefits? World Bank, Washington DC.

Eide, A. 2008. The right to food and the impact of liquid biofuels (agrofuels), *Food and Agriculture Organization of the United Nations.*

Food and Agricultural Organisation of the United Nations. 2015. The State of Food Insecurity in the World: Meeting the 2015 International Hunger Targets: Taking Stock of Uneven Progress.

Food and Agricultural Organisation. 2015. State of Food Insecurity in the World 2015: The FAO Hunger Map.

Food and Agriculture Organisation of the United Nations. 2016. State of Food Insecurity in the World. Food and Agriculture Organisation, Rome.

Herre, R. (*ed.*), 2010. German Investment Funds involved in Land Grabbing. FIAN Deutschland e.V., Köln, Germany.

Institute for Food and Development Policy. 2016. *Food First Backgrounder,* 22(2A): 1–4.

Karlsson, G. and Banda, K. (*eds.*), 2009. Biofuels for Sustainable Rural Development and Empowerment of Women: Case studies from Africa and Asia. Energia, Netherlands.

Knight, R.S. 2010. Statutory recognition of customary land rights in Africa: An investigation into best practices for lawmaking and implementation. Food and Agriculture Organization of The United Nations, Rome.

Kugelman, M. and Levenstein, S.L. *(eds.),* 2009. Land Grab? The Race for the World's Farmland. 2009. Woodrow Wilson International Center for Scholars, Washington, D.C.

Quesada-Aguilar, A. and Mata, G. Gender and National Biodiversity Strategies and Action Plans (NBSAPs). IUCN.

Rights and Resources Initiative 2015. Who Owns the World's Land? A global baseline of formally recognized indigenous and community land rights.

Secretariat of the Convention on Biological Diversity. 2011. Gender and Biodiversity: Living in Harmony with Nature.

Secretariat of the Convention on Biological Diversity. 2011. Gender and Biodiversity: Living in Harmony with Nature.

Stanford University Woods Institute for the Environment. The Impacts of Large-Scale Biofuel Use on Water Resources.

The National Academy of Sciences 2007. 'Water Implications of Biofuels Production in the United States'.

The Oakland Institute 2011. Understanding Land investment Deals in Africa: Country Report Mozambique. The Oakland Institute, Oakland, CA.

Twomey, H. 2014. Displacement and dispossession through land grabbing in Mozambique: The Limits of International and National Legal Instruments. Refugee Studies Centre Working Paper Series No, 101.

World Food Programme 2016. Hunger Statistics.

International Legal Documents

African Charter on Human and People's Rights. OAU Doc. CAB/LEG/67/3 rev. 5, 21 I.L.M. 58 (1982). Adopted 27 June 1981, entered into force 21 October 1986.

African Commission on Human and Peoples' Rights, Report of the African Commission's Working Group of Experts on Indigenous Populations/ Communities, adopted by the African Commission at its 34th Ordinary Session, November 2003.

Agenda 21. 1992. Adopted at the United Nations Conference on Environment & Development Rio de Janerio, Brazil, 3 to 14 June 1992. A/CONF.151/26/ Rev.1 (vol. I), annex II.

American Convention on Human Rights. *Adopted at an Inter-American Specialized Conference on Human Rights, San José, Costa Rica, 22 November 1969.* O.A.S. T.S. No. 36.

American Declaration of the Rights and Duties of Man. *Adopted at the Ninth International Conference of American States, Bogota, Colombia, 2 May 1948.* O.A.S. Res. XXX.

Commission on Human Rights. 16 April 2004. Resolution: 2004/28 on Prohibition of Forced Evictions. CHR Res 2004/28,E/CN.4/RES/2004/28.

Convention on Biological Diversity. 1992. Adopted at Rio de Janeiro on 5 June 1992. 1760 UNTS 79.

Convention on the Elimination of All Forms of Discrimination against Women. 18 December 1979. 1249 U.N.T.S. 13. Adopted by the UN General Assembly on 18 December 1979.

Convention on the Rights of Persons with Disabilities. Adopted by the General Assembly on 24 January 2007. A/RES/61/106.

Food and Agriculture Organisation. 13-17 November 1996. World Food Summit Plan of Action.

Human Rights Council. 27 March 2008. Resolution 7/14 on the Right to Food.

International Convention on the Elimination of All Forms of Racial Discrimination. 21 December 1965. 660 U.N.T.S. 195. (entered into force on 4 January 1969). International Covenant on Economic, Social and Cultural Rights. 16 December 1966. Adopted and opened for signature, ratification and accession by General Assembly Resolution 2200A (XXI), entered into force 3 January 1976.

Office of the United Nations High Commissioner for Human Rights. Principles and Guidelines for a Human Rights Approach to Poverty Reduction Strategies. HR/PUB/06/12.

Sub-Commission on Prevention of Discrimination and Protection of Minorities. 1998. Resolution 1998/15 on 'Women and the Right to Land, Property and Adequate Housing.

U.N. Commission on Human Rights. January 21, 2003. Report of the Special Rapporteur on the situation of human rights and fundamental freedoms of indigenous people, Rodolfo Stavenhagen, submitted in accordance with Commission resolution 2001/65 (Fifty-ninth session). U.N. Doc. E/CN.4/2003/90.

U.N. Commission on the Elimination of Racial Discrimination. 22 August 1997. Report of the Committee on the Elimination of Racial Discrimination: Annex V - General Recommendations XXIII. U.N. Doc. Supplement No. 18 (A/52/18).

U.N. High Commissioner for Human Rights. 12-30 March, 2007. Concluding Observations of the Human Rights Committee: Chile. U.N. Doc. CCPR/C/CHL/CO/5.

U.N. High Commissioner for Human Rights. 17 April, 2008. Concluding Observations of the Human Rights Committee: Panama. U.N. Doc. CCPR/C/PAN/CO/3.

U.N. High Commissioner for Human Rights. Dec. 12, 2008. Concluding observations of the Human Rights Committee: Nicaragua. U.N. Doc. CCPR/C/NIC/CO/3.

UN Committee on Economic Social and Cultural Rights. 11 August 2000. General Comment 14 - The Right to the Highest Attainable Standard of Health (Article 12). Adopted at the Twenty-second Session of the Committee on Economic, Social and Cultural Rights, on 11 August 2000. U.N. Doc. E/C.12/2000/4.

UN Committee on Economic Social and Cultural Rights. 12 May 1999. General Comment No. 12: The Right to Adequate Food (Article 11). Adopted at the Twentieth Session of the Committee on Economic, Social and Cultural Rights. U.N. Doc. E/C.12/1999/5.

UN Committee on Economic Social and Cultural Rights. 13 December 1991. General Comment 4: The Right to Adequate Housing: (Art. 11 (1) of the Covenant).

Adopted at the Sixth Session of the Committee on Economic, Social and Cultural Rights, on 13 December 1991. U.N. Doc. E/1992/23.

UN Committee on Economic, Social and Cultural Rights. 21 December 2009. General Comment no. 21, Right of everyone to take part in cultural life {art. 15(1) (a) of the International Covenant on Economic, Social and Cultural Rights}. Adopted at the Forty-third Session of the Committee on Economic, Social and Cultural Rights, on 21 December 2009. U.N. Doc. E/C.12/GC/21.

UN Committee on Economic, Social and Cultural Rights. General Comment 15 on the Right to Water (Adopted at the Twenty-ninth session of the Committee on Economic, Social and Cultural Rights, 2003). U.N. Doc. E/C.12/2002/11.

UN Human Rights Committee. 30 April 1982. General Comment No. 6: Article 6 (Right to Life). Adopted at the Sixteenth Session of the Human Rights Committee.

UN Human Rights Committee. 8 April 1994, CCPR General Comment No. 23: Article 27 (Rights of Minorities). CCPR/C/21/Rev.1/Add.5, para 3.2.

UN Human Rights Committee. 8 April 1994. CCPR General Comment No. 23: Article 27 (Rights of Minorities). CCPR/C/21/Rev.1/Add.5, para 7.

UN Human Rights Council. 15 July 2009. Report of the Special Rapporteur on the Situation of Human Rights and Fundamental Freedoms of Indigenous People. U.N. Doc. A/HRC/12/34.

UN Millennium Development Goals. 2000. Adopted at the UN Millennium Declaration. Goal 1: Eradicate Extreme Hunger and Poverty.

United Nations Conference on Environment and Development. 14 June 1992. Rio Declaration on Environment and Development. A/CONF.151/26.

United Nations Convention to Combat Desertification in Countries Experiencing Serious Drought and/or Desertification, Particularly in Africa. 1994. 1954 UNTS 3.

United Nations Declaration on the Rights of Indigenous Peoples. 13 September 2007. Adopted by the General Assembly by resolution No. 61/295.

United Nations General Assembly. 18 December 2009. Resolution adopted by the General Assembly, A/RES/64/159.

United Nations General Assembly. Resolution adopted by the General Assembly on 18 December 2009 (Sixty-fourth session). A/RES/64/172.

Universal Declaration of Human Rights. Adopted 10 December 1948. UNGA Res 217 A(III). Universal Declaration on the Eradication of Hunger and Malnutrition, Adopted on 16 November 1974 by the World Food Conference convened under General Assembly Resolution 3180 (XXVIII) of 17 December 1973 and endorsed by General Assembly Resolution 3348 (XXIX) of 17 December 1974.

Vienna Declaration and Programme of Action. 25 June 1993. Adopted by the World Conference on Human Rights in Vienna, UN Doc. A/CONF.157/23.

Regional and National Legal Documents

Commission of the European Communities. 2008. Proposal for a Directive of the European Parliament and of the Council on the promotion of the use of energy from renewable sources. COM (2008) 19 final 4-8.

Council Directive 2009/28/EC of 23 April 2009 promoting the use of energy from renewable sources and amending and subsequently repealing Directives 2001/77/EC and 2003/30/EC (2009) OJ L140/16.

Council of the European Union. 2007. Presidency Conclusions of the Brussels European Council (8/9 March 2007). 7224/1/10 Rev 1 Annex I.

Government of South Africa Department of Minerals and Energy. December 2007. Biofuels Industrial Strategy of the Republic of South Africa. *<http:// www.energy.gov.za/files/resources/renewables/biofuels_indus_strat. pdf(2).pdf>*

Government of South Africa. 2017. Energy.

Land Act 1999. Act No. 4 of 1999. Government of Tanzania.

Lei de Terras, 1997. Lei No. 19/1997. Government of Mozambique.

RSB Principles and Criteria for Sustainable Biofuel Production. 2010. RSB-STD-01-001.

Tribal Land Act (Chapter 32:02), Date of Original Text: 1968. Date of consolidation/reprint: 31 December 2008.

Village Land Act 1999. Act No. 5 of 1999. Government of Tanzania

Case Law

Centre for Minority Rights Development (Kenya) and Minority Rights Group International on behalf of Endorois Welfare Council v. Kenya. 2009. Comm. No. 276/03.

Delhi Water Supply and Sewage Disposal Undertaking v. State of Haryana (1996) 2 SCC 572 (Supreme Court of India).

Mary and Carrie Dann v. U.S. Inter-American Commission of Human Rights, Case 11.140, 2002. Report No. 75/02, OEA/Ser.L/V/II.117, doc. 5. Rev, para 1.

Maya Indigenous Communities of the Toledo Dist. v. Belize, Inter-American Commission for Human Rights. 2004. Case 12.053, Report No. 40/04, OEA/Ser.L/V/II.122 doc. 5 rev. 1, para 153.

Mayagna (Sumo) Awas Tingni Community v. Nicaragua. Merits, Reparations, and Costs, Inter-American Court of Human Rights (Judgment of 31 August 2001) [Series C, No. 79].

Saramaka People v. Suriname. Judgment of 28 November 2007. Preliminary Objections, Merits, Reparations and Cost. Inter-American Court of Human Rights [Series C, No. 174].

Subhash Kumar v. State of Bihar. 1991. 1 SCC 598 (Supreme Court of India).

The Matter of African Commission on Human and Peoples Rights v Republic of Kenya, 2013. eKLR. Application No. 006/2012. Order dated 15 March 2013.

The Social and Economic Rights Action Center for Economic and Social Rights v. Nigeria. 2001. Comm. No. 155/96.

Others

Coxon, C. 2007. EU Biofuels Policy Fuelling Evictions. *Action Aid*. *<http://www.actionaid.org/ guatemala/shared/eu-biofuels-policy-fuelling-evictions>*.

Ferrett, G. 2007. Biofuels "crime against humanity", *BBC* <*http://news.bbc.co.uk/2/hi/7065061.stm*>.

Ghana goes biofuel, despite global food crisis. 2008. *Ghana Business News*. <*https://www.ghanabusinessnews.com/2008/12/06/Ghana-goes-biofuel-despite global-food-crisis/*>.

Kennedy, J. 2013. Sierra Leone Farmers Evicted for Sugarcane Biofuel Plantations. *Business and Human Rights Resource Centre*.

Land Justice 4 West Africa. 2014. Land Grabbing in Northern Ghana: How Fair were the Deals Signed? *https://landjustice4wa.org/2014/04/01/landgrabbing-in-northern-ghana-how-fair-were-the-deals-signed/*

Land Justice 4 West Africa. 2014. When law fails to protect small farmers against land grabbers in Sierra Leone. *https://landjustice4wa.org/2014/06/03/when-law-fails-to-protect-small-farmers-against-land-grabbers-in-sierra-leone/*

Parakuiyo Pastoralists Indigenous Community Development Organization (PAICODEO). 2013. Report on The State of Pastoralists' Human Rights in Tanzania: Survey of Ten Districts of Tanzania Mainland 2010/2011. PAICODEO, Tanzania.

Rosenthal, E. 2007. World food stocks dwindling rapidly, UN warns. *New York Times*.

The Oakland Institute. 2012. Mapping the Land Grab in Africa. *https://www.oaklandinstitute.org/ mapping-land-grab-africa*

UN News Centre. 2005. Biofuels can cut poverty, provide energy and mitigate climate change – UN. <*http://www.un.org/apps/news/story.asp?NewsID=13971*>

14

Microbial Fuel Cell: A Promising Technology for Energy Generation and the Waste Water Treatment

BHAWANA SINGH[1*] AND MADHAV SINGH[2]

ABSTRACT

The recent energy crisis has been reinvigorated interest in microbial fuel cells (MFCs) technology as a mode of generating bio-electricity by using different kind of organic wastes and renewable biomass. However, the performance of microbial fuel cell is limited by irreversible reactions and processes occurring in cathode and anode compartments. In this chapter, different types of materials used for main parts of MFCs such as anode, cathode, membrane and various substrates are explored. This chapter also demonstrates their resulting performance, limitations as well as future potential materials and substrate. However, the generation of electricity and power density are relatively low or potential barriers to apply this technology practically at present, which can be improved with the improvement in current technology and knowledge.

***Key words*:** Microbial fuel cell, Electrode materials, Catalyst, Waste water, Micro-organisms

[1]Chair of Materials Processing, Faculty of Engineering Sciences, University of Bayreuth, 9544 (Aelterman, 2009) Bayreuth, Germany
[2]Institute of Nanotechnology, Karlsruhe Institute of Technology (KIT), 76344 Eggenstein-Leopoldshafen, Germany
**Corresponding author*: bhawana@iitdalumni.com

INTRODUCTON

Microbial fuel cells (MFCs) technology has become a potential renewable energy source for generating electricity from waste water containing bio-degradable organic materials such as sewerage waste, urine, industrial (*i.e.* brewery), animal, domestic waste, volatile acids, carbohydrates, proteins, alcohols, and recalcitrant materials like cellulose.

Power demand is increasing day by day as the global population is growing and is expected to increase upto 1.5 times to the current demand by 2035 according to the international energy agency projections (IEA) (Chu and Majumdar, 2012).

At present, most of the energy is derived from the use of limited fossil fuels sources. The consumption of fossil fuels has severe impact on human life through global warming and atmospheric pollution by emission of carbon dioxide (Rahimnejad, Ghoreyshi, Najafpour, Younesi and Shakeri, 2012). Alternative source of energy to meet future energy demands is nuclear energy, but with the recent incident which happened in Japan in 2011, there is a question of its safety and reliable implementation (Funabashi and Kitazawa, 2012). In current scenario, the treatment of wastewater consumes significant energy; microbial fuel cell technology can be used in a more economical way and has plethora of advantages over other kinds of bioenergy conversion processes (*i.e.,* gasification, anaerobic digestion and fermentation), *e.g.*, higher efficiency and no emission of environmental polluting gases such as SO_x, NO_x, CO_2 and CO. MFCs can operate under ambient environmental conditions and require no energy input for aeration.

In 1911, for the first time, the idea to produce electricity using microbes that oxidize organic molecules was given by British botanist Potter (Potter, 1911). The concept was not appreciated until 1999, for electricity generation and waste water treatment; recently, MFCs technology has attracted many scientists and researchers to develop it in a practical and cost-effective manner (Lovley, 2006; Rabaey and Verstraete, 2005; Min and Logan, 2004).

There are several critical parameters which determine the performance of MFCs such as microbial inoculum, the proton exchange membrane, the chemical substrate in the fuel and the cathode and anode materials.

MFCs could be used for powering the small portable electronic devices such as biosensors, performing fundamental research of microorganisms, screening bacterial strains and detection of toxicity in the waste water. MFCs have potential to be used use as a micro-power source in a wide variety of military mobile and wireless field applications (Qian, Baum, Gu and Morse, 2009; Choi *et al.*, 2011; Tommasi, Chiolerio, Crepaldi and Demarchi, 2014). Besides many advantages of this technology, some practical barriers such as low power and current density, high cost and high mass generation are still under research. In the last decade, to overcome these challenges, many research groups around the world have made remarkable efforts.

1. MICROBIAL FUEL CELL

MFC is electrochemical fuel cell which uses microorganisms to convert organic matter found in waste water into electricity. It consists of an anode and a cathode separated by proton exchange membrane (PEM) *e.g.,* Nafion and poly(tetrafluoroethylene) (PTFE) as shown in Fig. 1. Many different possible configurations of MFCs are shown in Figs. 2, 3 (Logan *et al.*, 2006). At anode, in anaerobic condition, the active biocatalyst converts the chemical energy into electrons and protons. The electrons are conveyed through the external circuit and the protons cross through PEM and get reduced to oxygen (*i.e.,* terminal electron acceptor) and form water at cathode (Rabaey and Verstraete, 2005; Min and Logan, 2004). Many inorganic metal ions, such as Mn and Fe, are reduced from higher oxidation state Mn(IV) to lower oxidation state Mn(II). The redox reactions in this process are catalyzed with the help of bacteria using (Rhoads, Beyenal and Lewandowski, 2005; Shantaram, Beyenal, Raajan, Veluchamy and Lewandowski, 2005) dissolved oxygen.

Many terminal electron acceptors such as oxygen, nitrate, sulfate, metal oxide like iron oxide and others are present which readily accept electrons and generate different products that can diffuse out of the cell. However, some microorganisms can transfer electrons exogenously (*i.e.,* outside the cell) such as *Geobacter* and *Shewanella,* which are known as exoelectrogens. Electron transfer takes place in three different ways:

1.1. Direct Electron Transfer

The outer membrane cytochrome (C-type) of microorganisms (*i.e., Geobacter* and *Shewanella* species) is involved in the direct transfer of

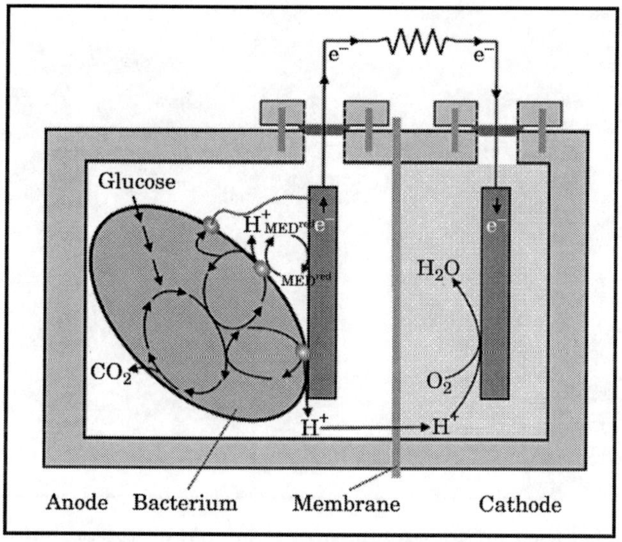

Fig. 1: A schematic diagram of the microbial fuel cell (MFC) (Logan *et al.*, 2006)

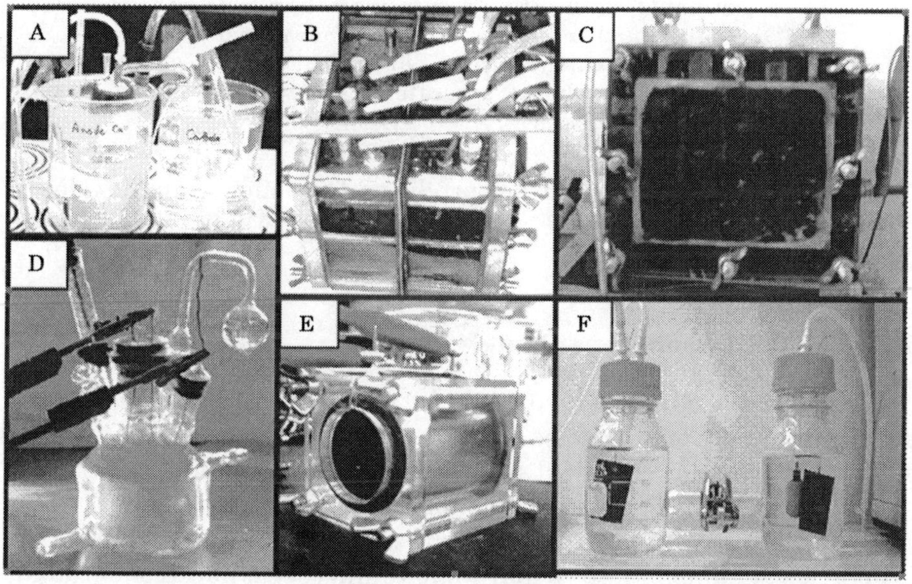

Fig. 2: MFCs setup used in research: (A) Salt-bridge system (B) Four batch type MFCs in which the chambers are separated by a membrane (C) Closed anode-cathode placement setup in which a continuous flow through anode occurred (D) Photoheterotrophic type MFC (E) Single-chamber setup where cathode is directly in contact with air and (F) H-type double chamber setup (Logan *et al.*, 2006).

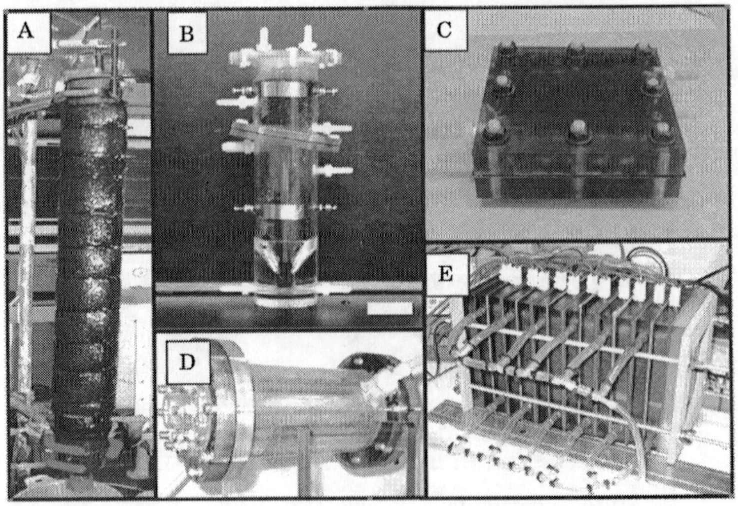

Fig. 3: MFCs used for continuous operation: (A) upflow, tubular type MFC with inner graphite bed anode and outer cathode (B) upflow, tubular type MFC with cathode above and anode below, the membrane is inclinated (C) flat plate design where a channel is cut in the blocks so that liquid can flow in a serpentine pattern across the electrode (D) single-chamber system with an inner concentric air cathode surrounded by a chamber containing graphite rods as anode and (E) stacked MFC, in which 6 separate MFCs are joined in one reactor block (Logan *et al.*, 2006).

electrons to the electrode surface produced from nicotinamide adenine dinucleotide (NADH) (Lies, Hernandez, Kappler, Mielke, Gralnick and Newman, 2005).

1.2. Electron Transfer through Mediators

The electrons are transferred from the outer membrane of bacteria to the electrodes. Some species such as *Shewanella* and *Pseudomonas* secrete a particular chemical species (*i.e.,* flavins, thionin, neutral red, pyocyanin) which act as shuttle or mediator molecules (Yang, Xu, Guo and Sun, 2012; Schröder, 2007). Good mediators should have certain properties such as low cost, non-toxic to microbes, non-biodegradable and high solubility in the anolyte. In addition, they should cross the cell membrane easily and possess a high electrode reaction rate (Ieropoulos, Greenman, Melhuish and Hart, 2005).

Some microbes can use microbial metabolites, which are known as "Endogenous mediators" such as humic acids, anthraquinone and the oxyanions of sulphur (sulphate and thiosulphate). All of these mediators have ability to transfer the electrons from inside the cell membrane to the anode (Lovley, 1993).

1.3. Electron Transfer through Nanowires

Some microorganisms (*i.e., Geobacter sulfurreducens* and *Shewanella oneidensis*) species use conductive appendages (*i.e.,* cellular outgrowth), which are 20 μm long and termed as "nanowires" for the transfer of electrons outside the cell (El-Naggar *et al.*, 2010; Gorby *et al.*, 2006; Malvankar *et al.*, 2011). Schaetzle *et al.* have reported another mechanism to transport electron by oxidizing the excreted catabolites by the bacteria (Schaetzle, Barrière and Schröder, 2009). The cell yield of the substrate converted by using these exoelectrogens into biomass is not well known. The Columbic efficiency ~85% is expected, indicating that only ~15% of the substrate is converted into biomass.

Microbial fuel cell can be operated at different temperatures, which depend on the tolerance of bacteria. Bacteria, such as *Geobacter sulfurreducens* and *Shewanella oneidensis*, in the room temperature range of 15–35°C, thermophiles (*i.e., Bacillus licheniformis* and *Bacillus thermoglucosidasius*) in the high temperature range of 45–65°C and psychrophiles at low temperatures <15°C, can be used to operate the MFCs (Logan, 2007). The major operating costs of the MFC for different types of waste water treatment come from waste water aeration, sludge treatment and waste water pumping. Besides these, it also depends on the ecology of the microbial communities that produce electricity as their role is mentioned above.

2. CHEMICAL CELL REACTIONS AND LIMITATIONS

As a feedstock, glucose and acetate molecules are used in MFCs to produce electrons, which travel through an external circuit and produce power. The cell reactions at the anode and cathode are mentioned below:

With sodium acetate molecules:

Anode: $$CH_3COO^- + 2H_2O \rightarrow 2CO_2 + 7H^+ + 8e^- \quad (1)$$

Cathode: $$O_2 + 4H^+ + 4e^- \rightarrow 2H_2O \quad (2)$$

With glucose molecules:

Anode: $C_6H_{12}O_6 + 6H_2O \rightarrow 6CO_2 + 24H^+ + 24e^-$ (3)

Cathode: $6O_2 + 24H^+ + 24e^- \rightarrow 12H_2O$ (4)

The theoretical half-cell potential can be thermodynamically calculated according to Nernst equation:

$$E_{half-cell} = E_o - \frac{RT}{nF}\text{In}Q \tag{5}$$

where E_0, R, T, Q are the standard half-cell potential, the gas constant, the temperature in Kelvin and the activity constant, respectively.

MFCs produce significantly different cell voltages ranging from 0.361 to 0.805 V and power output upon combining the same anode with different cathode materials (Logan *et al.*, 2006). This indicates that the MFCs performance mainly depend on the cathode materials. The Gibbs free energy (J) equation is used to calculate feasible energy produced by the electrochemical system.

$$\Delta G = -nFE_{emf} \tag{6}$$

where F is Faraday constant, n is number of electrons and E_{emf} is cell voltage (Worrell, 1997).

The generated overall cell electromotive force (emf), E_{emf} (V) can be calculated as:

$$E_{emf} = E_{cathode} - E_{anode} - \eta \tag{7}$$

where $E_{cathode}$ and E_{anode} are the half-cell potential at the cathode and anode, respectively and η is an irreversible loss term (*i.e.*, dominant Ohmic losses, activation losses and mass transfer losses) (Hoogers, 2002). Because of these irreversible losses, also called over-potential, the actual generated maximum voltage (~0.80) of an MFC is observed less than the predicted thermodynamic ideal voltage. These losses can be described as the voltage required to compensate for the current lost during the electrochemical reactions. The real operational voltage (V_{op}) can be calculated by subtracting the voltage losses associated with the cathode and anode chambers (Hayre, Cha, Colella and Prinz, 2005) from the standard voltage (E_{emf}) as follows:

$$V_{op} = E_{emf} - [(\eta_{act} + \eta_{ohmic} + \eta_{conc})_{cathode} + (\eta_{act} + \eta_{ohmic} + \eta_{conc})_{anode}] \quad (8)$$

where η_{act} is the activation loss due to reaction kinetics, η_{ohmic} is the Ohmic loss due to ionic and electronic resistance and η_{conc} is the concentration loss due to mass transport limitations. The above equations represents that these losses limit the overall performance of the MFC and can be improved by optimizing both the anode and cathode.

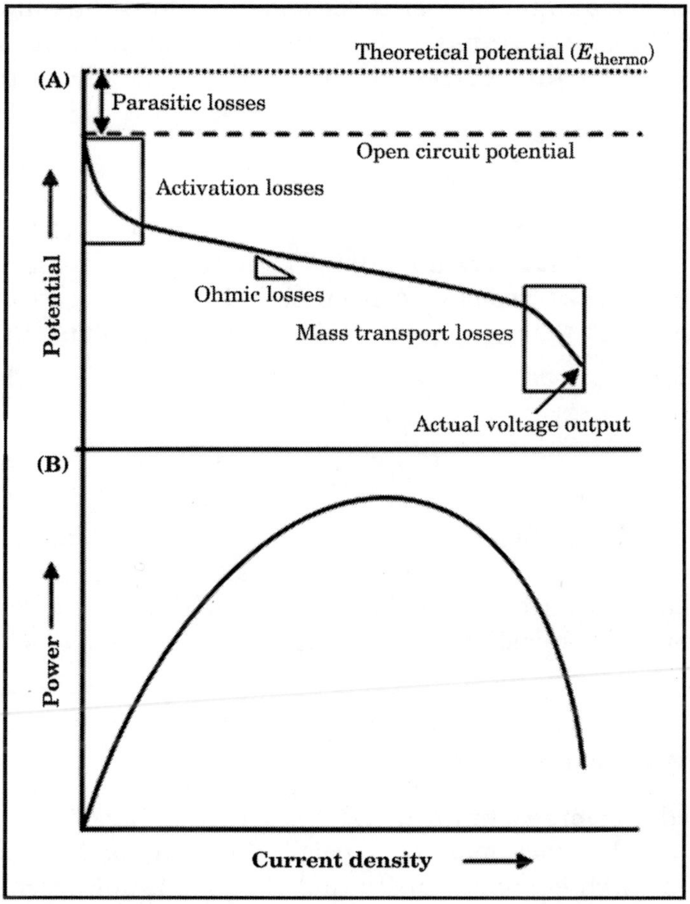

Fig. 4: (A) Schematic irreversible potential losses for a cathodic reaction (*i.e.,* activation, Ohmic, mass transport and parasitic loss) and (B) A typical power-current curve (Yazdi, Carver, Christy and Tuovinen, 2008).

2.1. Activation Losses

The dominant cathodic activation losses occur due to the reaction kinetics at low current densities and steadily increase when current density increases (Fig. 4) (Yazdi, Carver, Christy and Tuovinen, 2008). Kinetics performance can be improved by decreasing the activation barrier and increasing the temperature, oxidant concentration and reaction interface area.

2.2. Ohmic Losses

It is also known as internal resistance and represents the voltage which is lost during charge transport (*i.e.*, electrons and protons) and more pronounced at medium current densities (Fig. 4). It follows the Ohm's law:

$$\eta_{ohmic} = iR_{ohmic} \tag{9}$$

where i is the current (A) and R_{ohmic} is the Ohmic resistance (Ω) of microbial fuel cell. It is a combination of ionic and electronic resistance which includes the resistance from the electrode, electrolytes and interconnections. The ionic resistance of the electrolyte, R_{ion} can be expressed by

$$R_{ion} = \frac{l}{AK} \tag{10}$$

where l is the distance (cm), A is the cross-sectional area (cm^2) and K is the specific conductivity ($(\Omega\ cm)^{-1}$) of the electrolyte (Larminie and Dicks, 2003). Optimized electrode spacing, stable electronic connections and the use of low resistive membranes can significantly reduce Ohmic losses.

2.3. Mass Transport Loss

The degradation of reactants (*i.e.*, O_2) or accumulation of products (*i.e.*, H_2O) at cathode causes mass transport loss during the MFC operation. It occurs at high current density and the magnitude increases with increasing current density (Fig. 4).

3. ELECTRODE MATERIALS

MFCs are being constructed in diversity of configurations (Figs. 2, 3) and by using variety of materials. There are many operational parameters which should be optimized to achieve high power density such as different temperature range, pH, cost, high electrode surface area to enhance the electrode kinetics, reactor size, electron acceptor, cation exchange membrane and operation time, electrical conductivity, stability and durability of electrode materials.

3.1. Cathode Materials

The MFC cathode should have some properties such as high mechanical strength, catalytic property and high electronic and ionic conductivity. There is three-phase interface of the cathode in MFC; the catalyst layer, electrode base material and oxygen diffusion layer where oxidation–reduction reactions occurred.

Oxidation–reduction reactions (ORR) occurred in two different pathways in acidic medium at cathode: 4-electron pathway is more favorable than the 2-electron pathway, which is shown below:

4 electron pathway $O_2 + 4H^+ + 4e^- \rightarrow 2H_2O$ (11)

2 electron pathway $O_2 + 2H^+ + 2e^- \rightarrow H_2O_2$ (12)

$$H_2O_2 + 2H^+ + 2e^- \rightarrow 2H_2O \quad (13)$$

3.1.1. *Metal based composites*

Generally metal oxides with low cost such as MnO_2, PbO_2 and metal macrocyclic compounds such as iron phthalocyanines (FePc) and cobalt tetramethoxyphenylporphyrins (CoTMPP), are used as Pt-free cathode catalysts in MFCs. Carbon based materials, such as plane carbon paper, carbon cloth or graphite with high surface area which have poor catalytic property, coated with above mentioned catalyst are used as appropriate alternatives for MFCs applications (Zhou, Chi, Luo, He and Jin, 2011; Liew, Daud, Ghasemi, Leong, Lim and Ismail, 2014). Some limitations with platinum metal include its being highly expensive, sulfide poisoning, pH sensitivity; non-durable and non-sustainable. These issues have been the challenges to develop the alternative composite materials as substitute in electrodes in the future. Lead oxide, can be used as an alternative as a catalyst, it shows four times higher power densities (Morris, Jin, Wang, Zhu and Urynowicz, 2007) than the commercial Pt/C (10 wt.%). Palladium can be used as a cathode due to its excellent catalytic properties and low cost.

3.1.2. *Biocathode*

Microorganisms themselves act as the catalyst by making a biofilm over cathode. Biocathode can be divided into two groups (1) aerobic and (2) anaerobic. In aerobic biocathode, transition metals such as iron and manganese participate as electron mediators, oxygen is a terminal electron acceptor and hydrogen peroxide is an intermediate. In anaerobic biocathode, nitrates and sulfates act as terminal electron acceptors. The advantage of anaerobic group is that it prevents the loss of electron through oxygen, which could diffuse to the anode through PEM (Park and Zeikus, 2003). By using cyclic voltammetry it is also proved that graphite felt followed by carbon paper and stainless steel mesh facilitate the formation of biofilm over cathode (Zhang, Sun, Hu, Li and Xu, 2012). The advantages of using microorganisms at the cathode are that they can produce chemicals such as methane, ethanol and formic acid (Rabaey and Rozendal, 2010) and can help to detect the biological oxygen demand (BOD) in water (Kim, Chang, Gil, Park and Kim, 2003). The BOD indicates the amount of oxygen used to oxidize organic matter in waste water.

3.1.3. *Cathode without metal catalyst*

Chemically modified carbon based materials such as nitrogen doped carbon nano-tubes (CNTs), nitrogen doped graphene (GN) nano-sheets and graphite foam can be used as catalyst as they show higher electro-catalytic activity, lower over-potential and long-term stability for ORR compared to Pt catalyst (Qu, Liu, Baek and Dai, 2010; Feng, Yan, Chen and Wang, 2011; Ci *et al.*, 2012; Gong, Du, Xia, Durstock and Dai, 2009).

3.2. Anode Materials

The catalyst of anode should possess certain properties such as high surface area, suitable porosity, high electrical conductivity, appropriate mechanical strength, toughness and high stability under fuel cell operating conditions. Carbon materials (*i.e.,* graphite rod, graphite fiber brush wound around a titanium wire to increase microbial inoculation, carbon paper, carbon cloth, carbon felt and reticulated vitreous carbon) are the most widely used as anode because of their high electrical conductivity, chemical stability, specific surface area, low cost, biocompatibility and inert properties at ambient conditions (Fig. 5) (Wei, Liang and Huang, 2011; Zhou, Chi, Luo, He and Jin, 2011; Kumar, Sarathi and Nahm, 2013). The characteristic properties of anode materials can be improved by surface treatment such as heat treatment, ammonia treatment, and acid treatment, chemical and electrochemical oxidation or by addition of metallic ions such as Mn^{4+}, Mn^{2+} and Ni^{2+}. By treating the electrode surface with concentrated inorganic acids, protonation of functional groups over the electrode surface occurres, which increases the positive charges on the electrode surface. The combination of acid and heat treatment increases power production by 25%. All these methods facilitate the electron transfer between microbes and anode (Park and Zeikus, 2003; Lowy, Tender, Zeikus, Park and Lovley, 2006; Lowy and Tender, 2008).

3.2.1. *Metal-carbon nanotube/metal-graphene composites*

Some researchers have recently used CNTs and GN to coat the anode materials to support in low temperature operated fuel cells. The production cost of large amount of GN is much lower than that of CNTs. Due to the high surface area and conductivity of these materials, the power density of MFCs was higher than the uncoated anodes (Tsai, Wu, Lee and Shih, 2009; Zhang *et al.*, 2011). The rough surface of CNTs assists the bacteria to produce their nanowires which help them to adhere with each other and provide the electron transfer bridges. CNT-polyaniline (PANI) composites with high surface area increase the electro-catalytic property and adhesion with the bacterial cells. The metal oxides such as MnO_2, SnO_2 and TiO_2 coated on CNTs facilitate the electron transfer between microbes and anode, thus improve the MFCs performance (Kalathil, Hguyen, Shim, Khan, Lee and Cho, 2013; Mehdinia, Ziaei and Jabbari, 2014; Wen *et al.*, 2013).

3.2.2. *Polymer-carbon composites*

Conductive polymers should have some properties to act as a suitable fuel cell catalyst support such as high surface area for the attachment of bacterial cells, high porosity, high electrical conductivity, biocompatibility and high stability under fuel cell operating conditions. The most common examples of conducting polymers in literature are PANI coated CNTs and graphenes/polypyrrole (PPy), polythiophene (PTh) and their derivatives. They can also be used as proton conducting materials and can replace Nafion in the catalyst layer of fuel cell electrode with enhanced properties (Antolini and Gonzalez, 2009).

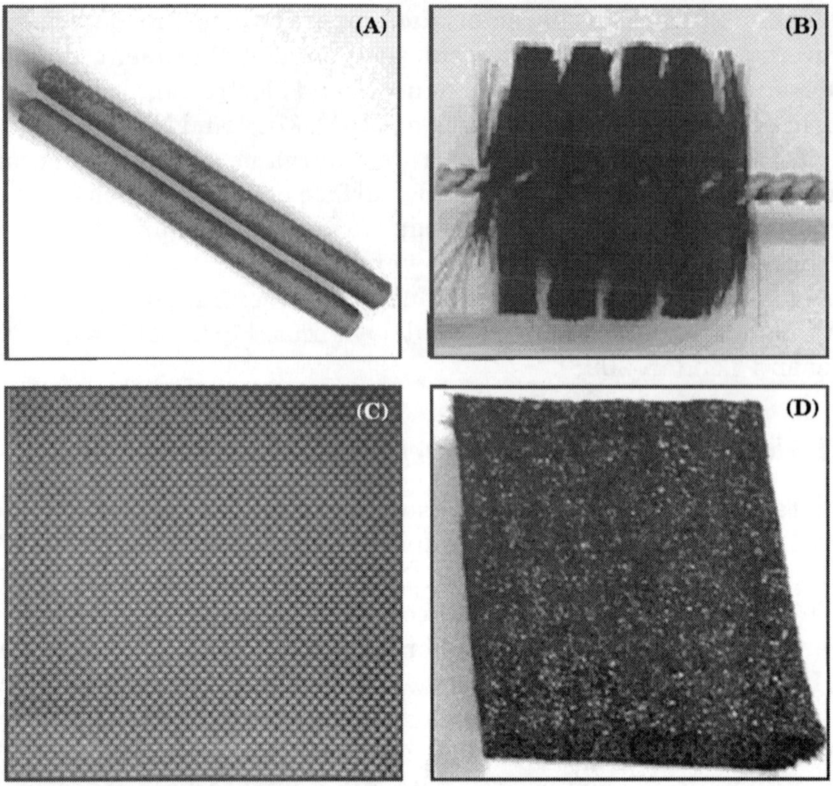

Fig. 5: Anode materials: (A) typical graphite rod electrode, (B) carbon fiber brush electrode, (C) carbon cloth electrode and (D) carbon felt electrode (Mustakeem, 2015).

4. CATION EXCHANGE MEMBRANE

Generally, Nafion membranes are used in MFCs. However, there are several problems associated with these membranes such as high cost, substrate loss, internal resistance, cation transport and accumulation rather than protons and fouling, oxygen leakage from the cathode to the anode. The main technological challenge to the use of Nafion membrane in MFCs is bio-fouling which normally occurs on the surface of membrane facing towards the anode by attachment of the bacteria and organic substrates. It can affect the power output and durability of the MFCs (Leong, Daud, Ghasemi, Liew and Ismail, 2013). High resistance membranes have low ion exchange capacity (IEC), reducing proton diffusion from the anode to the cathode, leading to low current and power densities. Microfiltration membranes which have high porosity and low resistance show lower MFCs performance because oxygen and substrate crossover through the membranes pores which reduce power density and columbic efficiency (Zhao, Slade and Varcoe, 2009). Carbon nano-fiber (CNF)/Nafion and activated carbon nano-fiber (ACNF)/Nafion nano-composite membranes show a reduction in pore size, roughness and porosity, which block the transfer of oxygen from cathode to anode and decrease the fouling problem in the membrane. With different membranes, the performance of MFCs follows the order ACNF/Nafion> CNF/Nafion> Nafion 117> Nafion 112 (Ghasemi, Shahgaldi, Ismail, Yaakob and Daud, 2012). The different roles of metal and CNT in case of anode, cathode and membrane are summarized in scheme 1 and 2 as shown below.

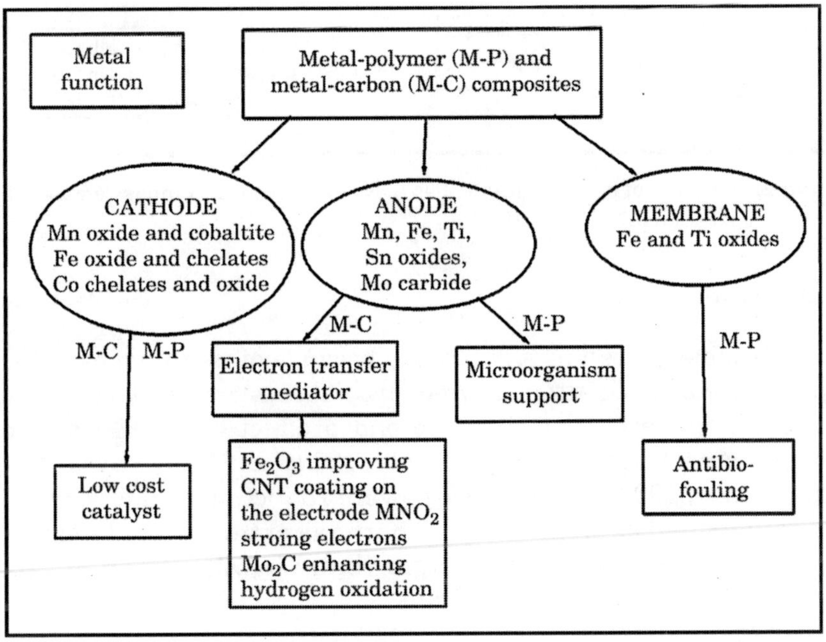

Scheme 1: The role of metals present in composite materials in MFCs (Antolini, 2015)

5. LIST OF SUBSTRATES USED IN MFCs

A great variety of substrates have been used so far as feedstock in operating the MFCs for electricity generation. However, the generation of electricity and power density are relatively low at present. It can be enhanced with the improvements in the current technology and knowledge.

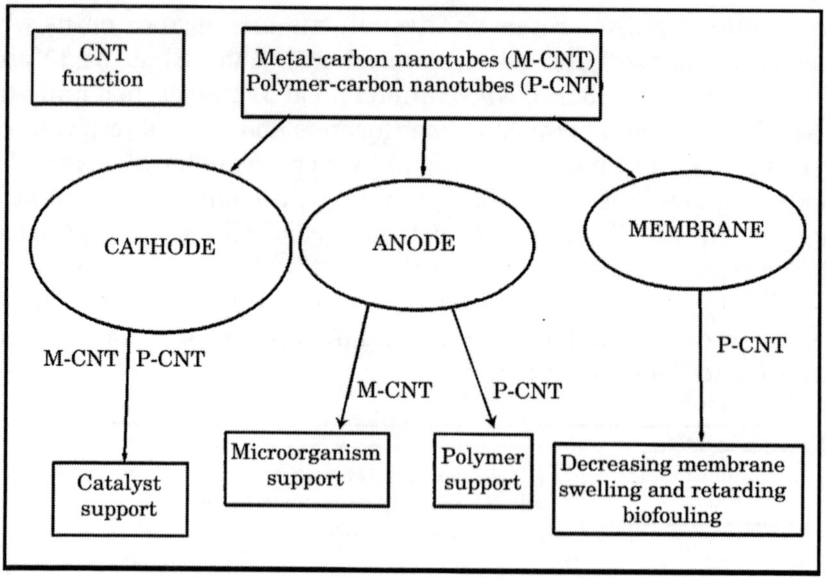

Scheme 2: The role of carbon nano-tubes (CNTs) present in composite materials in MFCs (Antolini, 2015)

5.1. Acetate

Acetate is mostly used as a substrate because of its inertness at room temperature towards alternative microbial conversions (fermentation and methanogenesis) and it is the end product of several metabolic pathways as carbon sources (Aelterman, 2009; Biffinger, Byrd, Dudley and Ringeisen, 2008). The electric power production by using acetate is more than 2-fold and one half of optimal external load resistance compared to protein-rich waste water (Liu, Liu, Zhang and Su, 2009).

5.2. Glucose

When the energy conversion efficiency (ECE) of acetate and glucose was compared in MFC, it was observed that with acetate, ECE was 42% while with glucose, it was only 3%. It indicates that the glucose generates low current and power density because glucose is a fermentable substrate and consumed by different competitive metabolism (*i.e.*, fermentation and methanogenesis) that cannot produce electricity (Chae, Choi, Lee, Kim and Kim, 2009).

5.3. Ligno-cellulosic Biomass

From agricultural residues, the abundance of ligno-cellulosic materials obtained as a promising feedstock for cost-effective energy production. It cannot be used directly by microorganisms. Firstly, it should be converted into monosaccharide and other low-molecular-weight compounds such as cellulose by hydrolysis (Ren, Ward and Regan, 2007; Catal, Li, Bermek and Liu, 2008).

5.4. Synthetic Wastewater

Well-defined composition is also used in MFCs to compare the electricity generation capacity as it is easy to control the loading strength, pH and conductivity.

5.5. Brewery Wastewater

Waste water from breweries is favorable because of its low strength and food-derived nature of the organic matter which has very low concentration of inhibitory substances (*i.e.,* ammonia in animal waste water) and high carbohydrate content (Feng, Wang, Logan and Lee, 2008). Although it has 10 times higher value of chemical oxygen demand (COD)/L compared to the domestic waste water. It causes reaction kinetic and mass transport losses which influence the performance of the MFC. It can be improved by increasing the concentration of brewery waste water and reaction temperature as well as the roughness of electrode, which provide more reaction sites (Wen, Wu, Cao, Zhao and Sun, 2009).

5.6. Others

Many researchers have tried other substrates also to generate the electricity and remove pollutants from waste water through MFCs, but there is requirement of considerable improvements to find the suitable bacterial community that is capable of utilizing these substrates. (1) Starch processing waste water which contains high amount of carbohydrates, sugars, protein and starch. (2) Azo dyes waste water, which is extensively present in effluent from dye-manufacturing industries and textile industries. Before discharge, the removal of these dyes is very important as they are very toxic and create severe environmental problems such as obstruction of light and oxygen transfer into rivers and sea which directly affect the aquatic life (Pant, Singh, Satyawali and Gupta, 2008). (3) In recent studies, some researchers have mentioned about photosynthetic microorganisms (*i.e, Rhodobacter sphaeroides*) and photosynthetic alga (*i.e., Chlamydomonas reinhardtii* produce hydrogen) and living plants for example rice paddy fields where the solar energy is converted into the electrical energy by MFCs called as "phototropic MFCs" (Rosenbaum, Schröder and Scholz, 2005; He, Kan, Mansfeld, Angenent and Nealson, 2009; Strik, Hamelers, Snel and Buisman, 2008; Cho, Donohue, Tejedor, Anderson, McMahon and Noguera, 2008). In recent years, the research has been focused on waste water from paper recycling plant, swine, meat processing plant as carbon and protein rich substrates, canning of fruits and vegetables, whey and dairy industry (Huang and Logan, 2008; Min, Kim, Oh, Regan and Logan, 2005; Heilmann and Logan, 2006).

6. LIMITATIONS AND FUTURE OUTLOOK

Despite several limitations in MFC's such as high cost of cation exchange membranes, bio-fouling and over-potentials, in recent years, the power generation from MFCs has improved considerably at small lab-scale systems. A major drawback of using MFCs is the start-up time which may vary from 4 to 103 days depending on many factors such as inoculum, high cost of precious metal catalyst (*i.e.*, platinum), electrode materials, reactor design, type of substrate and operating conditions (*i.e.*, temperature, pH, external loading etc.) and lack of buffer capacity of electrolytes. These factors are significant impediment in scaling up of MFCs for different kinds of waste water treatment. By using different types of electrode materials, catalyst and substrates and by optimizing the operational conditions, the existing limitations in MFCs operation could be overcome to achieve the sustainable and economical bio-energy in coming years.

Several other potential applications have also been suggested (Lovley, 2009) for using microbe-electrode technology in the form of implanted medical devices using blood sugar as a fuel, microbial transistors, circuits and electronic computing devices. Another lucrative application is the capture of carbon dioxide (CO_2) in biomass and its convertion it into useful compounds in a MFC (Cao *et al.*, 2009). The use of single chamber MFC with air cathode as biosensors to fix BOD and COD of waste water and obtain a good correlation between COD and the current output, which is another application of these systems

(Kim, Chang, Gil, Park and Kim, 2003; Di Lorenzo, Curtis, Head and Scott, 2009). The COD indicates that the organic matter should be fully oxidized whether it is biodegradable or not. The potential for hydrogen production can be determined by calculating the COD values. For example, one mole of COD means one mole of O_2 is needed for the reaction (*i.e.,* 1 mol COD = 2 mol-H_2 or 1 g of COD produces 0.125 g-H_2) (Logan, 2007). However, microbial fuel cell can also be designed to produce bio-hydrogen by generating the potential instead of electricity from the electrolysis of the organic matter by the bacteria, these cells are called microbial electrolysis cell (MEC) or bio-electrochemically assisted microbial reactor (BEAMR). Bio–Hydrogen can be accumulated for many applications such as low temperature polyelectrolyte membrane fuel cell.

7. CONCLUSIONS

This chapter summarized various electrode materials and substrates that have been used for improved power output as well as to make waste water treatment feasible. However, to increase the power output, many technological improvements such as materials, optimized operating conditions are needed to make this technology as cost-effective and reliable. When MFCs are scaled up for use in real waste water treatment plant, it is very important to know about the ecology of microbial communities that metabolize the waste or catalyze the reactions on biocathode, which would reduce the perceived risks and accelerate the adoption of MFCs.

APPENDIX OF ABBREVIATIONS

ACNF	Activated carbon nano fiber
BOD	Biological oxidation demand
BEAMR	Bio-electrochemically assisted microbial reactor
CoTMPP	Cobalt tetramethoxyphenylporphyrins
C	Carbon
COD	Chemical oxygen demand
CNTs	Carbon nano tubes
CNF	Carbon nano fiber
ECE	Energy conversion efficiency
FePc	Iron phthalocyanines
GN	Graphene

IEA	International energy agency
IEC	Ion exchange capacity
MFC/MFCs	Microbial fuel cells
M-P	Metal-Polymer
M-C	Metal-Carbon
M-CNT	Metal-Carbon nano tubes
MEC	Microbial electrolysis cell
NADH	Nicotinamide adenine dinucleotide
ORR	Oxidation reduction reaction
PEM	Proton exchange membrane
PTFE	Poly (Tetrafluoroethylene)
Pt	Platinum
PANI	Polyaniline
PPy	Polypyrrole
PTh	Polythiophene
P-CNT	Polymer-Carbon nano tube

BIBLIOGRAPHY

Aelterman, P. 2009. *Microbial fuel cells for the treatment of waste streams with energy recovery*, Gent University, Belgium.

Antolini, E. 2015. Composite materials for polymer electrolyte membrane microbial fuel cells. *Biosensors and Bioelectronics,* 69: 54–70.

Antolini, E. and Gonzalez, E. 2009. Polymer supports for low-temperature fuel cell catalysts. *Applied Catalysis A: General,* 365(1): 1–19.

Biffinger, J., Byrd, J., Dudley, B. and Ringeisen, B. 2008. Oxygen exposure promotes fuel diversity for *Shewanella oneidensis* microbial fuel cells. *Biosens. Bioelectron.*, 23(6): 820–826.

Cao, X., Hunag, X., Liang, P., Boon, N., Fan, M., Zhang, L. *et al.,* 2009. A completely anoxic microbial fuel cell using a photo-biocathode for cathodic carbon dioxide reduction. *Energy Environ. Sci.,* 2: 498–501.

Catal, T., Li, K., Bermek, H. and Liu, H. 2008. Electricity production from twelve monosaccharides using microbial fuel cells. *J. Power Sources,* 175(1): 196–200.

Chae, K., Choi, M., Lee, J., Kim, K. and Kim, I. 2009. Effect of different substrates on the performance, bacterial diversity and bacterial viability in microbial fuel cells. *Biores. Technol.,* 100(14): 3518–3525.

Cho, Y., Donohue, T., Tejedor, I., Anderson, M., McMahon, K. and Noguera, D. 2008. Development of a solar-powered microbial fuel cell. *J. Appl. Microbiol.,* 104(3): 640–650.

Choi, S., Lee, H., Yang, Y., Parameswaran, P., Torres, C., Rittmann, B. *et al.,* 2011. A ìL-scale micromachined microbial fuel cell having high power density. *Lab on a Chip,* 11: 1110–1117.

Chu, S. and Majumdar, A. 2012. Opportunities and challenges for a sustainable energy future. *Nature,* 488(7411): 294–303.

Ci, S., Wu, Y., Zou, J., Tang, L., Luo, S., Li, J. *et al.* 2012. Nitrogen-doped graphene nanosheets as high efficient catalysts for oxygen reduction reaction. *Chinese Sci. Bull.,* 57(23): 3065–3070.

Di Lorenzo, M., Curtis, T., Head, I. and Scott, K. 2009. A single-chamber microbial fuel cell as a biosensor for wastewaters. *Water Res.,* 43(13): 3145–3154.

El-Naggar, M., Wanger, G., Leung, K., Yuzvinsky, T., Southam, G., Yang, J. *et al.,* 2010. Electrical transport along bacterial nanowires from *Shewanella oneidensis* MR-1. *Proc. Natl. Acad. Sci.* USA, 107(42): 18127–18131.

Feng, L., Yan, Y., Chen, Y. and Wang, L. 2011. Nitrogen-doped carbon nanotubes as efficient and durable metal-free cathodic catalysts for oxygen reduction in microbial fuel cells. *Energy Environ. Sci.,* 4(5): 1892–1899.

Feng, Y., Wang, X., Logan, B. and Lee, H. 2008. Brewery wastewater treatment using air-cathode microbial fuel cells. *Appl. Microbiol. Biotechnol.,* 78(5): 873–880.

Funabashi, Y. and Kitazawa, K. 2012. Fukushima in review: A complex disaster, a disastrous response. *Bull. At. Sci.,* 68(2): 9–21.

Ghasemi, M., Shahgaldi, S., Ismail, M., Yaakob, Z. and Daud, W. 2012. New generation of carbon nanocomposite proton exchange membranes in microbial fuel cell systems. *Chemical Engineering Journal*, 184: 82–89.

Gong, K., Du, F., Xia, Z., Durstock, M. and Dai, L. 2009. Nitrogen-doped carbon nanotube arrays with high electrocatalytic activity for oxygen reduction. *Science,* 323(5915): 760–764.

Gorby, Y., Yanina, S., McLean, J., Rosso, K., Moyles, D., Dohnalkova, A. *et al.,* 2006. Electrically conductive bacterial nanowires produced by *Shewanella oneidensis* strain MR-1 and other microorganisms. *Proc. Natl. Acad. Sci.,* 103(30): 11358–11363.

Hayre, O., Cha, S., Colella, W. and Prinz, F. 2005. *Fuel Cell Fundamentals*, John Wiley & Sons, New York, USA.

He, Z., Kan, J., Mansfeld, F., Angenent, L. and Nealson, K. 2009. Self-sustained phototrophic microbial fuel cells based on the synergistic cooperation between photosynthetic microorganisms and heterotrophic bacteria. *Environ. Sci. Technol.,* 43(5): 1648–1654.

Heilmann, J. and Logan, B. 2006. Production of electricity from proteins using a microbial fuel cell. *Water Environ. Res.,* 78(5): 531–537.

Hoogers, G. 2002. *Fuel cell technology handbook,* CRC Press, London .

Huang, L. and Logan, B. 2008. Electricity generation and treatment of paper recycling wastewater using a microbial fuel cell. *Appl. Microbiol. Biotechnol.,* 80(2): 349–355.

Ieropoulos, I., Greenman, J., Melhuish, C. and Hart, J. 2005. Comparative study of three types of microbial fuel cell. *Enzyme and Microbial Technology,* 37(2): 238–245.

Kalathil, S., Hguyen, V., Shim, J., Khan, M., Lee, J. and Cho, M. 2013. Enhanced performance of a microbial fuel cell using CNT/MnO_2 nanocomposite as a bioanode material. *J. Nanosci. Nanotechnol.,* 13: 7712–7716.

Kim, B., Chang, I., Gil, G., Park, H. and Kim, H. 2003. Novel BOD (biological oxygen demand) sensor using mediator-less microbial fuel cell. *Biotechnol. Lett.,* 25(7): 541–545.

Kumar, G., Sarathi, V. and Nahm, K. 2013. Recent advances and challenges in the anode architecture and their modifications for the application of microbial fuel cells. *Biosens. Bioelectron.,* 43: 461–475.

Larminie, J. and Dicks, A. 2003. *Fuel Cell Systems Explained,* John Wiley & Sons, New York, USA.

Leong, J., Daud, W., Ghasemi, M., Liew, K. and Ismail, M. 2013. Ion exchange membranes as separators in microbial fuel cells for bioenergy conversion: A comprehensive review. *Renew. Sustain. Energy Rev.,* 28: 575–587.

Lies, D., Hernandez, M., Kappler, A., Mielke, R., Gralnick, J. and Newman, D. 2005. *Shewanella oneidensis* MR-1 uses overlapping pathways for iron reduction at a distance and by direct contact under conditions relevant for biofilms. *Appl. Environ. Microbiol.,* 71(8): 4414–4426.

Liew, K., Daud, W., Ghasemi, M., Leong, J., Lim, S. and Ismail, M. 2014. Non-Pt catalyst as oxygen reduction reation in microbial fuel cells: A review. *Int. J. Hydrog. Energy, 39*(10): 4870–4883.

Liu, Z., Liu, J., Zhang, S. and Su, Z. 2009. Study of operational performance and electrical response on mediator-less microbial fuel cells fed with carbon- and protein-rich substrates. *Biochem. Eng. J.,* 45(3): 185–191.

Logan, B., Hamelers, B., Rozendal, R., Schröder, U., Keller, J., Frequia, S. *et al.,* 2006. Microbial fuel cells: Methodology and technology. *Envio. Scie. & Techn.,* 40(17): 5181–5192.

Logan, B.E. 2007. *Microbial fuel cells,* John Wiley & Sons, New York, USA.

Lovley, D. 1993. Dissimilatory metal reduction. *Annu Rev. Microbial.,* 47: 263–290.

Lovley, D. 2006. *Bug juice*: Harvesting electricity with microorganisms. *Nat. Rev. Microbiol.,* 4(7): 497–508.

Lovley, D. 2009. Future shock from the microbe electric. *Microbial Biotechnol.,* 2(2): 139–141.

Lowy, D. and Tender, L. 2008. Harvesting energy from the marine sediment-water interface III: Kinetic activity of quinone- and antimony-based anode materials. *J. Power Sources,* 185(1): 70–75.

Lowy, D., Tender, L., Zeikus, J., Park, D. and Lovley, D. 2006. Harvesting energy from the marine sediment-water interface II: Kinetic activity of anode materials. *Biosens. & Bioelectronics.,* 21(11): 2058–2063.

Malvankar, N., Vargas, M., Nevin, K., Franks, A., Leang, C., Kim, B. *et al.,* 2011. Tunable metallic-like conductivity in microbial nanowire networks. *Nat. Nanotechnol.,* 6(9): 573–579.

Mehdinia, A., Ziaei, E. and Jabbari, A. 2014. Multi-walled carbon nanotube/SnO_2 nanocomposite: A novel anode material for microbial fuel cells. *Electrochim. Acta,* 130: 512–518.

Min, B. and Logan, B. 2004. Continuous electricity generation from domestic wastewater and organic substrates in a flat plate microbial fuel cell. *Environ. Sci. Technol.,* 38(21): 5809–5814.

Min, B., Kim, J., Oh, S., Regan, J. and Logan, B. 2005. Electricity generation from swine wastewater using microbial fuel cells. *Water Res.,* 39(20): 4961–4968.

Morris, J., Jin, S., Wang, J., Zhu, C. and Urynowicz, M. 2007. Lead dioxide as an alternative catalyst to platinum in microbial fuel cells. *Electrochemistry Commun,* 9(7): 1730–1734.

Mustakeem. 2015. Electrode materials for microbial fuel cells: Nanomaterial Approach, *Mater. Renew. Sustain. Energy,* 4(22): 1–11.

Pant, D., Singh, A., Satyawali, Y. and Gupta, R. 2008. Effect of carbon and nitrogen source amendment on synthetic dyes decolourizing efficiency of white-rot fungus *Phanerochaete chrysosporium. J. Environ. Biol.,* 29(1): 79–84.

Park, D. and Zeikus, J. 2003. Improved fuel cell and electrode designs for producing electricity from microbial degradation. *Biotechnol. Bioeng,* 81(3): 348–355.

Potter, M. 1911. Electrical effects accompanying the decomposition of organic compounds. *Proc. R. Soc. Lond. Ser. B Contain. Pap. Biol. Character,* 84(571): 260–276.

Qian, F., Baum, M., Gu, Q. and Morse, D. 2009. A 1.5 μL microbial fuel cell for on-chip bioelectricity generation. *Lab on a Chip,* 9: 3076–3081.

Qu, L., Liu, Y., Baek, J. and Dai, L. 2010. Nitrogen-doped graphene as efficient metal-free electrocatalyst for oxygen reduction in fuel cells. *ACS Nano,* 4(3): 1321–1326.

Rabaey, K. and Rozendal, R. 2010. Microbial electrosynthesis–revisiting the electrical route for microbial production. *Nat. Rev. Microbiol.*, 8(10): 706–716.

Rabaey, K. and Verstraete, W. 2005. Microbial fuel cells: Novel biotechnology for energy generation. *Trends Biotechnol.,* 23(6): 291–298.

Rahimnejad, M., Ghoreyshi, A., Najafpour, G., Younesi, H. and Shakeri, M. 2012. A novel microbial fuel cell stack for continuous production of clean energy. *J. Hydrogen Energy*, 37: 5992–6000.

Ren, Z., Ward, T. and Regan, J. 2007. Electricity production from cellulose in a microbial fuel cell using a defined binary culture. *Environ. Sci. Technol.*, 41(13): 4781–4786.

Rhoads, A., Beyenal, H. and Lewandowski, Z. 2005. Microbial fuel cell using anaerobic respiration as an anodic reaction and biomineralized manganese as a cathodic reactant. *Environ. Sci. Technol.,* 39(12): 4666–4671.

Rosenbaum, M., Schröder, U. and Scholz, F. 2005. Utilizing the green alga *Chlamydomonas reinhardtii* for microbial electricity generation: A living solar cell. *Appl. Microbiol. Biotechnol.*, 68(6): 753–756.

Schaetzle, O., Barrie‘re, F. and Schröder, U. 2009. An improved microbial fuel cell with laccase as the oxygen reduction catalyst. *Energy Environ. Sci.,* 2(1): 96–99.

Schröder, U. 2007. Anodic electron transfer mechanisms in microbial fuel cells and their energy efficiency. *Phys. Chem. Chem. Phys.,* 9(21): 2619–2629.

Shantaram, A., Beyenal, H., Raajan, R., Veluchamy, A. and Lewandowski, Z. 2005. Wireless sensors powered by microbial fuel cells. *Environ. Sci. Technol.,* 39(13): 5037–5042.

Strik, D., Hamelers, H., Snel, J. and Buisman, C. 2008. Green electricity production with living plants and bacteria in a fuel cell. *Int. J. Energy Res.,* 32(9): 870–876.

Tommasi, T., Chiolerio, A., Crepaldi, M. and Demarchi, D. 2014. A microbial fuel cell powering an all-digital piezoresistive wireless sensor system. *Microsyst. Technol.,* 20(4): 1023–1033.

Tsai, H., Wu, C., Lee, C. and Shih, E. 2009. Microbial fuel cell performance of multiwall carbon nanotubes on carbon cloth as electrodes. *J. Power Sources,* 194(1): 199–205.

Wei, J., Liang, P. and Huang, X. 2011. Recent progress in electrodes for microbial fuel cells. *Bioresour. Technol.,* 102(20): 9335–9344.

Wen, Q., Wu, Y., Cao, D., Zhao, L. and Sun, Q. 2009. Electricity generation and modeling of microbial fuel cell from continuous beer brewery wastewater. *Biores. Technol.,* 100(18): 4171–4175.

Wen, Z., Ci, S., Mao, S., Cui, S., Lu, G., Yu, K. *et al.,* 2013. TiO_2 nanoparticles-decorated carbon nanotubes for significantly improved bioelectricity generation in microbial fuel cells. *J. Power Sources,* 234: 100–106.

Worrell, J. 1997. Inorganic chemistry: An industrial and environmental perspective. *Journal of Chemical Education*, 74(12): 1399.

Yang, Y., Xu, M., Guo, J. and Sun, G. 2012. Bacterial extracellular electron transfer in bioelectrochemical systems. *Process Biochem.,* 47(12): 1707–1714.

Yazdi, H., Carver, S., Christy, A. and Tuovinen, O. 2008. Cathodic limitations in microbial fuel cells: An overview. *J. Power Sources,* 180: 683–694.

Zhang, Y., Mo, G., Li, X., Zhang, W., Zhang, J., Ye, J. *et al.,* 2011. A graphene modified anode to improve the performance of microbial fuel cells. *J. Power Sources,* 196(13): 5402–5407.

Zhang, Y., Sun, J., Hu, Y., Li, S. and Xu, Q. 2012. Bio-cathode materials evaluation in microbial fuel cells: A comparison of graphite felt, carbon paper and stainless steel mesh materials. *Int. J. Hydrog. Energy,* 37(22): 16935–16942.

Zhao, F., Slade, R. and Varcoe, J. 2009. Techniques for the study and development of microbial fuel cells: an electrochemical perspective. *Chem. Soc. Rev.,* 38(7): 1926–1939.

Zhou, M., Chi, M., Luo, J., He, H. and Jin, T. 2011. An overview of electrode materials in microbial fuel cells. *J. Power Sources,* 196(10): 4427–4435.

15

Sustainability and Biofuels Production-A Review

ABHISHEK NANDAN[1*] AND N.A. SIDDIQUI[1]

ABSTRACT

There is sudden increase in energy consumption in India which has resulted in shortage of fossil fuels such as oil, coal and petroleum products. Recession in India's economy leads to concerns about the availability of energy required for sustaining economic growth. In view of the fast depletion of petroleum products, the search for alternative fuels has become necessary, looking at huge demand of diesel for all sectors of economy, the biodiesel is being viewed as a substitute of diesel. The vegetable oils, fats from animal and oil are the fundamental stocks for biodiesel generation. Biodiesel fuel properties are practically very identical to diesel. In India, National Biodiesel mission begun in year 2003, which advanced the development and improvement of sustainable and eco-friendly fuel in India. This work provides the current status of biofuel production and sustainability in India.

***Key words*:** Sustainability, Biodiesel, Biofuel, Energy consumption

[1]Department of Health, Safety & Environmental Engineering, University of Petroleum and Energy Studies, Dehradun

**Corresponding author*: anandan@ddn.upes.ac.in

INTRODUCTION

Sustainable development today is the most politicized catchword of international developmental conferences and programs.

Sustainable development has emerged out of the fears of depleting natural resources and a subsequent slowing or even closing down of economic activities and production systems. It is the result of rapacious misuse of earth's precious and limited resource base by those few who had a control over production systems. The concept has emerged as a broad framework to debate and decide on desirable direction of change in social and economic systems, policies, programs and actions at the national, community or individual levels. It developed in the 1960s when people became aware of the detrimental effects of industrialization on the environment. Sustainability offers long term planning for productive techniques, industrial processes and equitable distribution policies for the exploitation of resources, such as, to name a few, coal, oil and water. Planning ensures their longer life span and broader user base so that the greatest number of people may benefit out of it for the longest possible time frame.

The emergence of the idea of sustainability also strikes at the indispensability of technological transformation towards energy saving devices, alternate and non-conventional systems for providing comfort to citizens without bringing down their quality of life. This has led to a total revolution in the way people and governments have started thinking and designing their developmental programs and projects. A new approach has emerged for the Grassroots governance which fuels growth by providing land, water and forests that constitute the three basic inputs to any form of industrialization. Thus sustainable development is also indicative of 'planning from below' in contrast to the ivory tower 'planning from top' in which grassroots ecosystems were driven by technological systems. As a result, the grassroots ecosystems started to wither away as they were notable to manage and cleanse the high amount of effluent discharges, pollution and resource overuse. In the next section, we introduce the various parameters that characterize sustainable development.

Sustainable bioenergy can possibly control carbon emissions from entering the environment–essentially by changing from fossil-based petroleum to bio-based fuels as our essential transportation fuel. Technology platforms that can convert biomass feedstock to bio-based fuels such as bioethanol and biodiesel are well established.

The question is the means by which to increase worldwide generation of bio-based fuels without applying distortionary forces that may be harmful to food security and the environment, particularly in developing countries.

In part, the answer may involve the selection of a wider array of biomass feedstock, including those available from the forestry, agricultural and waste recycling sectors. For instance, lignocellulose residues from woody biomass are low value fibers that are a more cost-effective biomass feedstock than first-generation feedstock used in bioethanol production. The first criteria in selecting a feedstock is that its production must not consume more fossil fuel carbon than it can displace, requiring that the production and harvesting practices must be monitored and certified to meet sustainability standards. The second criteria are that the feedstock conversion to biofuel will be done with advanced (bio)chemical processing technologies that meet rigorous fuel test standards, ensuring compatibility of biofuels with current and emerging engine designs.

Biofuels offer benefits in terms of GHG (greenhouse gas) reduction and fossil fuel replacement. However, concerns about the overall sustainability of biofuels have been raised in terms of competition with food production, water use and other resources to produce biomass. Also in terms of the release of stored carbon and impacts on biodiversity if land is cleared to grow energy crops.

1. SUSTAINABILITY & BIOFUELS PRODUCTION

The world is confronted with vital challenges sooner rather than later and it is accounted for that fossil fuel stores will be exhausted with in a century. This will be a remarkable change that will affect adversely on all anthropogenic exercises above all horticulture, industry and trade (Ryding, 1994). On account of this, it is significant to investigate renewable and financially suitable alternatives. It has been assessed that biomass could give around 25% of worldwide vital necessities and can likewise be a wellspring of significant chemicals, pharmaceuticals and nutraceuticals (Asgari *et al.*).

With the consumption and increment in costs of petrochemical products, the development of methods for producing biofuels has the capability of off-setting these applicable difficulties. Likewise, the growing urban populace represents a genuine risk to the earth because of highly polluted residential city wastewater. The utilization of microalgae is attractive since they can serve a double part of bioremediation of wastewater and also creating biomass for biofuel generation with corresponding carbon dioxide sequestration. Further, wastewater remediation by microalgae is an eco-accommodating procedure as the biomass delivered is reused (Mata *et al.*, 2010).

Recent analyses of the energy and greenhouse-gas performance of alternative biofuels have ignited a controversy that may be best resolved by applying two simple principles (MacLean and Lave, 2003). In a world seeking solutions to its energy, environmental, and food challenges, society cannot afford to miss out on the global greenhouse-gas emission reductions and the local environmental and societal benefits when biofuels are done right. However, society also cannot accept the undesirable impacts of biofuels done wrong (Brundtland *et al.*, 2012). Biofuels done right can be produced in substantial quantities. However, they must be derived from feed stocks produced with much lower life-cycle greenhouse-gas emissions than traditional fossil fuels and with little or no competition with food production (Mata *et al.*, 2010).

Residential wastewater streams have been as often as possible utilized as a promptly accessible and practical substrate for micro algal development for biomass generation and supplement evacuation. There are various benefits of utilizing microalgae for biodiesel creation (OLAWALE). Biodiesel produced from (MSW) in sterile landfills is typically connected with soil, surface water and groundwater pollution when the landfill is not legitimately developed (Mahjoub, 2013). At present, accumulation and treatment of land-fill leachates is a standout amongst the most problems that to be addressed encompassing the operation of landfills. It accordingly involves ecological and wellbeing hazards, therefore, should promptly be passed on far from its era source(s) and treated suitably before definite transfer (London, 2000).

The biofuels industry is positioned to undergo rapid growth. The attendant policy should anticipate and provide for a biofuels industry that meaningfully and positively addresses pressing sustainability and security challenges. Biofuels should receive policy support as substitutes for fossil energy only when they make a positive impact on four important objectives: energy security, greenhouse-gas emissions, biodiversity, and the sustainability of the food supply (Swanson and Bhadwal, 2009). Performance-based policies are needed that provide incentives proportional to the benefits delivered. Legislation that is vague could allow significant portions of the biofuels industry to develop along counter productive pathways (Tilman *et al.*, 2009). Complementary policies must directly target related goals, such as land-and water-efficient food production, reduced agricultural greenhouse gas emissions, and the prevention of habitat loss from land-clearing. The best available science, continually updated, should be used to evaluate the extent to which various biofuels achieve their multiple objectives, and policy should reward achievement (Whitten *et al.*, 2004).

Three steps should be taken: meaningful science-based environmental safeguards should be adopted, a robust biofuels industry should be enabled, and those who have invested in first-generation biofuels should have a viable path forward. In support of such policy, rigorous accounting rules will need to be developed that measure the impacts of biofuels on the efficiency of the global food system, greenhouse-gas emissions, soil fertility, water and air quality and biodiversity (Dauvergne and Neville, 2010). Accounting rules should consider the full life cycle of biofuels production, transformation, and combustion. Unless new technologies and life-styles are adopted globally over the coming decades, the massive projected increases in global energy and food consumption will greatly elevate atmospheric greenhouse-gas levels from fossil fuel combustion, land clearing, and livestock production and will create immense biodiversity loss from habitat destruction and climate change (Tester *et al.*, 2012). The quality of human life will be compromised. A central issue for the coming decades, then, is how the environmental impacts and potential benefits associated with meeting the global demand for food and energy can be internalized into our economic systems. This is a complex question that cannot be addressed with simplistic solutions and sound bites. It needs a new collaboration between environmentalists, economists, technologists, the agricultural community, engaged citizens, and governments around the world (Schmidheiny, 1992).

As the demand for energy continues to increase globally, fossil fuel usage will likewise continue to rise. There is still a plentiful supply of fossil fuels at reasonably low cost, although this is likely to change in the future, but more critically a rising use of fossil fuels is unlikely to be sustainable in the longer term principally due to the attributed increase in GHG emissions from using these fuels and the environmental impact of these emissions on global warming (Pittman *et al.*, 2011). There is therefore significant interest in identifying alternative renewable sources of fuel that are potentially carbon neutral. The majority of the current commercially available biofuels are bioethanol derived from sugar cane or corn starch or biodiesel derived from oil crops including soybean and oilseed rape (Nigam and Singh, 2011). Although biofuels have the potential to be environmentally beneficial compared to fossil fuels, there is some dispute as to whether these crop-based biofuels are economically competitive compared to fossil fuels (Naylor *et al.*, 2007). Furthermore, there is even more concern over the impact that the use of these crops for biofuels might have on food availability. Biofuels derived from the cultivation of algae have therefore been proposed as an alternative approach that does not adversely impact agriculture (Scott *et al.*, 2010).

1.1. Potential of Biofuel Production

One of the major attraction of microalgae as a biofuel feedstock is that they can be effectively grown in conditions which require minimal freshwater input unlike many plant-based biofuel crops. Also these utilize land, which is otherwise non-productive to plant crops, thus making the process potentially sustainable with regard to preserving freshwater resources. For example, microalgae could be cultivated near the sea to utilize saline or brackish water. There has therefore been significant interest in the growth of microalgae for biofuels under saline conditions (Speight and Singh, 2014).

However, another potentially sustainable growth medium for algal feedstock is wastewater. It has been appreciated for some years now that microalgae can be potentially utilized for low-cost and environmentally friendly wastewater treatment compared to other more commonly used treatment processes (Ullah *et al.*, 2014). The major problem with most wastewaters is the very high concentrations of nutrients, particularly total N and total P concentration as well as toxic metals, which require costly chemical-based treatments to remove them during wastewater treatment (Christenson and Sims, 2011). Total N and P concentrations can be found in ranges of 10–100 mg/L in municipal wastewater and >1000 mg/L in agricultural effluents. The ability of microalgae to effectively grow in nutrient-rich environments and to efficiently accumulate nutrients and metals from the wastewater, make them an extremely attractive means for sustainable and low cost wastewater treatment. However, it has also long been proposed that wastewater-grown algae could be used for energy production (Jorgensen and Johnsen, 2011).

There have been contrasting assessments to the economic viability of algal biofuels. A number of studies have argued that biofuel production from algae, particularly biodiesel production is both economically and environmentally sustainable, although there have been some sceptical views of the long term viability and economics of biofuels from algae (Greenwell *et al.*, 2009). One frequent criticism is that the use of fossil fuels in the biofuel production process, in the construction of algal growth facilities, supply of nutrients for algal growth, harvesting of algae and biomass processing, is not often considered in the evaluation of algal biofuel viability and would in fact give rise to a net negative energy output (Mata *et al.*, 2010). The use of wastewater resources may be a viable means to enhance the sustainability of algal biofuel production, both by providing a dual use process, an effective growth medium for algal cultivation, and freely available nutrient (particularly N and P) input (Rawat *et al.*, 2013).

1.2. Alternative Biofuel for Liquid

Liquid stimulates and alternative biofuel particles are fundamental to what makes a respectable fuel. The best fuel centers for this will be particles that are great with existing engines (begin engine for gas, weight begin engine for diesel fuel, and gas turbine for fly fuel) (Swarup *et al.*, 1992). There are a couple of appropriate factors to consider when arranging biofuels essentials substance, the start quality portrayed by octane or cetane number and the purpose of hardening, aroma, and harmful quality are basic parts to consider. In the next section, we will consider a couple biofuel contenders and their properties (Bajpai and Tyagi, 2006).

1.3. Gasoline And Its Alternatives

Fuel is an intricate blend of hydrocarbons including direct, extended, and cyclic alkanes (40–60%), aromatics (20–40%), and oxygenates. The carbon number of hydrocarbons in fuel differs from 4 to 12. Ethanol, the well-known added substance to gas, has an octane number of 129, yet its vitality content per gallon is around 70% of that of fuel (Schobert, 2013). Ethanol likewise has issues as a fuel inferable from high miscibility with water, which makes it hard to distil from the aging juices and to transport through existing pipelines. As of late, n-butanol has gotten more consideration as an option fuel added substance. Butanol has two a greater number of carbons than ethanol, which brings about a vitality substance of around 40% higher than that of ethanol (Antoni *et al.*, 2007). The octane number of butanol is 96, which is to some degree lower than that of ethanol yet is still practically identical to that of gas. Not at all like ethanol, butanol is less solvent in water than ethanol. It can likewise be utilized as an added substance to gas as well as a fuel independent from anyone else in customary motors. As a rule, with octane number increments the particle has methyl expanding and twofold bonds. Spread C4 and C5 alcohols are additionally viewed as potential fuel added substances (Demirbas, 2009).

1.4. Diesel and Its Alternatives

Diesel fuel is additionally an unpredictable blend of hydrocarbons including straight, expanded, and cyclic alkanes (75%) and aromatics (25%). The carbon number of hydrocarbons in petrodiesel fluctuates from 9 to 23, with a normal of 16. Biodiesel is for the most part made out of unsaturated fat methyl esters (FAMEs), and is for the most part derived from vegetable oil or animal fat (Pahl, 2008). The unsaturated fats in FAMEs by and large have an anchor length from 12 to 22, containing zero to two twofold securities. Biodiesel has a tantamount cetane rating and vitality substance to petrodiesel and extra favorable circumstances, for example, higher lubricity and lower emission of pollutants. Another hotspot for biodiesel would be isoprenoids, which are normally happening extended or cyclic hydrocarbons for the most part combined in plants (Eriksen, 2013). These as a rule have methyl branches, twofold bonds, and ring structures, which enhance the ease at lower temperatures yet bring down cetane evaluations (Kayhanian *et al.*, 2007).

1.5. Biojet Fuels/Jet Fuels

Biojet fuel is an extremely complex blend of hydrocarbons with a carbon number close to of 8–16 and around 25% (v/v) point of confinement of aromatics. Stream fuel is fundamentally the same as lamp oil or diesel fuel, yet requires a lower the point of solidification since it is utilized under harsh conditions (Smil, 2010). Straight or fanned hydrocarbons with medium carbon chain length delivered from the unsaturated fat or isoprenoid biosynthetic pathways are essential for biojet fuels. As of late, the utilization of isoprenoids as fly fuel has been explored, as they have low solidification which owing to their expanding and cyclic structure (Serrano-Ruiz *et al.*, 2012).

2. BIOMASS FEEDSTOCK FOR CELLULOSIC ETHANOL PRODUCTION: A NEW INDUSTRY

The economic competitiveness of cellulosic ethanol production is highly dependent on feedstock cost, which constitutes 35–50% of the total ethanol production cost. It depends on various geographical factors and the types of systems used for harvesting, collecting, preprocessing, transporting and handling of the material. Cellulosic ethanol is projected to be much more cost-effective, environmentally beneficial, and have a greater energy output to input ratio than grain ethanol (Solomon *et al.*, 2007).

Consequently, as the deployment of cellulosic ethanol biorefineries approaches, feedstock cost and availability are the driving factors that influence pioneer bio refinery locations and will largely control the rate at which this industry grows. Ethanol made biologically from cellulosic biomass, including agricultural and forestry residues, portions of municipal waste, and herbaceous and woody crops, is finally being widely recognized as a unique transportation fuel with powerful economic, environmental and strategic attributes (Wyman, 2007).

Initial scenarios were postulated to develop a pioneer dry feedstock supply system design case as a demonstration of the current state of technology. Based on this pioneer design, advanced scenarios were developed to determine key cost barriers, needed supply system improvements, and technology advancements to achieve government and private sector cost targets. Pioneer supply systems would start by using current infrastructure and technologies and be individually designed for biorefineries using specific feedstock types and varieties based on local geographic conditions. As the industry develops and cost barriers are addressed, the supply systems will incorporate advanced technologies that will eliminate downstream diversity and provide a uniform, tailored feedstock for multiple biorefineries located in different regions.

2.1. Cellulosic Feedstock and Its Advantage

Cellulosic feed stocks show great promise for technological advancement as an economically viable source for transportation fuel. They are abundant and can be converted into fuels and chemicals by either biochemical or thermo-chemical processes, but the best efficiency may come from integration of both the processes (Himmel *et al.*, 2007). In order to understand the issues impacting feedstock costs, it is helpful to first understand the processes for converting cellulosic feed stocks to ethanol and other fuels and products. The use of renewable energy sources is becoming increasingly necessary, if we are to achieve the changes required to address the impacts of global warming (McKendry, 2002).

Cellulosic biochemical conversion technologies, like grain ethanol processes, ferment sugars from the carbohydrate fractions of biomass into ethanol and other products. In grain ethanol processes, the fermentable glucose monomeric sugars are liberated from the grain starch, whereas, in cellulosic processes, the fermentable sugars are liberated from the cellulose and hemicelluloses cross-linked polymeric molecules that make up plant cell walls

Different thermo-chemical conversion processes that include combustion, gasification, liquefaction, hydrogenation and pyrolysis have been used to convert the biomass into various energy products (Goyal *et al.*, 2008). Economic development planners, commodity leaders and others are assessing their regions to define how and when their areas can become part of the industry, which is largely determined by the location of the bio refinery. Because of economic risk, the first, or pioneer, biorefineries will rely on existing feedstock types, growing conditions, and the grain commodity infrastructure as the basis of their feedstock supply systems, as exhibited by the bio refinery designs of the US Department of Energy commercial-scale facility. The supply systems for these facilities will be unique to each bio refinery as regional biomass variations, cropping practices, and equipment require optimization for relatively small, local areas. Thus, biorefineries will begin processing ethanol in limited areas and spread as feedstock production systems, based on advanced feedstock supply system technologies, begin supplying large enough quantities to enable them to be cost effectively replicated.

The large-scale problem is solved *via* Lagrange relaxation based decomposition method (Chen and Fan, 2012).

An integrated national supply network was proposed together with the collection, storage and transportation of raw materials and products. Finally, challenges and obstacles in legal context and policies implementation were elaborated, as well as infrastructure shortage and technology availabilities (Goh *et al.*, 2010).

Based on the average annual straw usage per animal and size of the Idaho dairy industry, an estimated 350000 to 450000 tons of straw were harvested annually for Livestock (Hess *et al.*, 2009). While this large crop residue biomass market provided an excellent feedstock baseline model, it also demonstrated that biorefineries were not the only potential large users of biomass resources. Duane Grant, a south-central Idaho straw broker, reported that the 2004 value of straw in the Idaho dairy market was $35 to $46 per ton delivered to the dairy (Hess *et al.*, 2009).

In a feedstock assembly system where a commodity grain is the focus of the harvest and collection operation, growers will be responsible for the timing and logistics associated with equipment and labor resources (Hess *et al.*, 2009). They will identify who will perform the labor (themselves or a custom operator), when the harvest will occur, and where the commodity will be sold.

They will also be responsible for baling and road siding the resulting straw. Because grain will remain the primary commodity, at least until the cellulosic bio refinery industry matures, the grain is assumed to bear all of the costs of the harvesting operation (Perlack *et al.*, 2005). Thus, the costs and logistics associated with the feedstock assembly system begin with the collection of the straw lying in the harvester-produced windrow, where it is allowed to dry to a moisture content of 15% or less.

3. BIOENERGY CROPS AND GENETICALLY MANIPULATED PLANTS AND SUSTAINABLE DEVELOPMENT

An energy crop is a plant grown as a low-cost and low-maintenance harvest used to make biofuels, such as bioethanol, or combusted for its energy content to generate electricity or heat (Murray, 2015). Energy crops are generally categorized as woody or herbaceous plants; many of the latter are grasses of the family Graminaceae.

Commercial energy crops are typically densely planted, high-yielding crop species which are processed to bio-fuel and burnt to generate power. Woody crops such as willow or poplar are widely utilised, as well as temperate grasses such as *Miscanthus* and *Pennisetum purpureum* (both known as elephant grass). If carbohydrate content is desired for the production of biogas, whole-crops such as maize, Sudan grass, millet, white sweet clover and many others can be made into silage and then converted into biogas (Basu, 2013).

Through genetic modification and application of biotechnology, plants can be manipulated to create greater yields, reduce associated costs and require less water (Gavrilescu and Chisti, 2005). High energy yield can also be realized with existing cultivars.

3.1. Types of Energy Crops

3.1.1. *Solid biomass*

Energy generated by burning plants grown for the purpose, often after the dry matter is pelletized. Energy crops are used for firing power plants, either alone or co-fired with other fuels (Sami *et al.*, 2001). Alternatively they may be used for heat or combined heat and power (CHP) production.

To cover the increasing requirements of woody biomass, short rotation coppice (SRC) were applied to agricultural sites. Within this cropping system fast growing tree species like willows and poplars are planted in growing cycles of three to five years. The cultivation of this cultures is dependent on wet soil conditions and could be an alternative for moist field sieds (Lowe and Butt, 2005). However, an influence on local water conditions could not be excluded. This indicates that an establishment should exclude the vicinity to vulnerable wetland ecosystems (Harris, 1967).

3.1.2. *Gas biomass (methane)*

Anaerobic digesters or biogas plants can be directly supplemented with energy crops once they have been ensiled into silage (Nordberg *et al.*, 2007). The fastest growing sector of German bio-farming has been in the area of "Renewable Energy Crops" on nearly 500,000 ha (1,200,000 acres) of land (2006). Energy crops can also be grown to boost gas yields where feedstocks have low energy content, such as manures and spoiled grain. It is estimated that the energy yield presently of bioenergy crops converted *via* silage to methane is about 2 GWh/km^2 (1.8×10^{10} BTU/sq mi). Small mixed cropping enterprises with animals can use a portion of their acreage to grow and convert energy crops and sustain the entire farms energy requirements with about one fifth of the acreage (Guthman, 2014). In Europe and especially Germany, however, this rapid growth has occurred only with substantial government support, as the German bonus system for renewable energy (Jacobsson and

Bergek, 2004). Similar developments of integrating crop farming and bioenergy production *via* silage-methane have been almost entirely overlooked in N. America, where political and structural issues and a huge continued push to centralize energy production has overshadowed positive developments (Murray, 2015)

3.1.3. *Liquid biomass*

3.1.3.1. *Biodiesel*

European production of biodiesel from energy crops has grown steadily in the last decade, principally focused on rapeseed used for oil and energy. Production of oil/biodiesel from rape covers more than 12,000 km^2 in Germany alone, and has doubled in the past 15 years (Saynor *et al.*, 2003). Typical yield of oil as pure biodiesel may be is 100,000 L/km^2 (68,000 US gal/sq mi; 57,000 imp gal/sq. mi) or more, making biodiesel crops economically attractive, provided sustainable crop rotations exist that are nutrient-balanced and preventative of the spread of disease such as Clubroot. Biodiesel yield of soybeans is significantly lower than that of rape (Ma and Hanna, 1999).

3.1.3.2. *Bioethanol*

Energy crops for biobutanol are grasses. Two leading non-food crops for the production of cellulosic bioethanol are switchgrass and giant miscanthus. There has been a preoccupation with cellulosic bioethanol in America as the agricultural structure supporting biomethane is absent in many regions, with no credits or bonus system in place (Visser, 2015). Consequently, significant private money and investor hopes are being pinned on marketable and patentable innovations in enzyme hydrolysis and the like.

Bioethanol also refers to the technology of using principally corn (maize seed) to make ethanol directly through fermentation, a process that under certain field and process conditions can consume as much energy as is the energy value of the ethanol it produces, therefore being non-sustainable. New developments in converting grain stillage (referred to as distillers grain stillage or DGS) into biogas energy looks promising as a means to improve the poor energy ratio of this type of bioethanol process (Lee and Shah, 2012).

3.2. Genetically Modified Crops

Genetically modified crops (GMCs, GM crops, or biotech crops) are plants used in agriculture, the DNA of which has been modified using genetic engineering techniques (Thomson, 2002). In most cases, the aim is to introduce a new trait to the plant which does not occur naturally in the species. Examples in food crops include resistance to certain pests, diseases, or environmental conditions, reduction of spoilage, or resistance to chemical treatments (*e.g.* resistance to a herbicide), or improving the nutrient profile of the crop. Examples in non-food crops include production of pharmaceutical agents, biofuels, and other industrially useful goods, as well as for bioremediation (Morin, 2008)

3.2.1. *Some of bioenergy crops include*

(1) Cellulosic crops including short rotation trees and shrubs such as *Eucalyptus* (*Eucalyptus* spp.), Poplar (*Populus* spp.), Willow (*Salix* spp.), and Birch (*Betula* spp.);
(2) Perennial grasses such as giant reed (*Arundo donax*), reed canary grass (*Phalaris arundinacea*), Switchgrass (*Panicum virgatum*).

ROLE IN SUSTAINABILITY

Bioenergy crops currently provide the only source of alternative energy with the potential to reduce the use of fossil transportation fuels in a way that is compatible with existing engine technology, including in developing countries (Morin, 2008). The use of organic waste and agricultural/forestry residues, and of lignocellulosic crops that could be grown on a wider spectrum of land types, may mitigate land and water demand and reduce competition with food (Ho *et al.*, 2014). Energy crops act as filter systems/ supplemental crops for pollution control as they remove pesticides and excess fertilizer from surface water before it pollutes groundwater or streams/rivers. They can protect a stream's bank and water from erosion, siltation, and chemical runoff and can still be harvested for energy.

Research has also shown that energy crops have increased soil stability, decreased surface water runoff, decreased transport of nutrients and sediment, and increased soil moisture, in comparison to traditional crops (Blanco-Canqui, 2010). They also require less fertilizers, herbicides and insecticides than traditional row crops, this reduction in herbicide and pesticide use reduces the potential for water pollution and other environmental problems. Another environmental benefit from the use of energy crops is a decrease in emissions. Unlike fossil fuels, plants grown for energy crops absorb the amount of carbon dioxide (CO_2) released during their combustion/use. Therefore, by using biomass for energy generation there is no net CO_2 generated because the amount emitted in its use has been previously absorbed when the plant was growing (Samal, 2015). It has been assessed the potential for CO_2 emission reduction by developing non grain based ethanol crops.

The results show that non-grain based bio-ethanol production can potentially reduce CO_2 emissions from the 2007 levels by 11 million tons and 49 million tons in 2015 and 2030, respectively (Baeyens *et al.*, 2015). Energy crops may also protect natural forests by providing an alternative source of wood, which can be grown on farm or pasture land that is no longer suitable for traditional row crops. Bioenergy-driven restoration of degraded ecosystems can also increase terrestrial carbon sequestration due to large biomass production and root residues as well as slowing decomposition of soil organic materials under no-till conditions (Saibi *et al.*, 2013).

CONCLUSIONS

Diminished reliance on fossil based powers has set the establishment for research on cleaner alternatives. All the more as of late research has concentrated on utilizing waste streams that have supplements in form of nitrogen and phosphorus as an asset. Giving ideal conditions to algae advances enhanced lipid generation. Subsequently the biomass gives double advantages of wastewater treatment and oil creation. In any case, as of now much accentuation is likewise being put on upgrading innovation for reaping of biomass and extraction of oil. The spent biomass can hence be utilized for the generation of bioethanol or bio-methane.

REFERENCES

Antoni, D., Zverlov, V.V. and Schwarz, W.H. (2007). Biofuels from microbes. *Applied Microbiology and Biotechnology*, 77: 23–35.

Asgari, N., Abdullah, A.M., Askari, A. and Enayati, N. Review on Algae as an alternative source of Biofuel.

Baeyens, J., Kang, Q., Appels, L., Dewil, R., Lv, Y. and Tan, T. (2015). Challenges and opportunities in improving the production of bio-ethanol. *Progress in Energy and Combustion Science,* 47: 60–88.

Bajpai, D. and Tyagi, V. (2006). Biodiesel: Source, production, composition, properties and its benefits. *Journal of Oleo Science,* 55: 487–502.

Basu, P. (2013). *Biomass gasification, pyrolysis and torrefaction: Practical design and theory*, Academic press.

Blanco-Canqui, H. (2010). Energy crops and their implications on soil and environment. *Agronomy Journal,* 102: 403–419.

Brundtland, G.H., Ehrlich, P., Goldemberg, J., Hansen, J., Lovins, A., Likens, G., Lovelock, J., Manabe, S., May, B. and Mooney, H. (2012). Environment and development challenges: The imperative to Act. *The Asahi Glass Foundation, Tokyo.*

Chen, C.-W. and Fan, Y. (2012). Bioethanol supply chain system planning under supply and demand uncertainties. *Transportation Research Part E: Logistics and Transportation Review,* 48: 150–164.

Christenson, L. and Sims, R. (2011). Production and harvesting of microalgae for wastewater treatment, biofuels, and bioproducts. *Biotechnology advances,* 29: 686–702.

Dauvergne, P. and Neville, K.J. (2010). Forests, food, and fuel in the tropics: The uneven social and ecological consequences of the emerging political economy of biofuels. *The Journal of Peasant Studies,* 37: 631–660.

Demirbas, A. (2009). Progress and recent trends in biodiesel fuels. *Energy Conversion and Management,* 50: 14–34.

Eriksen, D. (2013). Pathway optimization and engineering for biofuel production.

Gavrilescu, M. and Chisti, Y. (2005). Biotechnology–A sustainable alternative for chemical industry. *Biotechnology Advances,* 23: 471–499.

Goh, C.S., Tan, K.T., Lee, K.T. and Bhatia, S. (2010). Bio-ethanol from lignocellulose: Status, perspectives and challenges in Malaysia. *Bioresource Technology,* 101: 4834–4841.

Goyal, H.B., Seal, D. and Saxena, R.C. (2008). Bio-fuels from thermochemical conversion of renewable resources: A review. *Renewable and Sustainable Energy Reviews,* 12: 504–517.

Greenwell, H., Laurens, L., Shields, R., Lovitt, R. and Flynn, K. (2009). Placing microalgae on the biofuels priority list: A review of the technological challenges. *Journal of the Royal Society Interface*, rsif20090322.

Guthman, J. (2014). *Agrarian dreams: The paradox of organic farming in California*, Univ of California Press.

Harris, G.A. (1967). Some competitive relationships between *Agropyron spicatum* and *Bromus tectorum. Ecological Monographs,* 37: 89–111.

Hess, J.R., Wright, C.T., Kenney, K.L. and Searcy, E.M. (2009). Uniform-Format Solid Feedstock Supply System: A Commodity-Scale Design to Produce an Infrastructure-Compatible Bulk Solid from Lignocellulosic Biomass—Executive Summary. Idaho National Laboratory (INL).

Himmel, M.E., Ding, S.-Y., Johnson, D.K., Adney, W.S., Nimlos, M.R., Brady, J.W. and Foust, T.D. (2007). Biomass recalcitrance: Engineering plants and enzymes for biofuels production. *Science,* 315: 804–807.

Ho, D.P., Ngo, H.H. and Guo, W. (2014). A mini review on renewable sources for biofuel. *Bioresource Technology,* 169: 742–749.

Jacobsson, S. and Bergek, A. (2004). Transforming the energy sector: The evolution of technological systems in renewable energy technology. *Industrial and Corporate Change,* 13: 815–849.

Jorgensen, S.E. and Johnsen, I. (2011). *Principles of Environmental Science and Technology*, Elsevier.

Kayhanian, M., Tchobanoglous, G. and Brown, R.C. (2007). Biomass conversion processes for energy recovery. *Handbook of energy efficiency and renewable energy. Taylor and Francis Group*, 25: 4–25.8.

Lee, S. and Shah, Y.T. (2012). *Biofuels and Bioenergy: Processes and Technologies*, CRC Press.

London, L. (2000). Sustainable Landfill in the UK: A Review of Current Knowledge and Outstanding R&D Needs.

Lowe, C.N. and Butt, K.R. (2005). Culture techniques for soil dwelling earthworms: A review. *Pedobiologia,* 49: 401–413.

Ma, F. and Hanna, M.A. (1999). Biodiesel production: A review. *Bioresource Technology,* 70: 1–15.

Maclean, H.L. and Lave, L.B. (2003). Life cycle assessment of automobile/fuel options. *Environmental Science & Technology,* 37: 5445–5452.

Mahjoub, B. (2013). Plants for soil remediation. *Biomass for Sustainable Applications.*

Mata, T.M., Martins, A.A. and Caetano, N.S. (2010). Microalgae for biodiesel production and other applications: A review. *Renewable and Sustainable Energy Reviews,* 14: 217–232.

Mckendry, P. (2002). Energy production from biomass (part 1): Overview of biomass. *Bioresource Technology,* 83: 37–46.

Morin, X.K. (2008). Genetically modified food from crops: Progress, pawns, and possibilities. *Analytical and Bioanalytical Chemistry,* 392: 333–340.

Murray, C. (2015). *Losing ground: American social policy, 1950-1980*, Basic Books.

Naylor, R.L., Liska, A.J., Burke, M.B., Falcon, W.P., Gaskell, J.C., Rozelle, S.D. and Cassman, K.G. (2007). The ripple effect: Biofuels, food security, and the environment. *Environment: Science and Policy for Sustainable Development,* 49: 30–43.

Nigam, P.S. and Singh, A. (2011). Production of liquid biofuels from renewable resources. *Progress in Energy and Combustion Science,* 37: 52–68.

Nordberg, Å., Jarvis, +!., Stenberg, B., Mathisen, B. and Svensson, B.H. (2007). Anaerobic digestion of alfalfa silage with recirculation of process liquid. *Bioresource Technology,* 98: 104–111.

Olawale, T.D. The Design of a Pilot Plant That will Produce 5000m 3 Per Year of Biogas from Biomass a Final Year Design Project Presented to the Department of Chemical Engineering.

Pahl, G. 2008. *Biodiesel: Growing a new energy economy*, Chelsea Green Publishing.

Perlack, R.D., Wright, L.L., Turhollow, A.F., Graham, R.L., Stokes, B.J. and Erbach, D.C. (2005). Biomass as feedstock for a bioenergy and bioproducts industry: The technical feasibility of a billion-ton annual supply. DTIC Document.

Pittman, J.K., Dean, A.P. and Osundeko, O. (2011). The potential of sustainable algal biofuel production using wastewater resources. *Bioresource Technology,* 102: 17–25.

Rawat, I., Kumar, R.R., Mutanda, T. and Bux, F. (2013). Biodiesel from microalgae: A critical evaluation from laboratory to large scale production. *Applied Energy,* 103: 444–467.

Ryding, S.-O. (1994). *Environmental Management Handbook*, IOS Press.

Saibi, W., Brini, F., Hanin, M. and Masmoudi, K. (2013). Development of energy plants and their potential to withstand various extreme environments. *Recent Patents on DNA & Gene Sequences,* 7: 13–24.

Samal, A.P. (2015). *Characterization of Properties and Estimation of Power Generation Potentials of Residues of Some Woody Biomass Species.* National Institute of Technology, Rourkela.

Sami, M., Annamalai, K. and Wooldridge, M. (2001). Co-firing of coal and biomass fuel blends. *Progress in Energy and Combustion Science,* 27: 171–214.

Saynor, B., Bauen, A. and Leach, M. (2003). The potential for renewable energy sources in aviation. *Report commissioned for the DTI, Imperial College Centre for Energy Policy and Technology.*

Schmidheiny, S. (1992). *Changing course: A global business perspective on development and the environment*, MIT press.

Schobert, H. (2013). *Chemistry of fossil fuels and biofuels*, Cambridge University Press.

Scott, S.A., Davey, M.P., Dennis, J.S., Horst, I., Howe, C.J., Lea-Smith, D.J. and Smith, A.G. (2010). Biodiesel from algae: Challenges and prospects. *Current Opinion in Biotechnology,* 21: 277–286.

Serrano-Ruiz, J.C., Ramos-Fernández, E.V. and Sepúlveda-Escribano, A. (2012). From biodiesel and bioethanol to liquid hydrocarbon fuels: New hydrotreating and advanced microbial technologies. *Energy & Environmental Science,* 5: 5638–5652.

Smil, V. (2010). *Energy transitions: History, requirements, prospects*, ABC-CLIO.

Solomon, B.D., Barnes, J.R. and Halvorsen, K.E. (2007). Grain and cellulosic ethanol: History, economics, and energy policy. *Biomass and Bioenergy,* 31: 416–425.

Speight, J.G. and Singh, K. (2014). *Environmental Management of Energy from Biofuels and Biofeedstocks*, John Wiley & Sons.

Swanson, D. and Bhadwal, S. (2009). *Creating adaptive policies: A guide for policymaking in an uncertain world*, IDRC.

Swarup, R., Mishra, S. and Jauhari, V. (1992). *Environmental Pollution Ecology*, Mittal Publications.

Tester, J.W., Drake, E.M., Driscoll, M.J., Golay, M.W. and Peters, W.A. (2012). *Sustainable energy: Choosing among options*, MIT press.

Thomson, J.A. (2002). *Genes for Africa: Genetically modified crops in the developing world*, Juta and Company Ltd.

Tilman, D., Socolow, R., Foley, J.A., Hill, J., Larson, E., Lynd, L., Pacala, S., Reilly, J., Searchinger, T. and Somerville, C. (2009). Beneficial biofuels–the food, energy, and environment trilemma. *Science,* 325: 270–271.

Ullah, K., Ahmad, M., Sharma, V.K., Lu, P., Harvey, A., Zafar, M., Sultana, S. and Anyanwu, C. (2014). Algal biomass as a global source of transport fuels: Overview and development perspectives. *Progress in Natural Science: Materials International,* 24: 329–339.

Visser, W. (2015). *Sustainable frontiers: Unlocking change through business, leadership and innovation*, Greenleaf Publishing.

Whitten, S., Carter, M. and Stoneham, G. (2004). Market-based tools for environmental management. Proceedings of 6th Annual AARES Symposium.

Wyman, C.E. (2007). What is (and is not) vital to advancing cellulosic ethanol. *Trends in Biotechnology*, 25: 153–157.

Subject Index